Crafting a Sustainable Future Through Education and Sustainable Development

Javier Martínez-Falcó
University of Alicante, Spain

Bartolomé Marco-Lajara
University of Alicante, Spain

Eduardo Sánchez-García
University of Alicante, Spain

Luis A. Millan-Tudela
University of Alicante, Spain

A volume in the Practice, Progress, and Proficiency in Sustainability (PPPS) Book Series

IGI Global
PUBLISHER of TIMELY KNOWLEDGE

Published in the United States of America by
IGI Global
Engineering Science Reference (an imprint of IGI Global)
701 E. Chocolate Avenue
Hershey PA, USA 17033
Tel: 717-533-8845
Fax: 717-533-8661
E-mail: cust@igi-global.com
Web site: http://www.igi-global.com

Library of Congress Cataloging-in-Publication Data

Names: Martinez Falcó, Javier, 1997- editor. | Marco Lajara, Bartolomé, editor. | Sánchez García, Eduardo (Professor of Management), editor. | Millán-Tudela, Luis A., 1998- editor.
Title: Crafting a sustainable future through education and sustainable development / edited by Javier Martínez-Falcó, Bartolomé Marco-Lajara, Eduardo Sánchez-García, and Luis A. MillanTudela.
Description: Hershey, PA : Engineering Science Reference, [2023] | Includes bibliographical references and index. | Summary: "Crafting a Sustainable Future Through Education and Sustainable Development" serves as a comprehensive study guide, offering critical reflections and constructive critiques. It covers an array of relevant topics, ranging from artificial intelligence and big data to gender equality, game-based learning, and socio-technological innovation. By examining concepts like knowledge management, problem-solving skills, and innovative abilities, the book provides a robust foundation for understanding the intricate connections between education and sustainability"-- Provided by publisher.
Identifiers: LCCN 2023030272 (print) | LCCN 2023030273 (ebook) | ISBN 9781668496015 (h/c) | ISBN 9781668496053 (s/c) | ISBN 9781668496022 (eISBN)
Subjects: LCSH: Environmental education. | Sustainable development--Study and teaching. | Education--Aims and objectives.
Classification: LCC GE70 .C73 2023 (print) | LCC GE70 (ebook) | DDC 304.2071--dc23/eng20230919
LC record available at https://lccn.loc.gov/2023030272
LC ebook record available at https://lccn.loc.gov/2023030273

This book is published in the IGI Global book series Practice, Progress, and Proficiency in Sustainability (PPPS) (ISSN: 2330-3271; eISSN: 2330-328X)

British Cataloguing in Publication Data
A Cataloguing in Publication record for this book is available from the British Library.
All work contributed to this book is new, previously-unpublished material.
The views expressed in this book are those of the authors, but not necessarily of the publisher.
For electronic access to this publication, please contact: eresources@igi-global.com.

Practice, Progress, and Proficiency in Sustainability (PPPS) Book Series

Ayman Batisha
International Sustainability Institute, Egypt

ISSN:2330-3271
EISSN:2330-328X

MISSION

In a world where traditional business practices are reconsidered and economic activity is performed in a global context, new areas of economic developments are recognized as the key enablers of wealth and income production. This knowledge of information technologies provides infrastructures, systems, and services towards sustainable development.

The **Practices, Progress, and Proficiency in Sustainability (PPPS) Book Series** focuses on the local and global challenges, business opportunities, and societal needs surrounding international collaboration and sustainable development of technology. This series brings together academics, researchers, entrepreneurs, policy makers and government officers aiming to contribute to the progress and proficiency in sustainability.

COVERAGE

- Socio-Economic
- Intellectual Capital
- Eco-Innovation
- Global Business
- Environmental informatics
- E-Development
- Knowledge clusters
- Global Content and Knowledge Repositories
- Outsourcing
- Technological learning

IGI Global is currently accepting manuscripts for publication within this series. To submit a proposal for a volume in this series, please contact our Acquisition Editors at Acquisitions@igi-global.com or visit: http://www.igi-global.com/publish/.

The Practice, Progress, and Proficiency in Sustainability (PPPS) Book Series (ISSN 2330-3271) is published by IGI Global, 701 E. Chocolate Avenue, Hershey, PA 17033-1240, USA, www.igi-global.com. This series is composed of titles available for purchase individually; each title is edited to be contextually exclusive from any other title within the series. For pricing and ordering information please visit http://www.igi-global.com/book-series/practice-progress-proficiency-sustainability/73810. Postmaster: Send all address changes to above address.

Titles in this Series

For a list of additional titles in this series, please visit: http://www.igi-global.com/book-series/

5G, Artificial Intelligence, and Next Generation Internet of Things Digital Innovation for Green and Sustainable Economies
Patricia Ordóñez de Pablos (University of Oviedo, Spain) and Xi Zhang (Tianjin University, China)
Engineering Science Reference • © 2023 • 256pp • H/C (ISBN: 9781668486344) • US $280.00

Handbook of Research on Bioeconomy and Economic Ecosystems
Ana Pego (Universidade Nova de Lisboa, Portugal)
Business Science Reference • © 2023 • 502pp • H/C (ISBN: 9781668488799) • US $295.00

Handbook of Research on Safe Disposal Methods of Municipal Solid Wastes for a Sustainable Environment
Prateek Srivastava (Chitkara University Institute of Engineering and Technology, Chitkara University, India) Durgaprasad Ramteke (Fibre and Particle Engineering, Faculty of Technology, University of Oulu, Finland) Ankush Kumar Bedyal (Department of Physics, School of Sciences, Cluster University of Jammu, India) Monika Gupta (Chitkara Business School, Chitkara University, India) and Jasminder Kaur Sandhu (Chitkara University Institute of Engineering and Technology, Chitkara University, India)
Engineering Science Reference • © 2023 • 414pp • H/C (ISBN: 9781668481172) • US $295.00

Hydrogen Fuel Cell Technology for Mobile Applications
Raluca Andreea Felseghi (Department of Mechanical Engineering, Faculty of Automotive, Mechatronics, and Mechanical Engineering, Technical University of Cluj-Napoca, Romania)
Engineering Science Reference • © 2023 • 313pp • H/C (ISBN: 9781668467213) • US $240.00

IGI Global
PUBLISHER of TIMELY KNOWLEDGE

701 East Chocolate Avenue, Hershey, PA 17033, USA
Tel: 717-533-8845 x100 • Fax: 717-533-8661
E-Mail: cust@igi-global.com • www.igi-global.com

Table of Contents

Detailed Table of Contents

Carbon neutrality is considered one of the most important goals for countries and societies to achieve due to its anticipated benefits on the lives of all living organisms, as well as its central role in any attempt to achieve sustainable development goals. Therefore, shedding light on this topic and informing the next generation of its significance is essential. Educational institutions play a vital role in helping students improve knowledge, skills, and emotional resilience; they undoubtedly shape students' characters and enable them to face various challenges, particularly ones related to the environment. This chapter addresses carbon neutrality by defining it, outlining the goals related to it, and emphasizing its importance in protecting the planet. Additionally, the following discussion addresses the role of educational institutions, such as schools, in promoting awareness of carbon neutrality among stakeholders directly involved in education and in the local community. Finally, awareness-raising methods and class activities will be presented.

Worldwide guidelines since the 1980s have developed mechanisms with national and international economic funds for scientific and technological development that reduce greenhouse gas emissions (GHG) and mitigate environmental impacts. The objective of this chapter is to analyze the political, socioeconomic, and regulatory characteristics of Mexico as a country with a developing economy, as well as the actions that have made the country go backward in the energy transition, to make a proposal that serves as a guide. Among the precedents have been the Montreal Protocol and the Kyoto Protocol, which have sought to mitigate the use of greenhouse gases in industrialized countries and involve developing countries through various mechanisms that encourage the reduction of GHG, such as carbon credits. The UN (United Nations) continues to work with cutting-edge initiatives and programs such as the sustainable development goals (SDGs).

Higher education plays a crucial role in addressing sustainability challenges, enabling educators to equip students with the necessary knowledge, skills, and attitudes. However, many educators lack adequate training in teaching sustainability. This study used a quantitative approach to explore the impact of training on educators' ability to

The chapter presents the results of a spatiotemporal analysis of accident occurrence of the trolleybus mass transit system in Mexico City. Chi-square analysis for goodness of fit method was employed to assess the relationship between the temporal variables and the accident occurrence. The ArcGis software was used for the analysis of the spatial visualization. The employed data were for 2017. The main conclusion were a) accidents do not have a uniform distribution weekly and hourly (All lines and Line A), on weekdays and the seasons of the year for Lines B and C, and crashes occurring hourly for the case of lines D-G; b) there was not a relationship between weekdays and hourly accident occurrence when considering the seven lines altogether and Line A; c) the most-affected boroughs were those related to Cuauhtémoc (27%), Coyocan (21.6%), and GAM (20.5%); they accounted for 69.1% of the accidents; and d) the peaks of trolleybus accident occurrence depend on the lines, that is, Line F (one peak of hours), Lines B-E (two), and Lines A and G (three).

This chapter is based on researchers' experience of developing an undergraduate module (Level 4) in collaboration with students and colleagues supported by a curriculum investigation grant from the British Education Research Association (BERA). It presents a research-led module development initiative that was designed to engage and empower students by making them the co-creators of the curriculum which promotes practical examples of education towards sustainable development. This participatory research utilised to promote teacher's agency and students' voices. It empowered students as partners in co-creating curriculum. It also fosters teaching and learning that assists institutions and individuals to make informed choices that impact on individuals and communities to develop practice that promotes sustainability. This chapter reflects on the process of developing the module and learning from the process and outlines future directions embedding climate change mitigation and adaptation issues in different disciplines.

This chapter explores the significant role that artificial intelligence (AI) plays in advancing renewable energy technologies and promoting sustainability. It discusses how AI can address the challenges and complexities associated with renewable energy systems, improve their efficiency, and enable their seamless integration into existing power grids. The chapter also explores various AI applications in renewable energy generation, forecasting, grid optimization, energy management, and demand response. Additionally, it highlights the potential benefits of AI-driven solutions in accelerating the global transition to a sustainable energy future.

The improving the state of research ethics among the scientific community requires attention to the necessary infrastructure and quality in research. This chapter considers the training of ethical principles in the research process in curricula as a neglected aspect of research activities and it considers ethics and scientific integrity as the most important aspect of promoting academic excellence. The first step to prevent ethical abuse in research and to internalize scientific excellence in universities considers the formation of values or the formulation of an ethics and scientific integrity charter as the main solution. Therefore, this chapter first considers the reason for ethics and scientific integrity in research and tries to establish a link between these. Then it considers the position of ethics and scientific integrity in research in higher education centers. In the following, the higher education system is faced with two ethical discourses. Finally, it deals with standardization and drawing a vision of the ethics in the higher education system.

Food security is a fundamental human right. Enhancing agricultural practices, lowering food waste, and ensuring equitable resource allocation are among the actions taken to ensure food security. VGI has developed into a helpful tool for data collection, assisting in tracking resource availability, monitoring price fluctuations, mapping community food assets, and assisting with land management. VGI encourages fair pricing, capacity building, access to nutritious food, decision-making, locating food storage, and capacity building. VGI helps to address food waste, map crop production, compile production techniques, crowdsource food quality data, and track supply chains. Advanced technologies can be integrated with VGI to improve analysis and decision-making. Enhancing VGI programs and inclusivity can help people better understand and address issues related to food security.

As a developed and environmentally benign alternative for energy generation, renewable energy sources (RES) have attracted a lot of attention. The issue of energy quality, however, is one of the most difficult integration problems for RES. Voltage and frequency violations caused by waveform instability in the energy production from RES can harm equipment and lower the reliability of the power supply. In recent years, renewable energy sources like wind and solar power have drawn a lot of attention as environmentally friendly substitutes for traditional energy sources. Systems with artificial intelligence (AI) are becoming more and more common for use in automation, automated automation, and data analysis. In this research, the authors present a unique theory for an AI system that combines solar and wind energy to power AI applications. In order to improve the effectiveness and efficiency of AI systems, the suggested system makes use of the complementary properties of solar and wind energy, optimizing their generation and utilization.

The use of audiovisual content (short self-elaborated videos, movie extracts, and similar) in teaching has been adopted since the digitalization in education facilities took place. However, the inclusion of these materials should be assessed to check for the fit with the learning goals established in each learning environment. This chapter proposes a methodology to validate the adequacy and the level of congruence

Preface

In an ever-changing world marked by complex environmental, social, and economic challenges, the pursuit of sustainability has emerged as a paramount goal for societies across the globe. As we navigate the 21st century, it has become increasingly evident that the sustainable development of our planet requires concerted efforts from every individual, institution, and nation. At the heart of this transformative journey lies education—a potent catalyst for change and progress.

This book, *Crafting a Sustainable Future Through Education and Sustainable Development*, stands as a testament to the collective wisdom, knowledge, and passion of numerous researchers, educators, and practitioners who have come together to share their insights and innovative approaches to fostering sustainability through education. As editors at IGI Global publishing house, we are deeply honored to present this comprehensive collection of contributions, spanning a myriad of disciplines, all united under the common purpose of creating a more sustainable future for generations to come.

The volume begins by laying a strong foundation, delving into the fundamental principles of sustainability, its intricate link to education, and the dynamic interplay between the two. It explores how educational institutions can transform themselves into bastions of sustainability, instilling in their learners the values, skills, and mindsets necessary to confront global challenges and embrace a sustainable way of life.

Moving beyond theoretical frameworks, the chapters within this book showcase a wealth of practical applications and case studies from various regions and educational contexts. From innovative pedagogies that promote environmental stewardship to community-based initiatives that empower local communities to pursue sustainable development, these real-world examples illustrate the tangible impact of education as a potent force for positive change.

Moreover, *Crafting a Sustainable Future Through Education and Sustainable Development* also casts a spotlight on technology's role in advancing sustainability goals. By examining the transformative potential of digital learning, cutting-edge tools, and data-driven decision-making, the book underscores how technology can amplify educational efforts and expedite progress towards a more sustainable future.

In every chapter, the underlying thread of hope, resilience, and determination shines through. The collective expertise presented here sparks inspiration and demonstrates that change is possible when we unite in purpose and action. Whether you are an educator, policymaker, researcher, or concerned global citizen, the insights shared within these pages will undoubtedly enrich your understanding of the profound impact education can have in shaping a sustainable world.

As we embark on this journey together, let us embrace the notion that sustainability is not merely an aspiration, but a tangible destination within our reach. The path forward may be challenging, but with education as our steadfast guide and sustainable development as our lodestar, we have the power to craft a brighter, greener, and more equitable future for all.

We extend our heartfelt gratitude to all the contributors who have poured their expertise, passion, and dedication into this book. Your work exemplifies the spirit of collaboration and shared responsibility essential for addressing the global sustainability crisis. Special thanks to IGI-GLOBAL publishing house for their unwavering support in bringing this ambitious project to fruition.

Together, let us take the wisdom found within these pages and turn it into action, contributing to a sustainable future that we can proudly bequeath to generations yet unborn.

Javier Martínez-Falcó
University of Alicante, Spain

Bartolomé Marco-Lajara
University of Alicante, Spain

Eduardo Sánchez-García
University of Alicante, Spain

Luis A. Millan-Tudela
University of Alicante, Spain

Chapter 1
Carbon Neutrality in Schools: How We Can Achieve It?

Huda M. Aldayri
Ministry of Education, Oman

Abdullah K. Ambusaidi
Ministry of Education, Oman

Mohamed M. Shahat
https://orcid.org/0000-0002-9637-8192
Sultan Qaboos University, Oman & Aswan University, Aswan, Egypt

ABSTRACT

Carbon neutrality is considered one of the most important goals for countries and societies to achieve due to its anticipated benefits on the lives of all living organisms, as well as its central role in any attempt to achieve sustainable development goals. Therefore, shedding light on this topic and informing the next generation of its significance is essential. Educational institutions play a vital role in helping students improve knowledge, skills, and emotional resilience; they undoubtedly shape students' characters and enable them to face various challenges, particularly ones related to the environment. This chapter addresses carbon neutrality by defining it, outlining the goals related to it, and emphasizing its importance in protecting the planet. Additionally, the following discussion addresses the role of educational institutions, such as schools, in promoting awareness of carbon neutrality among stakeholders directly involved in education and in the local community. Finally, awareness-raising methods and class activities will be presented.

DOI: 10.4018/978-1-6684-9601-5.ch001

INTRODUCTION

Carbon neutrality has garnered considerable attention recently as a key measure in addressing climate change. According to the Intergovernmental Panel on Climate Change (IPCC), attaining carbon neutrality is critical for keeping global warming below the 2°C threshold and averting the catastrophic consequences of climate change (IPCC, 2018). The IPCC defines carbon neutrality as a state wherein human-induced greenhouse gas removals effectively balance human-induced greenhouse gas emissions over a specified timeframe (IPCC, 2018). Similarly, the International Energy Agency (IEA) defines net-zero carbon emissions as the point at which the volume of carbon emissions released equals the volume of carbon removed from the atmosphere. The IEA underscores the significance of reducing carbon emissions and bolstering carbon removal endeavors to achieve net-zero emissions (IEA, 2022).

Carbon neutrality is not easy to define, and various definitions exist depending on perspectives and context. The United Nations Framework Convention on Climate Change (UNFCCC) offers its perspective, defining carbon neutrality as harmonizing greenhouse gas emissions reduction with societal and economic needs (UNFCCC, n.d.). According to the UNFCCC, achieving carbon neutrality requires a multifaceted approach involving reducing emissions, embracing low-carbon technologies, and implementing offset mechanisms. The UK Government often frames carbon neutrality in terms of promoted goals; for example, their 'Ten Point Plan for a Green Industrial Revolution' outlines the target of transitioning to zero-emission vehicles by 2030 and achieving a carbon-neutral economy by 2050 (UK Government, 2020). The Science-Based Targets initiative (SBTi) is a collaborative effort between several organizations, including the United Nations Global Compact (UNGC), the World Resources Institute (WRI), the World Wide Fund for Nature (WWF), and the Carbon Disclosure Project (CDP), that provides a framework for companies, incentivizing them to establish and pursue scientifically grounded emissions reduction goals. According to the SBTi guidelines, net-zero emissions can be achieved by balancing the quantity of emitted greenhouse gases with the amount removed from the atmosphere (SBTi, 2021). The SBTi emphasizes the significance of emission reductions, renewable energy adoption, and the deployment of carbon removal technologies to attain net-zero emissions. There are also organizations, such as the UK-based Carbon Trust organization, that assist companies in reducing their carbon emissions; Carbon Trust define carbon neutrality as achieving net-zero greenhouse gas emissions by balancing emissions produced and removed from the atmosphere (Carbon Trust, n.d.). To facilitate this, they help companies reduce emissions, utilize renewable energy sources, and offset any remaining emissions through projects that decrease greenhouse gas emissions elsewhere. According to the Intergovernmental Panel on Climate Change (IPCC), zero carbon neutrality refers to a state without a general release of carbon dioxide or other greenhouse gases into the atmosphere (IPCC, 2018). This definition aligns with the broader objective of effectively eliminating net greenhouse gas emissions to address climate change.

Various researchers continue to fine-tune definitions and explore the complexities of achieving carbon neutrality. Shang and Lv (2022) defined it as the state where greenhouse gas emissions associated with a building or development throughout its entire life cycle are fully offset by removing an equivalent amount of greenhouse gases from the atmosphere. Wei et al. (2022) describe it as a specific situation where total greenhouse gas emissions are balanced by the removal of greenhouse gases, which can be partially achieved through renewable energy sources. Zhang and Chen (2022) define it as a state where net carbon dioxide emissions associated with a specific activity, process, or system are reduced to zero through emissions reduction measures and carbon offsetting techniques. Researchers from the University of Manchester offer a more comprehensive perspective, stating that achieving zero carbon neutrality involves striking a balance between greenhouse gas emissions and removals through a combination of measures such as reducing emissions, improving energy efficiency, deploying renewable energy sources, and implementing carbon sequestration methods (Magacho et al.,2023). This expanded definition emphasizes the need for a holistic and multifaceted approach to achieve zero carbon neutrality.

In addressing climate change, carbon neutrality assumes a pivotal role in helping find solutions and shape policy; however, the variety of published definitions point to this concept's intricate and multifaceted nature. They reveal the need for adopting a comprehensive and integrated approach that addresses the diverse factors driving emissions and facilitates the transition toward sustainable and low-carbon systems. Attaining carbon neutrality requires reducing carbon emissions, at least partially by adopting low-carbon technologies, and implementing offset mechanisms to compensate for any remaining emissions. A comprehensive understanding of the varied definitions and approaches to carbon neutrality is imperative for governments, businesses, and individuals to undertake the requisite measures in order to curtail their carbon footprint. While definitions may vary depending on the source, ultimately the goal is to achieve net-zero carbon emissions within any given domain.

Carbon Neutral Goals

Carbon neutrality has emerged as a critical objective in the battle against climate change. Governments and organizations worldwide are establishing targets to reach net-zero carbon emissions within the forthcoming decades. One prominent example is the European Union (EU), which has outlined its ambition to achieve net-zero greenhouse gas emissions by 2050 (European Commission, 2019). This objective forms a part of the EU's comprehensive climate strategy, which aims to reduce greenhouse gas emissions by at least 55% by 2030, relative to 1990 levels (European Commission, 2020). To realize these aspirations, the EU is implementing a range of measures including augmenting renewable energy sources, promoting energy efficiency, and investing substantial amounts in clean technologies.

China, the largest emitter of greenhouse gases globally, has committed to achieving carbon neutrality by 2060 (Xinhua, 2020). Given that China accounts for approximately 28% of worldwide carbon emissions (World Bank, 2021), the significance of this target cannot be overstated. Accomplishing carbon neutrality in China will necessitate substantial investments in renewable energy, electric vehicles, and other low-carbon technologies. Additionally, numerous companies worldwide are setting their own ambitious goals for carbon neutrality. For example, Microsoft has pledged to become carbon negative by 2030 and to offset all carbon emissions the company has generated since its inception in 1975 by 2050 (Microsoft, 2021).

Countries like the Sultanate of Oman have set ambitious targets to achieve carbon neutrality, in accordance with their commitments under the 2016 Paris Agreement on climate change. These countries recognize the urgent need to transition to a low-carbon economy to mitigate the impacts of climate change and protect the environment. Substantial investments in renewable energy, energy efficiency, and other measures to reduce emissions are required to achieve carbon neutrality. In 2020, Oman's Ministry of Energy and Minerals launched a National Energy Strategy with the stated goal of achieving sustainable and efficient use of energy resources. The strategy includes targets to increase the share of renewable energy in the electricity mix to 30% by 2030 and to 50% by 2040 (Ministry of Energy and Minerals, 2020). The government has also implemented policies to promote energy efficiency and conservation, such as the Energy Efficiency and Conservation Law and the Oman Energy Efficiency and Renewable Energy Program.

Oman has significant renewable energy potential, particularly in solar and wind power. The country's expansive desert landscapes and coastal regions offer favorable conditions for establishing large-scale solar and wind farms. As a member of the Middle East and North Africa (MENA) Hydrogen Alliance, Oman is actively engaged in initiatives promoting hydrogen production and utilization as a low-carbon energy solution. Notably, in 2019, Oman inaugurated its first utility-scale solar project, the Ibri II Solar Power Project, with a capacity of 500 MW. This initiative is expected to contribute to a reduction of over 340,000 tons of carbon emissions annually (Oman Observer, 2019). In addition to renewable energy, Oman is actively exploring diverse strategies to mitigate emissions; these include the adoption of carbon capture and storage (CCS) technology and the utilization of hydrogen as a clean fuel source. While Oman has made significant progress in transitioning to a low-carbon economy, further efforts are required to expedite emissions reduction and achieve the objective of carbon neutrality. Accomplishing this goal is going to require a coordinated endeavor across all sectors of the economy with collaboration and cooperation among the government, private sector, and civil society.

The importance of carbon neutrality goals in the fight against climate change cannot be emphasized enough. Governments, businesses, and organizations are actively setting ambitious targets to reach net-zero carbon emissions in the coming decades. The successful attainment of these objectives carries great significance in mitigating the devastating impacts of climate change. However, realizing these goals will require significant investments in renewable energy, energy efficiency, and other environmentally friendly technologies.

Amid the escalating challenges presented by climate change, the notion of carbon neutrality has emerged as an essential element in the global pursuit of sustainability. It directly and indirectly aligns with the three dimensions of sustainability – environmental, social, and economic (Rodrigues et al., 2022, Sharmin, 2023; Sen et al., 2022, Williams et al., 2019).

Carbon Neutrality and School Education

The concept of education for sustainable development is needed to enhance the knowledge and skills to safeguard our future. UNESCO argues that one of the most important jobs of education is to raise awareness and critical thinking in relation to environmental, social and economic sustainability (United Nations Educational, Scientific and Cultural Organization (UNESCO), 2022). They state that three aspects of learning - cognitive, behavioral, and emotional - need to be targeted, as detailed below:

- Knowledge: represented in understanding sustainability challenges and revealing innovative solutions and informed ideas.
- Behavioral aspect: represented in a series of actions that contribute to sustainable changes.
- Social and emotional aspect: represented in the formation and sustainability of positive values and attitudes towards the environment.

Education is, therefore, a powerful tool capable of enhancing students' knowledge and awareness, enabling them to deal with various problems that threaten life on the planet (UNESCO, n.d). Once students are able to understand climate change, they will have a much better chance of coming up with effective approaches to achieving carbon neutrality. Zhang et al. (2022) confirmed this and showed the impact of climate change knowledge on students' views and on their willingness to participate in carbon neutrality education. The study concluded that participants who were not involved in climate change-related activities were significantly less supportive of carbon neutrality education and were less willing to participate in relevant training courses. Hayles and De la Harpe (2010) found positive perceptions among students of architecture and environmental engineering in the US and a high level of awareness of sustainability issues after undergoing a module on sustainable design and construction strategies. The study also revealed that students have become more able to express their perspectives on sustainability and environmental issues. Aydin (2016) carried out a study that aimed to reveal the awareness levels of ninth graders about their own carbon footprints; the results showed a high level of awareness, with most believing that the most harmful behaviors to the environment are related to the home and transport.

As discussed, education has a pioneering role in moving society towards a net-zero carbon future. Through teaching practices, educational programs, and school activities, schools can hopefully equip students with the knowledge, skills, and behavior needed to achieve carbon neutrality.

CARBON NEUTRALITY AND THE SCHOOL ENVIRONMENT

A school's multiple buildings, specialist facilities and need for daily transport almost always constitute a source of carbon emissions; however, schools can serve as case studies that enable us to analyze and find sustainable ways to reduce these emissions (Chatterjee & Reynolds, 2008). The most likely major sources of emissions are represented in Figure 1.

Figure 1. Sources of carbon emissions in the school environment

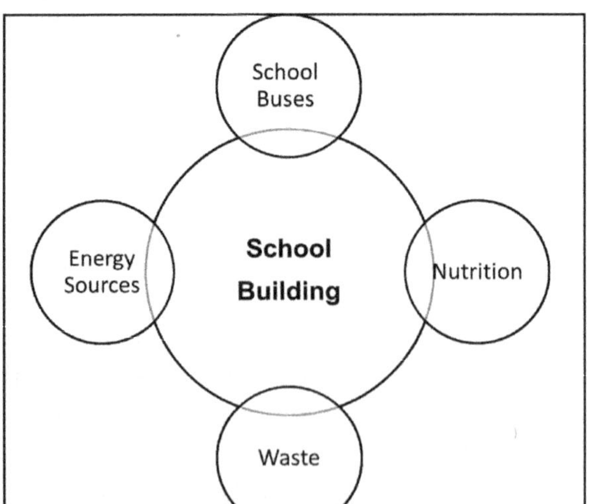

The figure shows multiple areas of significant carbon emissions in the school environment. Below they are each addressed in turn:

School Buildings

School buildings naturally require equipment and facilities to facilitate various educational tasks. However, some school buildings may, through emissions, pose a risk to the environment. Some buildings, of course, are more environmentally hazardous than others depending on age. It is an unavoidable fact that older buildings are responsible for more emissions due to being less fit for purpose and less energy efficient.

Efficient use of electricity and water in school buildings is a priority that factors in heavily to any calculation of sustainability; unfortunately, in Oman's extremely hot climate, air-conditioning and water use can be excessive and have negative impacts on the environment. In a study conducted by Gamarra et al. (2017), the Life Cycle Assessment (LCA) was employed to assess the environmental impacts associated with energy, water, and consumables consumption of two educational centers in Spain. The model consists of four phases: target and range determination, inventory analysis, impact assessment, and results interpretation. The results showed that the school needed to considerably increase their efficiency in the use of air conditioning and lighting in order to reduce greenhouse gas impacts. The study also showed that schools can reduce the overall demand for fossil energy per student to between 18-30% by using renewable sources of energy. On the other hand, Odell et al. (2021) presented a study that aimed at encouraging management in 13 schools to participate in an innovative two-year reduced-carbon training program in Australia. The study calculated how schools can effectively reduce carbon emissions and operational costs associated with buildings and infrastructure. By following the initiatives they suggested, most of which are simple and low cost, schools were able to reduce their energy expenditure by 15%. Care should be taken to use sustainable principles in a school's design. The design of school buildings constitutes a sustainable architectural, engineering and construction model which students and all stakeholders can learn and benefit from if it is developed using environmentally friendly guidelines (Bazzocch, et al., 2021).

School Buses

Oman's wide-ranging educational policies cover the provision of school bus systems to ensure convenient access and the safety of students. However, any transport system poses an inherent risk, especially with regard to the amount of carbon emissions produced. To alleviate this problem, schools can follow guidelines by from other countries; for example, the following recommendations come from the Ministry of Education, British Columbia (n.d):

- Encourage students and staff to walk or, for those living near the school, attend school by environmentally friendly means (electric scooter, bike, etc.)
- Organize a fun walking event to promote awareness of walking and the benefits to the environment that can arise from such activities.
- Design and distribute leaflets to bus and car owners about the need to switch off their cars while waiting to pick up or drop off students.

Nutrition

It is vital that administrators have systems in place to ensure the quality of food available in schools and make sure the materials used for preservation and packaging are safe. Food is a significant source of carbon emissions in the school environment, and therefore strict procedures must be followed by the school. A meta study by Dos Santos et al. (2022) analyzed 134 studies to determine which sustainable practices were employed and the most common sustainability recommendations related to school nutrition policies. The study presented recommendations related to purchasing sustainable food; and it offered guidelines on nutritional education focused on reducing food waste, using less plastic food packaging and consuming more organic products.

Waste

All the waste produced by schools, whether it be by the school administration, teachers or students, can lead to large amounts of carbon being emitted. Examples of waste include that from stationery, food, energy, water and many more. Therefore, schools should follow rigorous action to reduce the amount of waste such as checking the quantity and type of waste dumped in landfill as well as sustainable procedures such as recycling and reusing resources. The Ministry of Education in British Columbia, Canada has adopted several proposals to reduce the use of waste in school buildings as described below:

- Using e-mails in correspondence between teachers and management.
- Using technology to facilitate and record meetings rather than have in-person paper-based meetings.
- Using materials that can be used more than once (food containers, water preservation cans, meal packaging bags, etc.).

The dilemma of excess carbon emissions requires schools to review their policies and plans and develop a systematic carbon management plan. To do so, schools must review their current practices and periodically perform a thorough audit to identify the causes and possible solutions. This naturally requires schools to calculate the extent of their carbon footprint. A school's carbon footprint is the total net CO2 and other greenhouse gases from an institution's practices and processes. Thus, calculating it can direct individuals and institutions towards modifying actions that emit large amounts of carbon, i.e., modifying behaviors in the right direction (Growcom, 2008). This concept can be illustrated via a study by Bolstad, (2021) which aimed to build a model on how schools could use a footprint calculator to measure their carbon footprint in school. The study showed that there was interest within the school to use the concept of a carbon footprint to support students' learning. Teachers in the study successfully incorporated awareness raising of their school's carbon footprint into the curriculum. It was found that using the footprint as a basis for improvement could save up to 15% of the costs for the 10 participating schools in Auckland, New Zealand. The results of the study also showed benefits of using the carbon footprint calculator in terms of student learning, and the adoption of more environmentally friendly practices within their schools and communities.

THE ROLE OF SCHOOLS IN ACTIVATING CARBON NEUTRAL PRACTICES

Schools play an important role in achieving sustainability by contributing to students' knowledge and skills and instilling values that enable them as responsible citizens to protect the environment (National Governance Association, 2022). To meet environmental, social, and economic sustainability targets, schools have a three-dimensional role which consists of the following: the educational dimension - whereby schools impart knowledge, values, and skills to students; the social dimension – which involves institutions contributing to positive perceptions, not only for teachers and students, but also for families and communities; and lastly the cultural dimension – which refers to how cultural values and beliefs, such as valuing the environment, moderation and self-efficacy, can influence students to take sustainable environmental actions)Bazzocchi et al.,2021). A school is an integrated system and any practices that have a positive impact on the environment can be incorporated into all components of this system. The following Figure (2) illustrates this process:

The previous figure shows the multiple possible areas where a school's carbon neutrality practices can be integrated. They are grouped into three areas as described above - vision and values, sustainability, and change - each area is explained more fully in the table below)National Governance Association, 2022).

It is clear from Table 1 that a school's educational system has multiple components, which, when combined, allow for the comprehensive integration of carbon neutral

Figure 2. Areas where carbon neutrality practices can be incorporated in schools

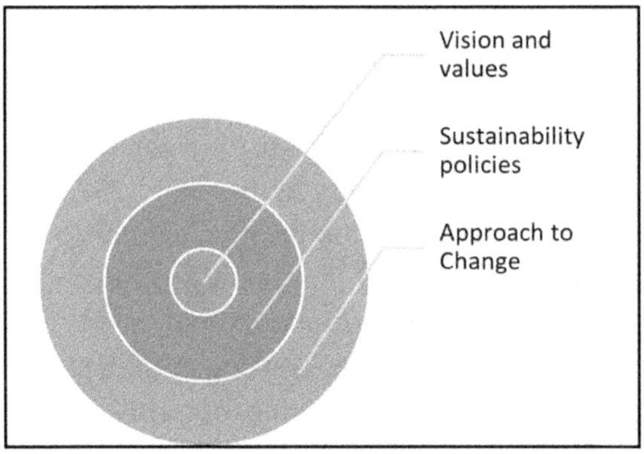

Table 1. Areas of inclusion in a school's carbon neutrality practices

Area	Questions a School Must Ask of Itself
Vision and values	- Does the school vision include the promotion of sustainability? - Did the school plan give details and place value on its sustainability goals?
Sustainability	Sustainability consists of four aspects: 1. Culture: the level of awareness of students and school staff about the dimensions of sustainability. In this regard, the school must consider the following questions: - Is there an environmental team at the school? - What is the degree of students' awareness of sustainability? - What practices and events has the school organized related to sustainability? - Do school students exhibit behaviors that support carbon neutrality practices (e.g., recycling, rationalizing water consumption, reducing electricity consumption, reducing waste)? 2. The school campus: the school building and the associated facilities and spaces. In this regard, the school must consider the following questions: - How does the school handle waste? - Does the school use modern technologies to reduce consumption (e.g., electricity, water, paper)? - Has the school established procedures to protect the school campus environment? 3. Society: the school's practices supporting zero-carbon neutrality in the community. In this regard, the school must consider the following questions: - Has the school implemented a plan to support carbon neutrality practices in partnership with the community? 4. The curriculum: the school's educational and training programs for students and teachers. In this regard, the school must consider the following questions: - What topics related to carbon neutrality has the school identified to educate students? - Has the school implemented training programs for teachers to enable them to understand carbon neutrality and its applications? - Are the appropriate educational tools required to teach subjects that address carbon neutrality available at school? - How can the school find innovative teaching methods to enhance its students' awareness of carbon neutrality practices?
Pioneering change	This area explores the approach to making changes to achieve carbon neutrality. It addresses the following points: - What are the roles and responsibilities of principals, teachers, and students in working towards carbon neutrality? - How can stakeholders, both within and outside the school, contribute to reducing the carbon footprint of the school?

practices. If done well, schools can be important incubators for fostering the next generation tasked with the protection and sustainability of the environment. Some major ways schools can lessen their carbon footprints are as follows:

1. Reduce Consumption (Electricity and Water)

To reduce consumption, schools need to consider other environmentally friendly alternatives. An example can be seen in Victorian schools participating in the Resource Smart schools' program, which is supported by the Victorian Government in Australia. The program has contributed to providing practical support to schools to allow them to reduce resource use and energy consumption rates. It has also provided educational programs with sustainability themes that have been integrated into the participating schools' curricula, as well as providing schools with mechanisms to raise community awareness off-campus in the participating 1300 schools. As a result, schools were able to save approximately $41 million in energy costs, plant 5.271.038 trees, and save more than 118,000 tons of greenhouse gas emissions (About Smart Resource Schools, 2019).

2. Engaging the Community on Sustainability

This can be done through initiatives implemented by the school in partnership with the community, for example, school parks can contribute to increasing community interaction; and students and teachers can be encouraged to engage in sustainability practices on and off the school campus (Rilla, 2013). An example of community engagement in action is the Green Schools Project - a comprehensive initiative developed by the Oman Ministry of Education, which targets all members of the school community including students, teachers, parents and administrators. It is not limited to work within the school but also asks members of the community to plan and partake in a range of environmentally friendly activities and projects (Ambusaidi et al., 2023). This initiative has several goals, including:

- Conservation of energy and natural resources.
- Reduction of water and sanitation waste.
- Enhancing the protection of natural environments.
- Improving indoor air quality.
- Preservation of fresh water.

The Ministry of Education of the Sultanate of Oman implemented the Green Schools Initiative in the school year 2022/2023 for a period of six months. It consisted of the following two aspects:

1. Istidamah Curriculum (Sustainability)

A curriculum entitled 'Istidama' was used; it consists of six themes: Waste Management, Climate Change, Renewable Energy, Afforestation, Water, and Carbon Neutrality. Teaching this curriculum included classroom activities, practical experiences, field visits, workshops and lectures by experts, and engaging parents. It also included some activities related to entrepreneurship and the potential use of environmental waste in recycling and reuse.

2. School Projects

School projects varied based on the six themes incorporated in the Istidama Curriculum. Schools developed 12 major projects, along with several accompanying sub-projects (Ministry of Education, 2023). Examples of these projects include:

1. Qantara: This project focuses on water management, particularly gray water (water from air conditioners, sinks, and filters). It serves as a link to other aspects of waste management, energy, and landscaping.
2. Wahj: This project centers around a protected house and an educational garden powered by solar energy.
3. Safia Green Garden: This project raises awareness about carbon neutrality by utilizing solar panels to create an attractive greenhouse. Solar energy is harnessed to operate a pump for irrigating fig plants and nurseries.

Furthermore, there are plans for a project involving a spherical fountain that will operate using solar energy, with the objective of tapping into sustainable resources within the school's natural surroundings.

It is hoped that through this Green Schools program schools can change their roles from traditional educational institutions to environmentally conscious model institutions that promote understanding of the importance of sustainability among those inside schools and in the surrounding community.

School Activities to Activate Carbon Neutral Practices

School activities of various kinds can encourage students to discuss, think critically, define problems and test solutions, and participate actively in protecting the environment. To assist educators in teaching the concept of carbon neutrality, some example activities are presented below that teachers could use in their classes:

Activity No. 1	
Activity Type:	Class-based individual activity
Activity Aim:	To allow students to understand the meaning/definition of carbon and carbon emissions in the school environment.
Activity Description:	What is meant by carbon? …………………………………………………………………………… …………………………………………………………………………… Draw a sketch of our school building and identify the areas of carbon emissions.
Activity No. 2	
Activity Type:	Class-based group activity
Activity Aim:	To allow students to think about the advantages and disadvantages of carbon in the environment
Activity Description:	Discuss the advantages and disadvantages of carbon in the environment:
Activity No. 3	
Activity Type:	Class-based group activity
Activity Aim:	To allow students to think about how the globe may look without carbon.
Activity Description:	The teacher asks his/her students to design postcards, specifying the source in each card (transport, home, school, buildings, food, etc.) Cards are then distributed to groups, and students are asked to discuss the specific topic and develop a plan to reach net-zero carbon in that area. Teachers can ask the following questions: ■ How can you reduce carbon emissions in future practices? ■ What should the government do to reduce emissions? ■ Other questions the teacher might think of.
Activity No. 4	
Activity Type:	Extracurricular individual activity
Activity Aim:	To allow students to calculate the consumption of electricity during a specific period of time.
Activity Description:	The teacher asks students individually to calculate the amount of electricity their family consumed during the last 6 months. Students will then develop an action plan to reduce consumption.
Activity No. 5	
Activity Type:	Extracurricular individual activity

Activity No. 1	
Activity Aim:	To train students how to measure the amount of consumed water and how to reduce waste.
Activity Description:	1. The teacher requests students to carry out a domestic activity aimed at measuring the amount of water consumed so that students understand the scale of their water consumption. The following tasks can be employed: ■ Read the water meter at home for a summer month and a winter month. ■ Calculate the amount of water consumed per month. ■ Calculate the amount of water consumed per individual (divide the total by the number of family members). 2. Design a poster with some tips that family members can follow to lessen water consumption.
Activity No. 6	
Activity Type:	Class-based group activity
Activity Aim:	To train students to think critically and innovatively.
Activity Description:	The teacher asks students to think about how to reduce the use of paper waste in the school environment by introducing a competition between classes. The teacher can ask the following questions: 1. How can we deal with used books? 2. How can paper be discarded in a safe manner? Come up with a way to take advantage of leftover paper inside the school.

TEACHING STRATEGIES TO ACTIVATE CARBON NEUTRAL PRACTICES

This section explores strategies for teaching topics related to carbon neutrality in order to provide guidance for teachers. Each strategy is presented along with examples to help aid understanding. Teachers can access the source material to gain a fuller understanding of theory underpinning the strategies.

1. Plateau Climbing Strategy

A teacher can apply the strategy of Plateau Climbing by following seven steps, as shown in figure 3 (Karkhi & Araiki, 2015):

Figure 3. Strategy steps

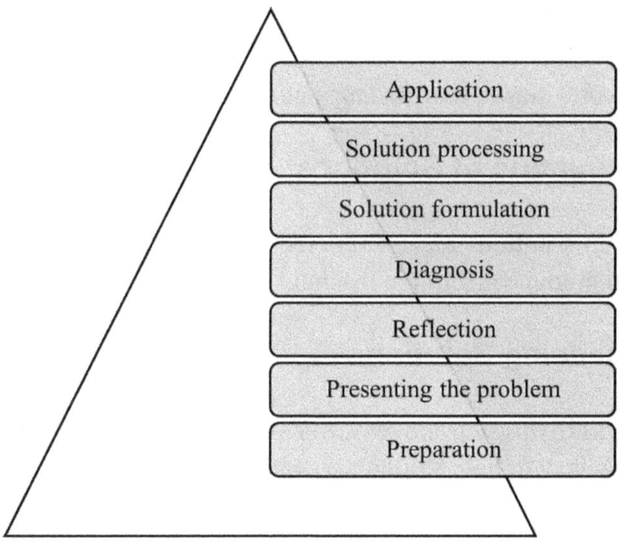

Step 1: Preparation

In this step, the teacher prepares the class for the topic (the school building is a source carbon emissions) by discussing with students what happens when there is no air conditioning (AC) in the classroom. They then gather as many answers as possible to prepare the ground for the topic to be discussed. Teachers should elicit the idea that despite the benefits of air conditioning, it is still a cause of carbon emissions.

Step 2: Presenting the Problem

The problem in this step is accurately presented and written on the board. Linking with the first step, the teacher explains to his students that despite the benefits of the AC, it is a source of carbon emissions. Subsequently, the problem is formulated as follows: "How do air conditioners constitute a source of carbon emissions? and can other areas of the school building pose a risk to the environment in terms of the amount of carbon they emit?"

Step 3: Reflection

The role of students here is to reflect on the problem using any means necessary (conceptual maps, illustrations, etc.). The aim is to encourage students to research the problem, clarify trends, and find possible solutions to address the problem.

Step 4: Diagnosis

The teacher diagnoses the students' findings and helps them delete or modify some of them.

Step 5: Formulation of Solutions

In this step, final solutions are formulated and commented on by students with supervision and guidance from the teacher.

Step 6: Processing Solutions

The teacher works in this step as a reviewer and may reformulate solutions in a more concise and easy way to help students to understand depending on needs.

Step 7: Application

In this step, it can be ascertained whether students are able to utilize their understanding by assigning them an application task (e.g., homework).

2. Functional Behavioral Analysis Strategy

This strategy involves carrying out a series of procedures to diagnose students' problematic behavior (in this case related to environmentally beneficial actions) in order to hypothesize about the causes of this behavior and suggest changes (Al-Shamri, 2019). The most important steps of this strategy are illustrated below in Figure 4.

Figure 4. Steps of functional behavioral change strategy

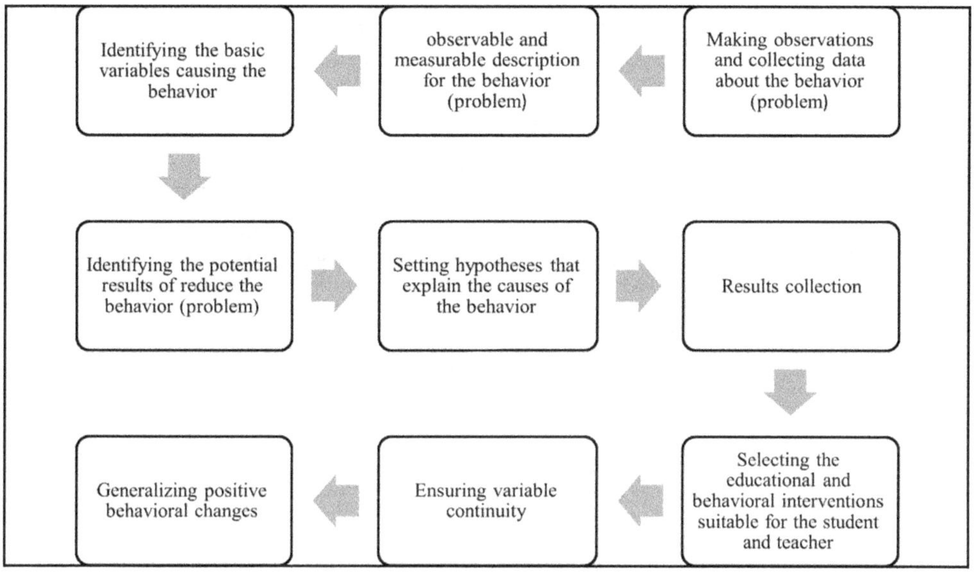

The teacher can use examples of the students' behaviors (use of plastic, throwing waste in the schoolyard, excess consumption etc.) and apply the previous steps to analyze students' behaviors and guide them to find appropriate alternatives.

3. Wall Construction

This is an active learning strategy which aims to help students work together. For this activity, the teacher asks his students to design a wall model, putting the most important actions at the bottom of the wall, and the less important ones nearer the top. This strategy can be adapted to address many issues related to carbon-neutral practices in the school environment, such as:

- Reducing paper consumption in the school.
- Promoting awareness of organic products.
- Planting trees around the school building.

4. Promoting Engagement in Environmental Activities

This strategy involves a teacher improving their students environmental awareness by following systematic steps (Table 2) that can be adapted depending on the situation) Ambusaidi et al., 2023):

Table 2. Steps to promote engagement in environmental activities

Steps	Procedure Example
Selection of Environmental Activity:	The teacher asks students to carry out a multi-stage environmental-awareness project or campaign. E.g., "Looking for the Environmental Thief" is a project that involves students searching for some sites in the school building to locate the 'environmental thief', which are sites that are sources of carbon emissions such as air conditioners, heating, buses and computers.
Data Collection:	The teacher asks students to gather information on the details of the environmental campaign through a) interviews with school stakeholders (staff, students), b) asking experts from the community, or c) using the Internet and the learning resource center to search for information.
Engagement in the Activity:	Students participate in the campaign. This can be done by announcing it on the school radio, conducting a peer training program to distribute tasks to the student team, creating media related to the promotion of their campaign (e.g., detecting the environmental thief), or any other activities that the teacher deems appropriate.
Assessment:	Students assess the impact of the campaign they participated in by making a survey or note card showing the degree of change in students' and teachers' behaviors in the school environment and propose some solutions (e.g., eliminating environmental thieves).

5. Claim, Evidence, and Reasoning Strategy (CER)

The teacher begins by asking students the question "How do we help the Sultanate of Oman reach net-zero carbon?". Students then answer the question in three phases (Gottschalk, 2021).

- Phase I: Claim: Student define their answers only, which must be concise and contain no in-depth explanations.
- Phase II; Evidence: Students demonstrate the validity of their first-stage answers by providing quantitative or qualitative evidence.
- Phase III: Reasoning: Students make use of questions like "why and how". At this stage, students interpret and analyze whatever conclusions are reached in the first and second stages.

Once students are familiar with the process, they can be given further questions, for example, "Here are some sources that help reduce carbon emissions. Using the CER strategy, think about how you can save your countries resources by creating an alternative energy plan, which includes energy efficient homes, electric cars, and renewable energy sources".

PERSPECTIVES FOR ACHIEVING CARBON NEUTRALITY

Based on the aforementioned information, a school can adopt several ideas to effectively implement carbon neutral practices within its premises. Some of these ideas include:

Zero Carbon Culture: An environmentally aware culture can be developed through scientific competitions related to carbon neutrality in the school environment. They can be done in the morning assembly or at any school events.

Carbon Officer: This idea involves choosing a student from the classroom on a weekly basis to be responsible for controlling the use of air conditioners, lighting, and waste. This will help raise awareness among peers of the importance of impactful behavior regarding the environment.

Zero Waste Campaign: This type of community campaign can be carried out in cooperation with government or private sector institutions, and hopefully involve the school's local community. The purpose is to strive for a waste-free society and enhance people's awareness about waste affecting the environment.

CONCLUSION

Given the significance of carbon neutrality to our future on this planet and based on the discussion in this chapter about suggested classroom activities and strategies, and successful local and global experiences, the authors suggest schools adopt the following recommendations to promote carbon neutrality practices within the school environment:

- Incorporate a specific carbon neutrality goal into the school's overall plan and actively work towards achieving it.
- Establish a permanent school group/committee dedicated to proactively achieving carbon neutrality in the school environment.
- Continuously monitor and maintain sources of school-related carbon emissions such as electrical appliances, water taps, and vehicles.
- Make concentrated efforts to raise awareness among all school staff and community members about the importance of reducing carbon emissions, adopting environmentally-friendly practices, and promoting sustainability.
- Foster partnerships with governmental and private community institutions to secure financial support and guidance, thus ensuring the sustainability of school practices and the attainment of carbon neutrality.

REFERENCES

About Resource Smart Schools. (2019). Available online: https://www.sustainability.vic.gov.au/School/About-ResourceSmartSchools

Ahmed, A. A., & Ahmed, A. E. A. A. (2021). Experimenting the strategy of climbing the plateau in teaching history to develop cognitive achievement, meta cognitive skills and mental alertness among first-grade secondary students. *Journal of Scientific Research in Education*, *22*(2), 431–461.

Al-Anbaki, A., & Al-Karkhi, A. (2015). The effect of the hill-climbing strategy on the achievement of fifth-grade literary students in modern and contemporary European history. *Diyala Journal*, (65), 343–371.

Alaa, M. (2021). Experimenting the Strategy of Climbing the Plateau in Teaching History to Develop Cognitive Achievement, Meta-cognitive Skills and Mental Alertness among First-grade Secondary Students. *Journal of Scientific Research in Education*, *22*(2), 431–461.

Ambusaidi, A., Al-Dairiya, H., & Al-Rashidi, H. (2023). *Sustainable Development in School Environment.* Al Maseera House for Publication and Distribution.

Bazzocchi, F., Ciacci, C., & Di Naso, V. (2021). Evaluation of environmental and economic sustainability for the building envelope of low-carbon schools. *Sustainability (Basel)*, *13*(4), 1702. doi:10.3390u13041702

Bolstad, R. (2021). Schools' Carbon Footprint Pilot Project: Final Report for Auckland Council. New Zealand Council for Educational Research.

Carbon Trust. (n.d.). *Carbon Neutral*. Retrieved from: https://www.carbontrust.com/what-we-do/assurance-and-labelling/carbon-neutral-verification

Chatterjee, L., & Reynolds, J. (2008). *Carbon Emissions from Schools: Where they arise and how to reduce them*. https://www.sdcommission.org.uk/data/files/publications/Publish_Schools_Carbon_Strategy.pdf

Dos Santos, E. B., da Costa Maynard, D., Zandonadi, R. P., Raposo, A., & Botelho, R. B. A. (2022). Sustainability recommendations and practices in school feeding: a systematic review. *Kompass Nutrition & Dietetics,* 2(2), 83-102. doi:10.1159/000526379

European Commission. (2019). *A European Green Deal.* Retrieved from https://ec.europa.eu/info/publications/communication-european-green-deal_en

European Commission. (2020). *EU climate action.* Retrieved from https://ec.europa.eu/clima/policies/strategies_en

Gamarra, A. R., Istrate, I. R., Herrera, I., Lago, C., Lizana, J., & Lechon, Y. (2017). Energy and water consumption and carbon footprint of school buildings in hot climate conditions. Results from life cycle assessment. *Journal of Cleaner Production, 195*, 1326–1337. doi:10.1016/j.jclepro.2018.05.153

Gottschalk, J. (2021). *Writing Strategies for Talent Development: From Struggling to Gifted Learners, Grades 3–8*. Routledge. doi:10.4324/9781003089247

Growcom, A. (2008). *Vegetable Industry Carbon Footprint Scoping Study*. https://www.daf.qld.gov.au/__data/assets/pdf_file/0003/59025/Hort-Fruit-Drought-Carbon-Report1.pdf

Hayles, C., & de la Harpe, B. (2010). Can Greening the Curriculum Influence Students' Perceptions and Awareness of Sustainability? In *W089-Special Track 18th CIB World Building Congress May 2010 Salford, United Kingdom* (p. 48). Academic Press.

Intergovernmental Panel on Climate Change (IPCC). (2018). *Global warming of 1.5°C*. Retrieved from https://www.ipcc.ch/sr15/download/

International Energy Agency (IEA). (2022). *Global Energy and Climate Model documentation*. Retrieved from https://iea.blob.core.windows.net/assets/2db1f4ab-85c0-4dd0-9a57-32e542556a49/GlobalEnergyandClimateModelDocumentation2022.pdf

Magacho, G., Espagne, E., Godin, A., Mantes, A., & Yilmaz, D. (2023). Macroeconomic exposure of developing economies to low-carbon transition. *World Development, 167*, 106231. doi:10.1016/j.worlddev.2023.106231

Microsoft. (2021). *Microsoft to be carbon negative by 2030*. Retrieved from https://blogs.microsoft.com/blog/2020/01/16/microsoft-will-be-carbon-negative-by-2030/

Ministry of Education. (2023). *Green Schools Project*. Sultanate of Oman Ministry of Energy and Minerals-Oman. Retrieved from https://www.mem.gov.om/en-US/Sectors/Electricity/Pages/National-Energy-Strategy.aspx

Ministry of Education, British Columbia. (n.d.). *Sustainable Schools: Best Practices Guide*. https://www2.gov.bc.ca/assets/gov/education/kindergarten-to-grade-12/teach/teaching-tools/environmental-learning/sustbestpractices.pdf

National Governance Association. (2022). *Environmental sustainability: a whole school approach A guide for governing boards*. https://www.nga.org.uk/getmedia/42a15743-7a17-4dfb-b15a-804e5f26b5a2/nga-environmental-sustainability-20211104.pdf

Odell, P., Rauland, V., & Murcia, K. (2020). Schools: An untapped opportunity for a carbon neutral future. *Sustainability, 13*(1), 1-24. doi:10.3390/su13010046

Oman Observer. (2019). *Oman inaugurates 500MW solar project in Ibri*. Retrieved from https://www.omanobserver.om/article/1070002/business/oman-inaugurates-500mw-solar-project-in-ibri

Öz-Aydin, S. (2016). The Awareness of Turkish High School Students about Carbon Footprint and the Effects of the High School Biology Curriculum on This Awareness. *International Electronic Journal of Environmental Education*, *6*(2), 29–53. doi:10.18497/iejee-green.56932

Rilla, C. (2013). *School garden design as catalyst for environmental education and community engagement: Los Angeles unified school district case studies*. University of Southern California.

Rodrigues, R., Pietzcker, R., Fragkos, P., Price, J., McDowall, W., Siskos, P., Fotiou, T., Luderer, G., & Capros, P. (2022). Narrative-driven alternative roads to achieve mid-century CO2 net neutrality in Europe. *Energy*, *239*, 121908. doi:10.1016/j.energy.2021.121908

Science-Based Targets initiative (SBTi). (2021). *Science-Based Targets: Scaling Urgent Corporate Climate Action Worldwide.* Retrieved from https://sciencebasedtargets.org/resources/files/SBTiProgressReport2021.pdf

Sen, G., Chau, H. W., Tariq, M. A. U. R., Muttil, N., & Ng, A. W. (2022). Achieving sustainability and carbon neutrality in higher education Institutions: A review. *Sustainability (Basel)*, *14*(1), 222. doi:10.3390u14010222

Shang, W.-L., & Lv, Z. (2023). Low carbon technology for carbon neutrality in sustainable cities: A survey. *Sustainable Cities and Society*, *92*, 104489. doi:10.1016/j.scs.2023.104489

Sharmin, S., Yabar, H., & Richards, D. (2023). Green energy optimization in dinajpur, Bangladesh: A path to net neutrality. *Sustainability (Basel)*, *15*(2), 1336. doi:10.3390u15021336

The United Nations Framework Convention on Climate Change (UNFCCC). (n.d.). *What is carbon neutrality?* Retrieved from https://enb.iisd.org/negotiations/un-framework-convention-climate-change-unfccc#:~:text=The%20UN%20Framework%20Convention%20on,interference%20with%20the%20climate%20system.%E2%80%9D

Tronchin, L., Manfren, M., & Nastasi, B. (2018). Energy efficiency, demand side management and energy storage technologies–A critical analysis of possible paths of integration in the built environment. *Renewable & Sustainable Energy Reviews*, *95*, 341–353. doi:10.1016/j.rser.2018.06.060

UK Government. (2020). *Ten Point Plan for a Green Industrial Revolution*. Retrieved from https://assets.publishing.service.gov.uk/government/uploads/system/uploads/attachment_data/file/936567/10_POINT_PLAN_BOOKLET.pdf

UNESCO. (2022). *Education for Sustainable Development: A Roadmap*. https://en.unesco.org/themes/education-sustainabledevelopment

Wei, Y. M., Chen, K., Kang, J. N., Chen, W., Wang, X. Y., & Zhang, X. (2022). Policy and management of carbon peaking and carbon neutrality: A literature review. *Engineering (Beijing), 14*, 52–63. doi:10.1016/j.eng.2021.12.018

Williams, J. H., DeBenedictis, A., Ghanadan, R., Mahone, A., Moore, J., Morrow, W. R. III, ... Torn, M. S. (2019). The technology path to deep greenhouse gas emissions cuts by 2050: The pivotal role of electricity. *Science*, *365*(6458), 71–76. doi:10.1126cience.1208365 PMID:22116030

World Bank. (2021). *CO2 emissions (kt).* Retrieved from https://data.worldbank.org/indicator/EN.ATM.CO2E.KT?locations=CN

Xinhua. (2020). *China aims to hit peak emissions before 2030, carbon neutrality before 2060.* Retrieved from http://www.xinhuanet.com/english/2020-09/23/c_139388712.htm

Zhang, J., Tong, Z., Ji, Z., Gong, Y., & Sun, Y. (2022). Effects of Climate Change Knowledge on Adolescents' Attitudes and Willingness to Participate in Carbon Neutrality Education. *International Journal of Environmental Research and Public Health, 19*(17), 10655. doi:10.3390/ijerph191710655

Chapter 2
Development of a Methodology for Educational Management Entailing Government, Economic Sectors, and Educational Institutions for Sustainable Development

María E. Raygoza L.
Universidad Politécnica de Baja California, Mexico

Roxana Jiménez-Sánchez
Universidad Politécnica de Baja California, Mexico

Jesús Heriberto Orduño-Osuna
https://orcid.org/0009-0004-4850-7481
Universidad Politécnica de Baja California, Mexico

Diego Ramon Bonilla G.
Universidad Autónoma de Baja California, Mexico

Abelardo Mercado-Herrera
Universidad Politécnica de Baja California, Mexico

Carlos Morales
Universidad Politécnica de Baja California, Mexico

Rafael Ortiz
Universidad Politécnica de Baja California, Mexico

Ivette Cota-Rivera
Universidad Politécnica de Baja California, Mexico

Guillermo Limón-Molina
https://orcid.org/0000-0002-5445-8928
Universidad Politécnica de Baja California, Mexico

Fabian N. N. Murrieta-Rico
https://orcid.org/0000-0001-9829-3013
Universidad Politécnica de Baja California, Mexico

DOI: 10.4018/978-1-6684-9601-5.ch002

ABSTRACT

Worldwide guidelines since the 1980s have developed mechanisms with national and international economic funds for scientific and technological development that reduce greenhouse gas emissions (GHG) and mitigate environmental impacts. The objective of this chapter is to analyze the political, socioeconomic, and regulatory characteristics of Mexico as a country with a developing economy, as well as the actions that have made the country go backward in the energy transition, to make a proposal that serves as a guide. Among the precedents have been the Montreal Protocol and the Kyoto Protocol, which have sought to mitigate the use of greenhouse gases in industrialized countries and involve developing countries through various mechanisms that encourage the reduction of GHG, such as carbon credits. The UN (United Nations) continues to work with cutting-edge initiatives and programs such as the sustainable development goals (SDGs).

INTRODUCTION

Since the late 19th century, there has been a significant increase in the amount of Greenhouse Gases emitted into the atmosphere, mainly caused by the global growth in energy consumption. The current era of globalization has exacerbated the problem of climate change, affecting all ecosystems. The overexploitation of natural resources, such as coal, oil, and natural gas, to meet the growing needs of the industrial and transportation sectors has caused ecological imbalances that must be addressed through a government strategy for sustainable development (UNECE, 2021).

The anthropogenic footprint has been the leading cause of the current environmental impact on the planet, and socioeconomically developed countries have implemented various strategies to mitigate the environmental impact of anthropogenic. The government of emerging economies is required to implement measures and relevant public policies on climate change and environmental education (Lin Zhang, 2022).

To address significant global challenges such as education, world leaders, through the Union Nations (UN), have developed Sustainable Development Goals (SDG), these goals integrate internationally significant indicators related to water, environment, education, etc. Emerging markets have not prioritized the UN's SDG for a guideline strategy focused on achieving these goals (United Nations, 2023). The two main challenges focused on are this chapter are environmental protection and quality education.

Based on SDG 4, which focuses on quality education, countries should focus their attention on public policies with the objective of integrating educational policies and the stakeholders involved.

Public and Education Policies are high on the agenda of governments around the world. The government is focused on the results and implications for economic prosperity and social citizenship (Ran, J., & Nedovic-Budic, Z, 2018).

This chapter aims to review and position public and educational policies as the central axis to drive environmental culture and sustainable development, taking Mexico as a country with an emerging economy as the central case of this study, and developing a general methodology for the integration of policies and other important instruments for a cultural transition that involves all economic sectors, Countries with emerging markets Mexico has addressed these issues at government plans and different educational levels, but the results have not been sufficient to monitor and meet the goals and strategies. Currently, sustainable development continues to be complex for government and society.

Referring to the problems currently detected in emerging markets countries, is a requirement of public policies that integrate the efforts of the different internal organizations, to issue better proposals for laws and guidelines that contribute to economic growth through the involvement of stakeholders. of the nation, so the unification of a methodology that serves as the basis for the integration of public policies is a tool that can help to improve management for sustainable development. The systems of integration of public policies with objectives for education are complex, from the perspective of social, cultural, and economic development, looking for a multidirectional interaction, good development of these policies with intelligent reforms proposes a base for economic development (OECD, 2018).

Bell and Stevenson propose a model to explain the multidirectional interactions between the development stage of a policy and its "implementation", in which the policy shapes educational institutions down to the local level, where external factors are included in the implementation process such as an integral part (Bell, L. and H. Stevenson, 2015). Authors such as Cairney and Weible, consider the practical formulation of public policies complex, as well as leading to their implementation due to the little integration of the axes benefited in conjunction with objectives not aligned with national policies (Cairney, P. and C. Weible, 2018).

A basis for national growth is to maintain hierarchical structures and integrate and align public policies with the educational policies of a country through methodological procedures to maintain order in internal agencies and other national structures, so public police management is a solid strategy for the functioning of a country (United Nations, 2018).

Currently, many countries have not issued sufficient government policies and laws that support reference frameworks for penalizing acts with environmental impacts caused by various productive sectors and generators of conventional energy use. In the educational system, the curricula require more relevant content for the development of sustainable technological innovation and awareness programs about caring for the environment, with strategies that involve initiatives to reduce GHG emissions and the environmental footprint from a global perspective.

Changes in society require joint work between the educational, governmental, and economic sectors. This cultural change towards the care and protection of the environment must begin from the initial educational levels to develop awareness on this issue, relevant procedures for the efficiency of natural resources, and laws and regulations that reinforce the educational strategies implemented, aligning a country towards sustainable development (International Labour Organization, 2021).

Due to the geopolitical circumstances of developing countries in environmental matters, it is pertinent to carry out reviews and comparisons with the educational ones of developed countries for the implementation of strategies and initiatives that include the management of public and educational policies, as well as a greater study.

The link between education, government, society, and industry to face climate change is the main measure to generate programs with international scope. Government institutions and all the actors involved must develop inclusive public policies for a timely intervention, addressing issues related to ecology, energy efficiency, renewable energy sources, climate change and the loss of ecosystems and water resources, with development programs aligned with social awareness and the culture of environmental protection, where the public and private sectors work together with government strategies.

BACKGROUND- INTEGRATION OF DOMESTIC POLICIES AND EDUCATIONAL POLICIES AS A REFERENCE FRAMEWORK FOR SUSTAINABLE DEVELOPMENT

The incorporation of a triad for the efficient management of public policies between public, private, and educational institutions, which collaborate with society and the government through programs that integrate internal institutional plans and programs with economic support, and the promotion of Research and Development (R&D) projects, and scholarships at all academic levels for the development of sustainable strategies and projects with established plans for monitoring and continuous improvement by government institutions in productive sectors would achieve a cultural transition that executes the strategies and goals delivered in the planning of public policies.

Governments, in conjunction with education, should focus efforts on establishing a transparent methodology for compliance with national indicators aligned with international ones through continuous training for all economic sectors. The integration of economic, environmental, and social issues should be considered when implementing the agenda and the SDGs. To do this effectively, greater inter-ministerial cooperation is needed. Horizontal and vertical policy integration tries to deal with the knowledge silos that form between national ministries and tries to bring ministries together to discuss policies and their potential effects on all aspects of sustainable development to find balanced solutions (Mulholland, 2018).

Public policies, due to their interaction, can be called vertical and horizontal policies, called "horizontal" if the interactions occur at the same level of government or "vertical" if they relate to two different levels of government (Saint-Germain, 2007). The 2010 ESDN (Economic, Social, Political, Natural/Environmental) report frames the vertical and horizontal integration of public policies as a development platform. Mexico, like other countries with emerging economies, has maintained a vertical political system, which implies leaving out from lobbying social actors and academics, shocking at not obtaining the desired percentage of development in government perspectives, so it is necessary to adjust towards a mixed system of public policies. Policies can be stratified by different techniques, by their topic of interest, their objective, as well as by their hierarchical structure. Table 1 shows their hierarchical division.

Table 1. Scheme of policies according to the hierarchical system that is disseminated and promulgated in different political contexts

Coerce	Horizontal Coordination	Vertical Coordination
Internal coordination within the government	Coordination between agencies or public sectors	coordination between subordinate agencies
External coordination between the government and the public and private sector	Coordination with civil society organizations and the private sector	Coordination: a) Upwards with international organizations b) Down with the local and regional government

Source: Christensen and Laegrein (2007)

The authors Christensen and Laegrein discuss "whole-of-government" initiatives as a reaction to the negative effects of New Public Management reforms such as structural devolution, "single-purpose organizations," and performance management but also as a reaction to a more insecure world. The authors examine what a "whole-of government" approach means and explore how this concept might be interpreted in analytical terms. The structural approach is contrasted with a cultural perspective (Christensen & Lægreid, 2007).

Mexico, as part of the countries affiliated with the UN, aligned itself with the indicators issued by this organization. This decree leads to the creation mentions the National Development Plan 2019-2024 of Mexico, within this is the commitment to promote sustainable development, considered an essential factor of well-being, incorporating the academic sector as an actor (Secretaría de Gobernación, 2019). Given the complexity of executing these goals, countries with emerging economies can take as an instrument the integration of public policies that alienate national economic objectives and budgets as well as the commitment of all stakeholders.

Figure 1 shows the place that Mexico is giving to education as one of the most relevant goals within the development plan. and its alignment with the UN 2030 agenda.

The foundation of the integration of public policies is based on the study of various authors who affirm the need for various instruments to work together. As there is agreement that policy instruments need to work together to best address complex public problems, policy instruments have also focused on identifying the best possible combinations of tools to achieve this goal. These combined instruments can produce more effective governance than when employees separately (Cejudo & Michel, 2019).

Figure 1. Hierarchy of education in the development plan of Mexico, with information based on the Regional Observatory for Development Planning

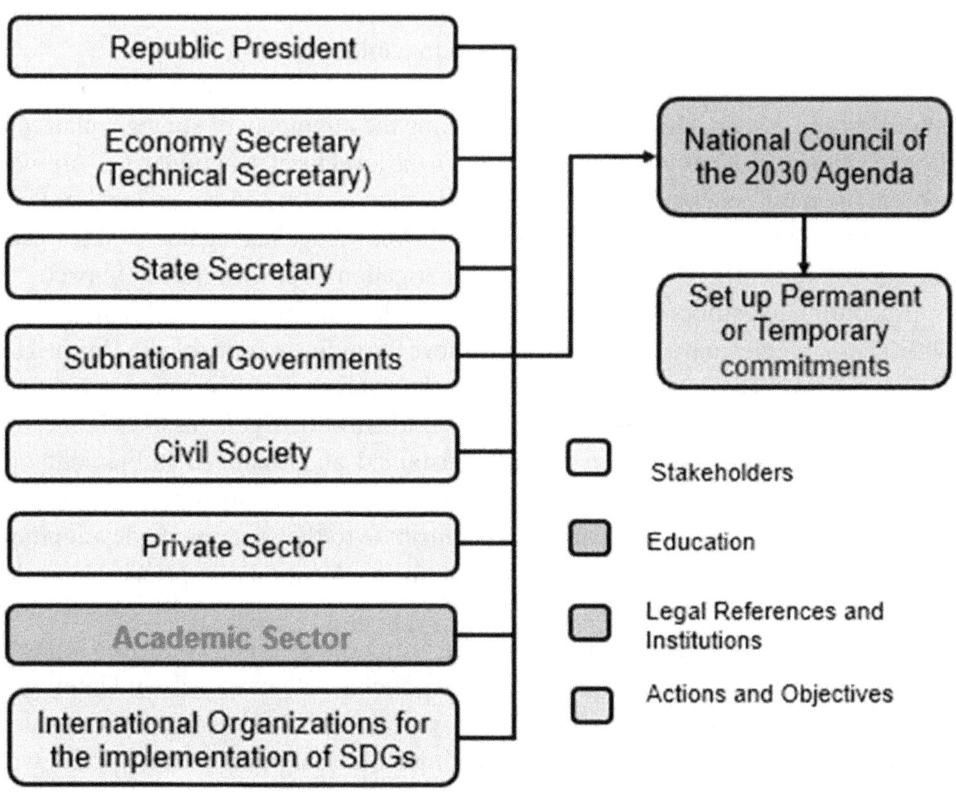

Since the instruments within a policy mix are not isolated from each other, but constantly interacting (Howlett, 2017), such interaction should lead them to complement and reinforce, rather than counteract Therefore, to be effective. The instruments of a policy mix must be coherent, which means that its components fit together because they share a set of ideas or objectives". In other words, a coherent set of instruments means that those who are part of the same mix can contribute, reinforce, or improve the chances of achieving their individual goals (Cejudo and Michel 2017).

Solving complex problems has long been a concern for governments and academics. Different theoretical approaches have been put forward to advance new ways to tackle them, such as network governance, joint government, and political integration (Candel, J., & Biesbroek, 2017). Common to these approaches is the underlying idea that, with different policy instruments.

Regarding quality education and the environment, governments must focus on pre-education levels to timely promote cultural change in what it implies to conserve the environment for sustainable development. The strategies for the achievement of the planned objectives must be articulated between all educational levels, taking advantage of the experience of universities and institutes with environmental research, ecology, and technological development. The commitment of industry, society, and government is essential for this sustainable transition.

As mentioned above, the indicators of the UN should be an axis to integrate quality education and sustainable development, seeking the alignment of strategic plans and the execution of projects for the countries from the level local to national. Government institutions must generate agreements that develop public and educational policies that prioritize pre-established global goals, with knowledge management that impacts social development, economic growth, and mitigation of environmental impact.

According to authors and experiences through quality educational management with local strategies and plans aligned to achieve the indicators established by the UN. This development must be simultaneous with the awareness of all economic, political, and social sectors, making educational interventions the main generators of innovative ideas in environmental matters, with a sustained and financed implementation through national and international projects, strategically contributing solutions to problems, and alternative measures that contribute to the sustainable development of emerging economies. Governments must place education as a priority in policy and practice (United Nations, 2023).

The relevance of studying education from preschool to higher education is based, focusing countries on public and educational policies with proposals and intelligent instruments such as the relationship with the private sector (The industry as a source of economic growth), the sources of financing and fiscal policies in countries with emerging economies, in which education is the basis of all changes, mainly the eradication of poverty and the promotion of sustainable development. UNESCO is the only UN agency with a mandate to cover all aspects of education and provide global leadership (UNESCO, 2023).

Quality education must be developed with the integration of the cognitive and the affective and the instructive and educational, as essential psychological and pedagogical requirements in the formation of their personality. This goal is universal, applicable to all countries and not just developing countries, and serves as a guide for a global transition towards sustainable development (UNESCO, 2022).

Examples of World Organizations and Institutions With a Focus on Sustainable Development and Education

The recognition of interdependencies, trade-offs, and synergies among the various goals, and their integration into policy design, is recognized as critical for going forward toward sustainable development. is important for policy to reflect links between education and other goals. The science-policy interface has a crucial role to play in this regard. One of the critical roles of the UN system is to provide a platform for the science-policy interface at the international level. Through their activities, specialized UN agencies, as well as UNDP (United Nations Development Program) and the World Bank provide a conduit for conveying evidence from science to decision-makers.

In Latin American countries, as emerging economies, with education with high vulnerability is reflected in the little cultural and scientific development, this lag in the classrooms contributes as a barrier to delay the goals of sustainable development, for this first, to achieve universal primary education, greater attention should be paid to excluded children, especially from rural areas, who have been left behind. The developed countries of the European Union managed strategies to establish ministries whose main goal is sustainable development. Table 2 shows some countries with ministries and institutions, as well as the topics these institutions assigned as a priority.

The institutions or ministries cited in Table 2 aim to facilitate the sharing of experiences, including successes, challenges, and lessons learned, with a view to accelerating the implementation of the agenda. Also seeks to strengthen policies and institutions of governments and to mobilize multi-stakeholder, support, and partnerships for the implementation of the SDG's.

Regarding the lessons, Mexico should take the success stories, specifically in education, with complementary actions such as the adoption of the European system mentioned in the previous table, where all the ministries are involved in the main objectives of the SDGs, this would be supported to develop benchmarking and efficiently migrate to sustainable development guidelines. Countries with emerging economies such as Mexico, India, and Brazil, have also promoted proposals and ministries for sustainable development. Table 3 shows some of these proposals and their current impacts.

Table 2. Information on the specific national ministry or institution that leads the implementation of 2030

Country	Ministry/Institution	Topic	Impact
Belgium	Federal Institute for Sustainable Development	Constitution establishes that every federal entity is pursuing the objectives of sustainable development, in its social, economic, and environmental dimensions	Fair access and equitable distribution of natural resources, beyond generations
Estonia	The Government Office of Estonia - Mainly the Ministry of Environment and Ministry of Foreign Affairs	All the ministries are involved via the inter-ministerial Working Group on Sustainable Development.	Formation of inter-ministerial committees on Climate Change and Biological Diversity. -National Council for Sustainable Development
Finland	Prime Minister's Office	The Ministry of the Environment and the Ministry for Foreign Affairs are also very much involved in implementation.	National Commission on Sustainable Development. - Roadmap with measures that Finland is taking to meet the objectives and defines a schedule to achieve them
Germany	Federal Chancellery	All ministries are closely involved, national ministries are actively involved: under the coordination of GSG, an "Interministerial Coordination Network for SDGs"	A future in which nature and climate are protected, fewer people suffer deprivation, and social integration is maintained

Source: Own development with information from ESDN (2018), (European Environment Agency, 2023), (European Environment Agency, 2023), (Government and ministries, 2023) & (Deutschland.de, 2023)

Table 3. Information on the specific national ministry or institution that leads the implementation of 2030

Country	Ministry/Institution	Topic	Impact
Mexico	Office of the federal executive power.	The federal government, in conjunction with state and municipal governments	Promote the development of people, communities, and cultures to achieve a global quality of life, health, and education.
Brazil	constitutional powers such as the legislative, with Congress, the executive power, with the Presidential Palace of the Planalto; and the judicial, with the Federal Supreme Court.	The Brazilian government has established a National SDG Commission including 32 representatives from civil society and the government	Promotes the development of people, communities, and cultures to achieve an adequate and equitable global level of quality of life, health, and education.
India	Federal Republic, with a president, Indian national parliament, and a prime minister.	Governments and local authorities	Reduction of poverty and food insecurity improvements in education, health, and infrastructure

Source: Own development with information from ESDN (2018), (Gobierno de Mexico, 2023), (Wbcsd, 2023) & (UN Sustainable Development Group, 2023)

It is important to mention that among the countries that invest the most financing in education and have the highest ranking in educational matters, are among others Japan, Denmark, Sweden, Belgium, and Finland, among others, so a benchmarking of their educational system implemented in these countries would be relevant for social, sustainable, and economic development.

The educational system in Latin American countries must have academic guidelines relevant to the social, cultural, economic, and geographical context, and fully train young people and adults, with the skills, abilities, and skills necessary to insert themselves favorably in the labor market, contribute to personal progress and the sustainable development of their communities and the country. The students in the classrooms know and competencies, in companies, linking educational theory and practice early and simultaneously Education will be the main driving force behind inclusive growth and will provide tools and practical knowledge for children and youth to adopt healthy and sustainable lifestyles, with social and environmental commitment.

Education in developing countries must seek to link all multidisciplinary areas of knowledge and stay at the forefront, for example, tools such as artificial intelligence (AI).

AI refers to the field of computer science that focuses on creating intelligent machines capable of performing tasks that typically require human intelligence. It involves the development of algorithms and models that enable machines to learn from data, reason, make decisions, and solve complex problems (Weng, 2015).

AI has made important contributions to the field of education; AI plays a vital role in tracking the mental steps of learners, such as self-regulation, monitoring, and explanation.

AI in the education sector aims to create a more personalized, effective, and inclusive learning environment that supports teachers and students by providing personalized content, adaptative learning paths, and oriented-content feedback, as well as providing accommodation and support to students with disabilities, speech recognition, text-to-speech tools, and translation services, providing equal opportunities for all learners Currently, (Luckin, 2016).

UNESCO in 2019 has published the Beijing Consensus on Artificial Intelligence and Education, this document proposes recommendations on how to obtain a better alignment of AI to the 2030 education agenda. The Beijing Consensus comes after the 2015 Qingdao Declaration, which stated that emerging technologies must be harnessed to strengthen education systems, through which UNESCO Member States committed to Effectively capitalize on emerging technologies towards the achievement of SDG 4 (UNESCO, 2019).

According to this consensus, the systematic integration of AI in education opens opportunities for innovative teaching and learning methods while effectively tackling significant educational challenges.

The economic development for emerging economies is currently among the biggest challenges, sustainability and development are apparently opposed, the current efforts to face the Planetary Crisis, technology and in particular Artificial Intelligence is little explored. Eco-innovation actions focus mainly on smart transport, smart use of energy and water, and waste recycling, but do not consider the necessary evolution of behaviors and approaches.

In sustainability issues, there are currently different applications, for a timely application of AI, it is necessary to seek to maintain an ethical involvement, development of public policies that promote its application in the branches of knowledge that support the environment and sustainability and reduce the carbon emissions with green projects.

The forums and congresses allowed for generating interest in the development of environmental education and established as a priority the regional exchange of educational experiences for the best use of the natural and cultural environment of the region, to build a new educational-environmental profile (Zabala, I., & García, M. 2008).

National Programs and Public Financing Initiatives That Contribute to the Sustainable Development of Mexico Within the Educational System

Developing countries, aligning themselves with international educational initiatives and policies, created programs for the development of science and technology, the main institutions and economic funds allocated to education that Mexico implemented are the following:

On December 27, 1970, the National Council of Science and Technology CONACYT program was created by the Law of the Congress of the Union, with reform in 1974, as a decentralized public agency of the federal government, forming part of the education sector (MEXICO, 2023), with the primary objective of consolidating the system to address the needs and requirements of the country, seeking to increase the quality of life, with the commitment to publicly invest in culture, science, technology, and innovation, CONACYT does not specifically focus on sustainability education, but it has financed projects focused on sustainable education.

The most important program restructuring for technological development in Mexico took place in 2023, involving CONAHCYT (from Spanish: Consejo Nacional de Humanidades, Ciencias y Tecnologías), with a focus on highlighting the social sciences as part of the holistic development of individuals.

- The Professional Development Program for Teachers PRODEP (from Spanish: Programa de Desarrollo Profesional para Docentes), aims to strengthen the quality of education in higher education institutions in the country. This program provides financial support for research projects and technological development related to sustainability and environmental education (PRODEP, 2023).

- The Faculty Improvement Program PROMEP (from Spanish: Programa de Mejoramiento del Profesorado) seeks to enhance teaching competencies and improve the quality of academic bodies, with the option to present projects related to sustainability (MEXICO, Faculty Improvement Program, PROMEP, 2023).
- Program for Incentives to Innovation and Research Development PEI (from Spanish: Proyecto Educativo Institucional), provides financial support for research and technological development projects in various areas, with a focus on linking them to businesses, aiming to generate diverse products, processes, and services that have positive impacts (CONAHCYT, 2023).
- The National Association of Universities and Higher Education Institutions ANUIES (from Spanish: Asociación Nacional de Universidades e Instituciones de Educación Superior) is a civil association that aims to promote continuous improvement in the services provided to the currently associated public and private institutions. It seeks recognition from the social sector and fosters collaboration, including academic exchange among the member institutions. This aims to have an impact on regional and national development, promoting integration and the advancement of the higher education system (SUPERIOR, 2023).
- The Ministry of Public Education SEP (from Spanish: Secretaría de Educación Publica) offers various programs, including the Federal Government's scholarship portal. These scholarships cover basic education, upper secondary education, higher education, postgraduate studies, culture, sports, and teaching. The intention is to expand opportunities for study, personal growth, and professional development, catering to the specific profiles of interested individuals. The General Directorate of International Relations (from Spanish: Dirección General de Relaciones Internacionales) of SEP (DGRI-SEP) aims to promote and enhance cooperation and exchange activities in the fields of science, culture, education, technology, and the arts with other countries. This fosters stronger ties with international organizations and ministries of education from other countries (SEP, 2023).

In addition to the aforementioned options, students and professionals in various fields have other international possibilities to further their development in their areas of interest. Consider the following alternatives.

- Funding from the government of the United Kingdom is provided for one-year master's degrees at any British university through the CHEVENING scholarships, focusing on topics such as energy, climate change, sustainable development, human rights, justice, and security, among others. These scholarships also involve research benefits for the British government.

- The German Academic Exchange Service (DAAD) serves as an organization that promotes the international exchange of students and scientists through scholarships funded by universities and student communities in Germany, thereby consolidating support for developing countries in university construction and contributions related to educational, scientific, and development policies.
- The OeAD agency for education and internationalization, working on behalf of the Federal Ministry of Education, Science, and Research, is committed to advising, promoting, and connecting students and educational institutions. It is an agency of the Republic of Austria.
- The Japan Student Services Organization (JASSO) provides scholarships for studying abroad in Japan.
- The La Caixa Foundation offers national and international scholarships for postgraduate studies in Europe, North America, and the Asia-Pacific region.
- Additionally, there are the Federal Supplemental Educational Opportunity Grants (FSEOG), which provide similar grants to the Pell Grants. These grants are funded by the school at 25%, while the federal government contributes 75% (GRANTS, 2023).

According to the previously mentioned national and international funding programs for continuous training scholarships at the undergraduate, master's, doctoral, and cultural levels, various advantages and significant opportunities are offered, including the following: Equity in terms of equal opportunities for any student who wishes to continue their development, regardless of their socioeconomic background, with the same quality education for all, fostering innovation and strengthening educational development in collaboration with the government, society, and industry.

These are not international funding programs and scholarships directly focused on sustainable development, but they allow exchange students to learn about programs and technologies from developed countries in the field of sustainable development and energy efficiency. This enables technological transfer to countries with emerging economies.

SYSTEM INTEGRATION OF THE ECONOMIC SECTORS, EDUCATION SYSTEM, AND GOVERNMENT FOR SUSTAINABLE DEVELOPMENT IN DEVELOPING COUNTRIES

Public policies for quality governance should be the basis for development, as the authors mentioned in the background state, with relevance and directing efforts to guidelines in conjunction with policy integration, these efforts should be directed to the development of R&D projects in the areas of renewable technologies and energy efficiency and the actor s involved, with education as basis for the development of various innovation projects regarding all areas of knowledge, for the environment (Garavita, 2008). Environmental education also entails practice in decision-making and self-formulation of a code of behavior about issues concerning environmental quality (UNESCO, 1990).

According to the UN guidelines, in reference to issues for sustainable development, education should be the main follow-up focus for government development plans of countries with emerging economies, where they teach, generate knowledge, technologies and raise awareness among institutions public and private to safeguard the environment with solid strategies of consumption, production, protection of marine and terrestrial ecosystems, where the main goal is not production for economic purposes but the quality and dignity of life, Figure 2 presents a diagram of the interactions of education with other Sustainable Development Goals (SDGs) based on a teaching-learning process and cultural change. Actions to achieve one of the SDGs have an impact on the others, along with an integration of public policies that will have a positive impact on their achievement.

The search for sustainability has been an important issue for public (education), private (industry), and governmental organizations, but at the same time is complex, sustainability based on the concept is organized into three categories: Economic, Environmental, and Social. Integrated management systems can be a way to induce sustainability, therefore the government must opt for the integration of models that contribute to sustainable development (Nadae, Carvalho, & Monteiro, 2019).

Educational plans and programs must be constantly updated, with educational goals based on human development, science, and technology, promoting a sustainability approach. The continuing education of the professional should not only be in universities but also keep the industrial sector involved as one of the main sources of waste and polluting emissions. The government and the industrial sector must link to work together to modernize and equip educational establishments of all types, levels, and educational modalities so that students are formed as free, responsible, active, and competitive individuals with R&D projects, pertinent to different economic sectors.

Sustainable development ought to mean the creation of a society and an economy that can come to terms with the life-support limits of the planet. The institutions of the social and economic order have yet to respond in any meaningful way, but a beginning has been made, and that beginning has brought with it some institutional changes, for these changes indicate both the nature of the response and the impediments to further progress. These early changes also clarify what further institutional adjustments will be desirable to continue the transition, what they might look like, and how they might be achieved.

Figure 2. Interaction of the main SDGs with the education guideline as the basis of sustainable development (own development)

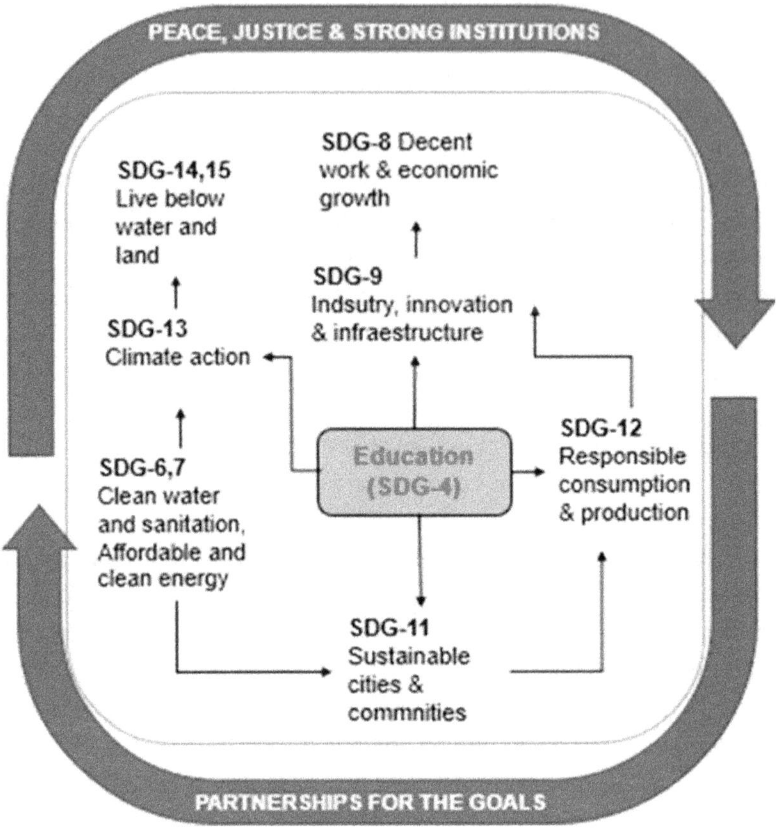

In the area of educational policy issues, different governmental and social elements must be integrated into their planning to consolidate a policy integration model, where educational, fiscal, and environmental policies and national and international funding sources for research and technological development are linked.

Inputs Required to Public Policy Development to Drive to Management System With Methodological Framework Integral

- Linking and alignment of relevant stakeholders in the issuance of public policies by corresponding topic current and previous development plan
- Methodological background for the preparation of the development plan
- Government program
- Sector plans, projects, policies, and programs derived from the development plan

- Annual budget
- Regulations related to the national planning system
- Existing mechanisms or instances of work between different public institutions and different levels of government

Adding to the points described in the previous list, AI must be incorporated into the agenda of the development plans of countries with emerging economies for the creation of public policies that promote the industry, universities, and research centers for the development of renewable technologies and energy efficiency, the use of this tool as support in educational programs, linking the various multidisciplinary areas, from data analysis to a projection of energy potential from the local, national, to the global level, to achieve a diversification of the energy matrix that integrates a greater percentage of friendly clean energy and reduces the current high consumption of fossil fuels. These projects must involve civil society, non-governmental institutions, and different economic sectors, mainly the industrial and transport sectors as the largest energy consumers. Universities, civil society, and industry in general must work together with interested parties to develop R&D projects that reduce the high GHG emissions and global warming generated by the generation of electricity and fuels, and industry (Sarpong & Derrick Boakye, 2023).

AI will transform business practices and industries and has the potential to address major societal problems, including sustainability (Mercier-Laurent, 2021). Degradation of the natural environment and the climate crisis are exceedingly complex phenomena requiring the most advanced and innovative solutions and support the derivation of culturally appropriate organizational processes and individual practices to reduce the natural resource and energy intensity of human activities. As the author Nishant mentions "the true value of AI will not be in how it enables individuals and society to reduce its energy, water, and land use intensities. Rather, the true value of AI will be realized at a higher level, in how it facilitates and fosters effective environmental governance" (Kumar, Rao, Kaiwartya, Kaiser, & Padmanaban, 2022).

Energy is the basis of economic development, so it must continue to grow in line with techniques such as AI, for efficient processes and to meet the goals of providing clean energy and satisfying energy demand using energy resources efficiently through optimal energy sizing in energy systems, with the drawbacks of the variable nature of energy resources. Anticipating lower energy costs with the application of hybrid systems for constant distribution and meeting the requirements of the growing global energy demand with the least negative impact on the environment; the electricity industry is one of the biggest polluters due to GHG emissions.

The debate on the effects on the economy continues, the analysis in this area must be from the perspective of intelligent decisions, social governance, work, innovation, and industry, among other relevant variables (Qin, Xu, Wang, & Skare,

2023). The development plans of educational programs and policies in conjunction with governments must manage, through fiscal policies, public and private, national, and international financing sources that promote the development of R&D projects linked to AI, in green technologies. To incubate high technology projects with the scope of technology transfer from developed countries.

As referenced in this chapter, quality education is essential for the international requirements and guidelines for sustainable development, in the current situation that occurs in developing countries where, although they have not yet reached the high levels of GHG emissions presented by developed countries due to the high consumption of electrical energy and industrial production, it is necessary to take pertinent and preventive measures in terms of the environment that go hand in hand with the socioeconomic development of countries with emerging economies. Based on the investigation of cases of countries with progress in sustainable development, this chapter proposes that governments implement models with integration of public policies, and fiscal and educational policies linking higher education institutions and industry, with the objective of reaching sources of financing. That promotes technological development with a socioeconomic and cultural vision for sustainable development. The Figure 3 presents a methodological proposal represented in an interactive diagram for a comprehensive policy system and stakeholder involvement, which begins with the linkage of the education system, civil society, and tax policy to obtain financing sources that are used for the development of R&D projects with a vision of sustainable development (Sarpong & Derrick Boakye, 2023).

In an era of economic globalization and computerization, public institutions around the world must provide a higher quality public service in response to citizen demands. However, different authorities have faced unprecedented challenges in meeting these expectations, contributing to growing public discontent and suspicion towards the public sector. Public management practitioners and scholars are increasingly realizing the need to develop innovative approaches and governance capacity to meet the difficult challenges posed by the increasing complexity of economic and political governance. Consequently, reform and innovation programs and innovative solutions. Such reforms and innovation are expected to lead to better governance (Burns & Zhiren, 2012). The Beijing consensus recommends that governments, in accordance with their legislation, public policies, and practices, consider implementing lessons in response to the opportunities and challenges related to education, the following are some of the most important recommendations:

Proposal of benefits obtained by efficient integration of public policies:

- Increase investment in sustainable development.
- Design sectoral and derivative programs based on the vision of the 2030 agenda.
- Guide the work of the State Councils of the 2030 Agenda.

- Accompany the legislatures for the creation of sustainable laws and strengthen the working Committees of the National Council of the 2030 Agenda.
- Maintain a territorial approach and implement it in the municipalities.
- Proposals for instruments, applications, and contributions for sustainable development using technological tools and algorithms with AI.
- Technological applications and AI algorithms proposed for public and educational policy initiatives that contribute to sustainable development.
    - Planning AI in education policies as decision-making.
    - AI empowers social governance.
    - Education.
    - AI for education management and delivery
    - Eco-innovation actions.
    - Smart transportation and fuels innovation.
    - Water, and waste recycling.
    - Smart use of energy, renewable energy grid, and stability and energy generation.
    - Reduction of energy generation, transportation, distribution, and power demand.
    - Dimensioning of renewable resources.
    - Reduction of global carbon (CO_2) emissions.

The educational system must educate in a way that is relevant to the social, cultural, economic, and geographical context, and comprehensively form personal progress for the sustainable development of communities and the country.

Public policies were "defined by their effectiveness and the level of integration and coherence with other policies". This concern with policy integration has forced practitioners and academics to rethink many underlying principles about this relatively old but little understood concept in political science. In practice, integrative strategies such as integration and coordination efforts were long thought to be capable of overcoming domain-specific constraints (Howlett, M. P. & Saguin, K. 2019). Along with these other elements proposed for the methodology of this chapter are green funds, fiscal policies along with environmental ones, as well as almost all environmental funds, which up to now have not managed to be effective or efficient policy tools. environmental (Grammk, 2021).

Good practices for integrating environmental public expenditure management and public finance into the public policy system can provide additional guidance for the reform of public environmental funds in emerging economies.

Following the guidelines proposed by authors as Grammk, Saguin, and Howlett regarding the integration of public policies, the following Figure 3 was prepared where it is proposed that countries with emerging economies implement in their development plans the integration and linkage between them policies such as: public, educational, fiscal policies for to drive towards sustainable development, using different instruments such as: grants, civil society, financing sources, green technologies, updating like other countries in energy efficiency and AI.

Figure 3. Proposal for an interaction diagram between public, educational, fiscal policies, stakeholders, and government tools for sustainable development (own development)

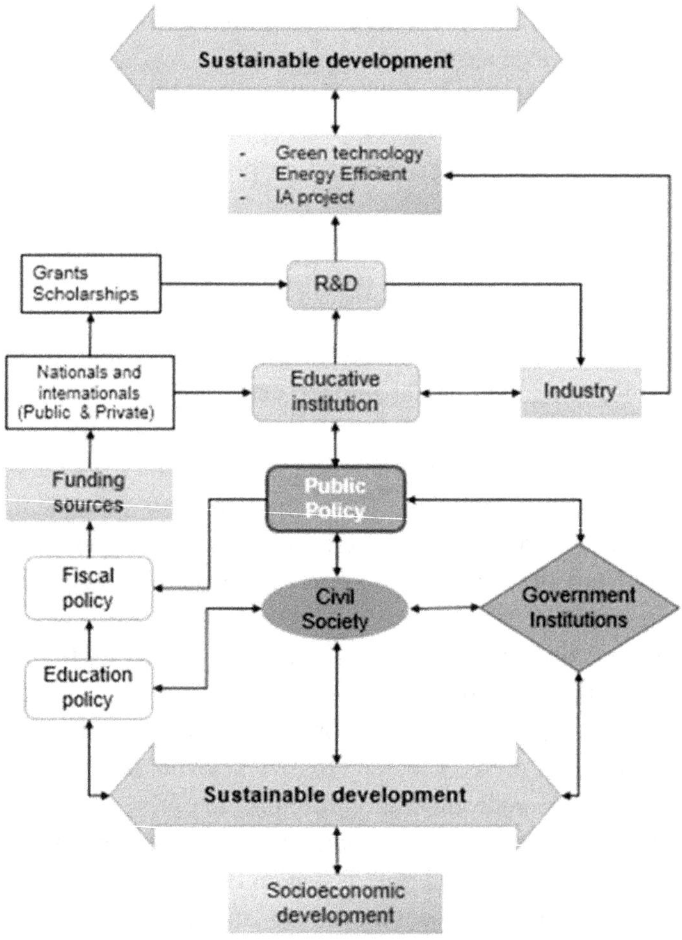

Explanation of the Methodology Based on the Interactive Diagram of Integration of Policies and Stakeholders

- Incorporation of a public policy model in the development plans of countries with emerging economies.
- Alignment of national and local goals with international goals
- Alignment of the internal organizations of the countries at the local to the national level, where the norms and permits that civil society collaborates within the elaboration of the national goals, seeking local economic development without leaving out the care of the environment.
- Development of public policies for sustainable development together with civil society migrate from a vertical public political structure to a mixed political structure.

Development of public policies elaborated together with the stakeholders where the educational institutions take leadership for the promotion of high technology projects in the environmental environment.

- Link between educational institutions, civil society, industry, and sources of financing for the development of joint R&D projects relevant to caring for the environment, the energy potential of localities for energy diversification, and transition to energy efficiency and renewable energy.
- Involvement of the quality government for the issuance of guidelines and programs that manage national and international financing sources for R&D projects with relevance in green technologies, energy efficiency, and AI.
- Evaluation and promulgation of fiscal policies that allow subsidizing and penalizing energy users, in order to use economic resources collected from educational institutions that develop R&D projects, under the scheme of scholarships and subsidies.
- Development of guidelines with transparency that can manage financing sources and subsidies focused on sustainable development.
- Issue regulations with tax penalties and generate economic funds that can be managed by research centers to promote R&D projects.
- Evaluate results of the integration of public policies on sustainable and socioeconomic development from a transparent and quality governance

The methodology integral system for sustainable development must have interaction between the integration of policies, government instruments, and grants, with equity and transparency of economic funds will provide a solid foundation for the development of countries with emerging economies. This methodological proposal of this chapter is based on arguments of authors mentioned in this chapter who affirm that policies should not be addressed separately as they were executed in previous political models, since the results show that they do not achieve the expected benefits or benefit the actors involved. The issue of integration of public policies is a proposal that several countries such as the United States of America, Germany, and others from the European Union are beginning to implement as a measure to obtain joint benefits.

CONCLUSION AND FUTURE RESEARCH DIRECTIONS

The current governance system of the same countries works in isolation without joint guidelines in programs of sustainable development, the main vector for sustainable development is the environmental market, globalization must be implicit in regional economic growth, and natural diversity for the energy matrix, to stop global warming that grows accelerated.

Government policies should implement recommendations of developed countries and make benchmarking them in national development plans with emerging economies, considering and adopting future renewable energy, designing the development plan, and policy through proposals for governance regulations and laws in matters of sustainability, from industry to rural areas. Evaluate if the goals, objectives, and contributions included in the national institutions are aligned with the internationals.

To establish guidelines for economic and sustainable development, the programs must be monitored by civil society and educational institutions with groundwork with local developers, with aims in an economic system compatible with sustainable development.

Quality education is the only solid foundation to achieve sustainable development, which contributes to the development of multidisciplinary knowledge, using modern tools such as AI, with the impetus of financing sources from private institutions and economic grants from developed countries. Educational programs must recognize the impact on the environment as one of the biggest current problems that affect all biodiversity and natural resources, such as the health of individuals, quality education programs promote and drive radical changes in the behavior of society will be achieved through the awareness of all the instances involved, such as public, private and government institutions, integrating objectives and values in the political and economic systems that reduce poverty and social lag.

To mitigate the global warming problem, changes are required to mitigate the current educational lag and global warming in conjunction with quality education, regulations on GHG emissions and soil and water pollutants must be established by governments and revolutionize the industrial and energy market as the main polluting sources.

Reversing the impact on the environment, as one of the biggest current world problems, must be born within the governments of all countries. A strategy that opens the possibility of economic and social development for countries with emerging economies is to implement a methodology that integrates policies from the preparation of the development plan of the governments.

REFERENCES

Bell, L., & Stevenson, H. (2015). Towards an analysis of the policies that shape public education: Setting the context for school leadership. *Management in Education*, *29*(4), 146–150. doi:10.1177/0892020614555593

Burns, J. P., & Zhiren, Z. (2012). Performance Management in the Government of the People's Republic of China: Accountability and Control in the Implementation of Public Policy. *OECD Journal on Budgeting*.

Cairney & Weible. (2018). Practical lessons from policy theories. *Policy and Politics*.

Candel, J., & Biesbroek, R. (2016). Toward a processual understanding of policy integration. *Policy Sciences*, *49*(3), 211–231. doi:10.100711077-016-9248-y

Cejudo, G. M., & Michel, C. L. (2019). *Instruments for policy integration*. Public Administration Department.

Christensen, T., & Lægreid, P. (2007). The Whole-of-Government Approach to Public Sector Reform. *JSTOR Collection,* 1059-1066.

CONAHCYT. (2023). *Consejo nacional de humanidades ciencia y tecnologias*. Obtenido de Consejo nacional de humanidades ciencia y tecnologias: https://conahcyt.mx/conahcyt/areas-del-conahcyt/uasr/desarrollo-regional/fondos/programa-de-estimulos-a-la-innovacion/

Deutschland.de. (2023, July 19). https://www.deutschland.de/es/desarrollo-sostenible

ECOES. (2023). *Espacio en comun de educacion superior*. Obtenido de http://www.ecoes.unam.mx/presentacion.html

European Environment Agency. (2023, July 19). *European environment agency*. Retrieved from European environment agency: https://www.eea.europa.eu/themes/sustainability-transitions/sustainable-development-goals-and-the/country-profiles/belgium-country-profile-sdgs-and

Gobierno de Mexico. (2023, July 19). Retrieved from Gobierno de Mexico: https://e.economia.gob.mx/guias/conocer-los-objetivos-de-desarrollo-sostenible-y-el-plan-de-acciones-en-mexico/

Government and Ministries. (2023, July 19). *Prime Minister's office*. Retrieved from Prime Minister's office: https://vnk.fi/en/sustainable-development

Gramk, C. (2021). *Green fiscal policies, An armoury of instruments to recover growth sustainably*. United Nations- ECLAC Office Brasilia.

Howlett, M., Vince, J., & del Rio, P. (2017). Policy integration and multi-level governance: Dealing with the vertical dimension of policy mix designs. *Politics and Governance*, *5*(2), 69–78. doi:10.17645/pag.v5i2.928

Howlett, M. P., & Saguin, K. (2019). *Policy capacity for policy integration: Implications for the sustainable development goals*. Lee Kuan Yew School of Public Policy Research Paper, (18-06).

International Labour Organization. (2021). *Guidelines for a just transition towards environmentally sustainable economies and societies for all*. ILO.

Kumar, K., Rao, R. S., Kaiwartya, O., Kaiser, S., & Padmanaban, S. (2022). *Sustainable Developments by Artificial Intelligence and Machine Learning for Renewable Energies*. Elsevier.

Luckin, R., Holmes, W., Griffiths, M., & Forcier, L. B. (2016). *Intelligence Unleashed: An argument for AI in education*. Academic Press.

Mercier-Laurent, E. (2021). Can Artificial Intelligence Effectively Support Sustainable Development? In *Artificial Intelligence for Knowledge Management: 8th IFIP WG 12.6 International Workshop, AI4KM 2021, Held at IJCAI 2020, Yokohama, Japan, January 7–8, 2021, Revised Selected Papers 8* (pp. 144-159). Springer International Publishing.

Mexico. (2023). *Creación del Consejo Nacional de Ciencia y Tecnología*. Obtenido de Creación del Consejo Nacional de Ciencia y Tecnología: https://www.gob.mx/epn/articulos/creacion-del-consejo-nacional-de-ciencia-y-tecnologia#:~:text=El%20Consejo%20Nacional%20de%20Ciencia,personalidad%20jur%C3%ADdica%20y%20patrimonio%20propio

Mexico. (2023). *Programa de Mejoramiento del Profesorado, PROMEP*. Obtenido de Programa de Mejoramiento del Profesorado, PROMEP: https://www.gob.mx/sep/documentos/programa-de-mejoramiento-del-profesorado-promep-eed

Mishra, B. K. (2023). *Handbook of Research on Applications of AI, Digital Twin, and Internet of Things for Sustainable Development*. IGI. doi:10.4018/978-1-6684-6821-0

Mulholland, E. (2018). *The Implementation of the 2030 Agenda and the SDGs in Europe: Overview and Updates*. ESDN.

Nadae, J., Carvalho, & Monteiro, M. (2019). Integrated management systems as a driver for sustainability: the review and analysis of the literature and the proposition of the conceptual framework. São Paulo: Associação Brasileira de Engenharia de Produção.

OECD. (2018). *Education policy implementation: A literature review and proposed framework*. Directorate for Education and Skills.

PRODEP. (2023). *Programa para el Desarrollo Profesional Docente, para el Tipo Medio Superior*. Obtenido de Programa para el Desarrollo Profesional Docente, para el Tipo Medio Superior: http://prodep.desarrolloprofesionaldocente.sems.gob.mx/

Qin, Y., Xu, Z., Wang, X., & Skare, M. (2023). Artificial Intelligence and Economic Development: An Evolutionary Investigation and Systematic Review. *Journal of the Knowledge Economy*, 214–219. doi:10.100713132-023-01183-2

Ran, J., & Nedovic-Budic, Z. (2018). Designing an Information Infrastructure for Policy Integration of Spatial Planning and Flood Risk Management. *International Journal of E-Planning Research*, *7*(1), 53–85. doi:10.4018/IJEPR.2018010104

Republic, Ministry of the Environment of the Czech. (2023, July 19). *Ministry of the Environment of the Czech Republic*. Retrieved from Ministry of the Environment of the Czech Republic: https://www.mzp.cz/en

Saint-Germain, P. M. (2007). *Faculty Integrating Technology*. Retrieved from home.csulb.edu: https://home.csulb.edu/~msaintg/ppa590/intro.htm

Secretaria de Gobernación. (2019). *Plan Nacional De Desarrollo 2019-2024*. Presidencia de la Republica.

SEP. (2023). *Secretaria de Educcion Publica*. Obtenido de https://becas.sep.gob.mx/

Superior. (2023). *ANUIES*. Obtenido de ANUIES: http://www.anuies.mx/anuies/acerca-de-la-anuies/mision-vision-y-objetivos-estrategicos

Tlili, A. H., Huang, R., Chang, T.-W., Nascimbeni, F., & Burgos, D. (2019). Open educational resources and practices in China: A systematic literature review. *Sustainability (Basel)*, *11*(18), 4867. doi:10.3390u11184867

UN Sustainable Development Group. (2023, July 19). *India*. Retrieved from India: https://unsdg.un.org/es/un-in-action/india

UNECE. (2021). *Sustainable Use of Natural Resources*. United Nations.

UNESCO. (2019). Retrieved from Artificial intelligence in education: https://www.unesco.org/en/digital-education/artificial-intelligence

UNESCO. (2021). *Interagency Consultation Meeting on Co-operation and Co-ordination in Environmental Education, 3rd, Paris, 1990. Contacto, 4*.

UNESCO. (2022). Education in Latin America and the Caribbean at a crossroads, Regional monitoring report SDG4 - Education 2030. France. UNESCO.

UNESCO. (2023). *Education transforms lives*. Retrieved from UNESCO EDUCATION: https://www.unesco.org/en/education

United Nations. (2018). *Working together: Integration, institutions and the sustainable development goals*. Department of Economic and Social Affairs.

United Nations. (2023a). *Sustainable Development Goals*. Retrieved from United Nations Web site: https://www.un.org/sustainabledevelopment/education/

United Nations. (2023b). *Sustainable Development Goals: 17 Goals to Transform our World*. Retrieved from https://www.un.org/en/exhibits/page/sdgs-17-goals-transform-world

Wbcsd. (2023, July 19). *Brazil - CEBDS*. Retrieved from Brazil - CEBDS: https://sdghub.com/brazil-cebds/

Weng, J. (2015). Brain as an Emergent Finite Automaton: A Theory and Three Theorems. *International Journal of Intelligence Science*, *5*(2), 112–131. doi:10.4236/ijis.2015.52011

Zabala, I., & García, M. (2008). Historia de la Educación Ambiental desde su discusión y análisis en los congresos internacionales. *Revista de Investigacion*, *32*(63), 201–218.

Zhang, M. X. (2022). Globalization, Green Economy and Environmental Challenges: State of the Art Review for Practical Implications. *Frontiers in Environmental Science, 10*.

ADDITIONAL READING

Dearlove, J. (2011). *The politics of policy in local government: the making and maintenance of public policy in the Royal Borough of Kensington and Chelsea*. Cambridge University Press.

Enríquez-Díaz, J., Castro-Santos, L., & Puime-Guillén, F. (Eds.). (2021). *Financial management and risk analysis strategies for business sustainability*. IGI Global. doi:10.4018/978-1-7998-7634-2

Fields, Z., & Atiku, S. O. (2019). Collective green creativity and eco-innovation as key drivers of sustainable business solutions in organizations. In Green Business: Concepts, Methodologies, Tools, and Applications (pp. 415-439). IGI Global. doi:10.4018/978-1-5225-7915-1.ch022

Kumar, K., Rao, R. S., Kaiwartya, O., Kaiser, S., & Padmanaban, S. (Eds.). (2022). *Sustainable developments by artificial intelligence and machine learning for renewable energies*. Academic Press.

Mishra, B. K. (2023). *Handbook of Research on Applications of AI, Digital Twin, and Internet of Things for Sustainable Development*. IGI. doi:10.4018/978-1-6684-6821-0

Palmer, J. (2002). *Environmental education in the 21st century: Theory, practice, progress, and promise*. Routledge. doi:10.4324/9780203012659

Sarpong, D., Boakye, D., Ofosu, G., & Botchie, D. (2023). The three pointers of research and development (R&D) for growth-boosting sustainable innovation system. *Technovation*, *122*, 102581. doi:10.1016/j.technovation.2022.102581

Schwachula, A. (2019). Sustainable Development in Science Policy-Making. In Sustainable Development in Science Policy-Making. Transcript Verlag.

Tripathi, S. L., Dubey, M. K., Rishiwal, V., & Padmanaban, S. (Eds.). (2021). *Introduction to AI Techniques for Renewable Energy System*. CRC Press. doi:10.1201/9781003104445

Weng, J. (2015). *Brain as an emergent finite automaton: A theory and three theorems. International Journal of Intelligence Science*. Springer.

KEY TERMS AND DEFINITIONS

Anthropogenic: Is referring to environmental change caused or influenced by people, either directly or indirectly.

Artificial Intelligence: This is the branch of computer science that deals with the simulation of intelligent behavior in computers. It involves the study and development of intelligent agents capable of perceiving their environment, reasoning, learning, and taking appropriate actions to achieve specific goals.

Civil Society: The connections that exist between different individuals and institutions. A narrower concept of civil society would be the set of organizations that provide the infrastructure for those connections, from charities to trade to social enterprises.

Energy Market: This is the combination of diverse primary energy sources used to satisfy the energy needs in a geographic region or for a system and complex systems that influence and are affected by various sectors such as transportation, industry, electric power, and heat, and fuel products.

Environmental Education: This is the process of educating our citizens about matters and issues that affect our environment - our homes, villages, citizens, farmlands, water supplies, forests, and weather in order to improve our environmental awareness, and skills and engender a change of attitude towards our environment.

Fiscal Policy: Describes changes to government spending and revenue behavior in an effort to influence the economy. By adjusting its level of spending and tax revenue, the government can affect economic outcomes by either increasing or decreasing economic activity.

Grant: The grant data is collected at scheme programmers and awards level where the scheme is related to an overarching policy area and grant award is given out to meet the objectives of the grant scheme.

Renewable Energy: Energy from renewable resources that are naturally replenished on a human timescale, examples of renewable energy sources include wind power, solar power, and bioenergy.

Research and Development (R&D): This is creative and systematic work undertaken in order to increase the stock of knowledge including knowledge of humankind, culture, and society, and to devise new applications of available knowledge.

Sustainable Development: Development that meets the needs of the present without compromising the ability of future generations to meet their own needs.

Chapter 3
Empowering Educators:
Training and Professional Development for Sustainability

Preksha Yadav
https://orcid.org/0000-0001-6316-8837
JAIN University (Deemed), Bangalore, India

Sonia Singh
JAIN University (Deemed), Bangalore, India

ABSTRACT

Higher education plays a crucial role in addressing sustainability challenges, enabling educators to equip students with the necessary knowledge, skills, and attitudes. However, many educators lack adequate training in teaching sustainability. This study used a quantitative approach to explore the impact of training on educators' ability to teach sustainability. A survey of 170 educators assessed their sustainability-related knowledge, skills, and attitudes before and after a training program. Results revealed that the training program positively influenced educators' knowledge, skills, and attitudes. Participants scored significantly higher on sustainability knowledge tests than non-participants. Additionally, the training bolstered educators' self-reported confidence in teaching sustainability. These findings underscore the effectiveness of training and professional development in empowering higher education educators to teach sustainability. Educators can better equip students to tackle sustainability challenges effectively.

DOI: 10.4018/978-1-6684-9601-5.ch003

INTRODUCTION

Background and Rationale

The world is facing a number of sustainability challenges, including climate change, environmental degradation, and social inequality. Higher education has a critical role to play in addressing these challenges by preparing students with the knowledge, skills, and attitudes they need to be part of the solution.

One of the most important ways that higher education can contribute to sustainability is by providing training and professional development for educators. This training should help educators to develop the knowledge and skills they need to effectively teach sustainability. It should also help them to feel empowered to create a more sustainable future.

There is a growing body of research on the importance of training and professional development for sustainability educators. For example, a study by Tilbury (2007) found that educators who had received training in sustainability were more likely to integrate sustainability into their teaching. A study by Wals (2007) found that professional development programs can help educators to develop a more holistic view of sustainability. And a study by Sterling (2001) found that training and professional development can help educators to become more critical thinkers and to engage students in more meaningful learning.

Education plays a pivotal role in fostering a sustainable future by empowering individuals to make informed decisions and take responsible actions. As the frontline facilitators, educators hold the key to shaping the mindset and behaviors of the next generation towards sustainability. However, research suggests that many educators lack the necessary training and professional development opportunities to effectively incorporate sustainability concepts into their teaching practices. According to the UN Decade of Education for Sustainable Development (DESD, 2005-2014), the current educational landscape faces several obstacles to the effective integration of sustainability education. Firstly, traditional teacher training programs often do not prioritize sustainability, resulting in a knowledge gap among educators about its importance and practical implementation. Secondly, there is a lack of standardized and comprehensive sustainability education guidelines and curricula, leading to inconsistencies in how sustainability is taught across different schools and regions. Furthermore, the lack of capacity building and training in the sense that professional development needs, at all levels including leadership needs to be addressed and focus on building upon existing actors and expertise including NGOs, private sector, and civil society needs to be emphasized.

Rationale

Empowering educators with adequate training and professional development in sustainability will bridge the knowledge gap and equip them with the necessary tools and resources to effectively teach sustainability concepts. This will enhance their confidence in addressing sustainability issues in the classroom and allow them to act as role models for sustainable practices. Effective sustainability training can help educators develop a deep understanding of the interconnectedness between ecological, social, and economic systems. This, in turn, will enable them to foster critical thinking and systems-oriented approaches in students, encouraging the development of sustainable mindsets from an early age. (Weik et al., 2011)

Well-trained educators can create interactive and engaging sustainability-focused lessons, inspiring students to become active participants in sustainability initiatives within and beyond the school environment. By instilling a sense of ownership and responsibility, students are more likely to apply sustainable practices in their daily lives. Empowering educators with sustainability focused training will promote the integration of sustainability principles across various subjects, not just limited to environmental studies. This interdisciplinary approach allows students to recognize the relevance of sustainability in real-world scenarios and fosters a holistic understanding of its applications. (Jones et al., 2010)

A school with sustainability-conscious educators can develop a culture that values environmental stewardship, community engagement, and ethical decision-making. This positive school culture will influence not only students but also staff, parents, and the wider community. Empowering educators with sustainability training aligns with international efforts to achieve sustainability goals, such as the United Nations' Sustainable Development Goals (SDGs). By investing in educators, societies can accelerate progress towards a more sustainable and equitable future. (UNESCO, 2017).

Benefits of Empowering Higher Education Educators

There are a number of benefits to empowering higher education educators to teach sustainability. UNESCO (2023) defines education for sustainable development as an education that equips learners with the knowledge, skills, values, and behaviors to contribute to sustainable development.

First, it can help to improve the quality of education. When educators are trained and supported in incorporating sustainability into their teaching, they are better able to help students understand the challenges of sustainability and develop the skills they need to be part of the solution.

Second, empowering educators can help to create a more sustainable campus. When educators are aware of sustainability issues and how they can be addressed, they can make changes to their own practices that can have a positive impact on the environment. They can also encourage students to make similar changes.

Third, empowering educators can help to raise awareness of sustainability issues in the community. When educators talk about sustainability in their classes and research, they help to educate others about the importance of sustainability. This can lead to more people taking action to address sustainability challenges.

Research Objectives

The primary purpose of this research paper is to examine the significance of empowering higher education educators in sustainability. The objectives are as follows:

1. To explore the role of educators in promoting sustainability within higher education.
2. To assess the need for specialized training and professional development in sustainability.
3. To identify best practices in sustainability-focused educator training.
4. To analyze the impact of empowered educators on students' sustainability knowledge, attitudes, and behavior.

LITERATURE REVIEW

The world is facing a number of sustainability challenges, including climate change, environmental degradation, and social inequality. Higher education has a critical role to play in addressing these challenges by preparing students with the knowledge, skills, and attitudes they need to be part of the solution.

There is a growing body of literature on the impact of training and professional development on educators' ability to teach sustainability. A 2017 study by the Sustainability Consortium found that educators who participated in a training program on sustainability were more likely to include sustainability content in their teaching (The Sustainability Consortium, 2017). The study also found that educators who participated in the training program reported feeling more confident in their ability to teach sustainability.

Another study, conducted by the National Wildlife Federation, found that educators who participated in a professional development program on sustainability were more likely to incorporate sustainability into their teaching and to use active learning methods (National Wildlife Federation, 2017). The study also found that educators who participated in the professional development program reported feeling more confident in their ability to teach sustainability.

The Need for Training and Professional Development for Sustainability Educators

There are a number of reasons why training and professional development are essential for sustainability educators. First, sustainability is a complex and multifaceted issue. It requires educators to have a deep understanding of the different aspects of sustainability, such as environmental science, economics, social justice, and ethics (Tilbury, 2007; Wals, 2007; Sterling, 2001).

Second, sustainability education often requires a cross-disciplinary approach. This means that educators need to be able to integrate sustainability into their teaching across a variety of disciplines (Wals, 2007). For example, an economics professor might teach a course on sustainable development that integrates economic principles with environmental and social justice concerns.

Third, sustainability education can be controversial. There are often different viewpoints on how to address sustainability challenges. This can make it difficult for educators to present a balanced and objective view of sustainability (Sterling, 2001). For example, a professor of environmental studies might be teaching a course on climate change and there are different viewpoints on the causes and solutions to climate change.

Specialized training and professional development for educators in sustainability are essential for several other reasons:

- **Promoting Systems Thinking:**

Sustainability education requires a systems thinking approach, where educators can help students analyze and understand the interconnections and interdependencies between different components of complex systems. Specialized training can enhance educators' ability to facilitate systems thinking among students, fostering a holistic understanding of sustainability challenges (Wiek et al., 2011).

- **Experiential and Active Learning**

Specialized training enables educators to develop and implement experiential and active learning methodologies that engage students in real-world sustainability challenges (Brown et al.,2010). These approaches empower students to become active agents of change and apply their knowledge to address sustainability issues in practical ways.

- **Aligning with Global Initiatives**

Many countries and international organizations have endorsed sustainability-focused education as a critical component of addressing global challenges. Specialized training ensures that educators align their teaching practices with these initiatives, contributing to broader sustainability goals (Ohman, 2008)

- **Empowering Educators as Change Agents**

Educators play a crucial role in shaping the values and attitudes of the next generation. Specialized training empowers educators to become change agents, promoting sustainability principles not only in their classrooms but also in their educational communities and beyond (Mezirow, 2000).

- **Keeping Up with Evolving Research and Practices**

Sustainability is a rapidly evolving field with ongoing research and innovative practices. Specialized training provides educators with up-to-date knowledge and resources, allowing them to stay informed about the latest developments in sustainability education (Chalkley, 2006).

- **Enhancing Student Learning Outcomes**

According to Sterling (2001) and Burns (2011), effective sustainability education can positively impact students' critical thinking, problem-solving, and decision-making skills. Specialized training ensures that educators can design and implement instructional strategies that enhance student learning outcomes related to sustainability.

- **Inspiring Future Leaders**

Specialized training in sustainability empowers educators to inspire and motivate their students to become future leaders and change-makers in sustainability, driving positive environmental, social, and economic impact (Scott et al., 2012).

Different Types of Training and Professional Development for Sustainability Educators

There are a variety of different types of training and professional development that can be offered to sustainability educators. Some common types of training and professional development include:

- **Sustainability Courses and Workshops**

These courses and workshops provide educators with the knowledge and skills they need to teach sustainability. For example, the University of California, Berkeley offers a number of sustainability courses and workshops for educators, such as "Sustainability in Higher Education" and "Teaching Sustainability" (Tilbury, 2007).

- **Mentorship Programs**

Mentorship programs pair experienced sustainability educators with new sustainability educators. This allows the new educators to learn from the experienced educators and get support as they develop their sustainability teaching skills. For example, the Sustainability Institute offers a mentorship program for sustainability educators, called the Sustainability Educator Mentorship Program (Wals, 2007).

- **Research Opportunities**

Research opportunities allow educators to explore sustainability issues in more depth. This can help them gain a deeper understanding of sustainability and develop new teaching materials. For example, the Sustainable Development Institute at the University of New Hampshire offers a number of research opportunities for sustainability educators, such as the Sustainable Development Research Fellowship (Sterling, 2001).

- **Online Resources**

There are a number of online resources available to sustainability educators. These resources can provide educators with information on sustainability, teaching methods, and professional development opportunities. For example, the Association for the Advancement of Sustainability in Higher Education (AASHE) offers a number of online resources for sustainability educators.

The Importance of Empowering Sustainability Educators

Empowering sustainability educators is essential for the success of sustainability education. When educators feel empowered, they are more likely to be confident in their teaching abilities and to be able to effectively engage students in sustainability learning (Tilbury, 2007).

There are a number of ways to empower sustainability educators, such as:

- Providing them with opportunities to learn from each other: This can be done through networking events, professional development workshops, and online forums. For example, the Association for the Advancement of Sustainability in Higher Education (AASHE) hosts a number of networking events and professional development workshops for sustainability educators.
- Giving them the freedom to experiment with different teaching methods: This allows educators to find teaching methods that work best for them and their students. For example, the Sustainability Institute offers a number of resources for educators on how to use different teaching methods in sustainability education, such as the Teaching Sustainability Toolkit.
- Providing them with the resources they need: This includes providing them with access to sustainability materials, funding for sustainability projects, and time to plan and implement sustainability education. For example, the Sustainable Development Institute at the University of New Hampshire offers a number of resources for sustainability educators, such as the Sustainability Education Resource Center.

Role of Educators in Advancing Sustainability

The role of educators in advancing sustainability is crucial for creating a more sustainable future. Educators play a vital role in shaping the attitudes, behaviors, and values of students towards environmental, social, and economic sustainability.

- **Fostering Sustainability Literacy**

Research has emphasized on the importance of educators in fostering sustainability literacy among students with respect to the effectiveness of incorporating sustainability-related content and themes across various subjects in the curriculum. Educators who integrate sustainability concepts into their teaching practices can promote critical thinking, problem-solving, and systems thinking skills that are essential for understanding and addressing sustainability challenges (Boeve-de Pauw et al., 2021).

- **Cultivating Values and Attitudes Towards Sustainability**

Efforts have been made to understand the role of educators in cultivating values and attitudes towards sustainability. Tilbury & Mochizuki (2021) explored the influence of educators' personal beliefs, commitment to sustainability, and role modeling behaviors on students' sustainability-related attitudes and behaviors. They highlighted the significance of educators' pedagogical approaches in fostering pro-environmental attitudes and ethical decision-making.

- **Promoting Experiential Learning and Active Engagement**

Stern & Powell (2022) found that the educators who incorporate hands-on experiences, outdoor learning, and service-learning projects can enhance students' connection to nature, sense of place, and responsibility towards the environment. Active engagement in sustainability initiatives empowers students to become change agents and active participants in promoting sustainable practices.

- **Educator Professional Development for Sustainability**

Corcoran et al., (2021) have explored the impact of sustainability-focused training, workshops, and networking opportunities on educators' capacity to address sustainability topics effectively. They found that Professional development helps equip educators with the knowledge, skills, and confidence to integrate sustainability into their teaching practices.

- **Embedding Sustainability in Educational Institutions**

Barth et al., (2021) in their study explored the role of leadership, curriculum development, institutional policies, and partnerships in promoting sustainability initiatives within schools and universities. Furthermore, studies have examined the influence of sustainable campus practices and organizational culture on advancing sustainability education

- **Interdisciplinary Collaboration and Systems Thinking**

Recent studies have focused on the importance of interdisciplinary collaboration and systems thinking in advancing sustainability. Educators who collaborate across disciplines can provide holistic perspectives on sustainability challenges and solutions. Integrating systems thinking in the curriculum can help students understand the interconnectedness of social, economic, and environmental issues (Stibbe, 2021).

How empowered educators can create an impact on students' sustainability knowledge, attitudes, and behavior?

The impact of empowered educators on students' sustainability knowledge, attitudes, and behavior can be significant and far-reaching. When educators are equipped with specialized training and professional development in sustainability, they can play a crucial role in shaping the sustainability outcomes of their students in the following ways:

- **Enhanced Sustainability Knowledge**

Empowered educators possess a deeper understanding of sustainability concepts and principles. They can effectively convey complex sustainability topics to students, leading to increased sustainability knowledge and awareness among learners. Students exposed to knowledgeable educators are more likely to comprehend the interconnectedness of environmental, social, and economic issues and understand the importance of sustainable practices (Blake et al., 2013).

- **Cultivation of Pro-Sustainability Attitudes**

By demonstrating their commitment to sustainability and modeling sustainable behaviors, educators influence students' perceptions and attitudes towards sustainability. Positive role modeling and passionate teaching can inspire students to adopt environmentally responsible attitudes and ethical decision-making, which can extend beyond the classroom setting (Sleurs, 2008).

- **Promotion of Sustainable Behavior**

Students can be engaged in experiential learning and active participation in sustainability initiatives. Through hands-on projects, field experiences, and service-learning opportunities, students are encouraged to apply their knowledge to real-world sustainability challenges. This active engagement can motivate students to adopt sustainable behaviors and take actions that contribute to a more sustainable future (Leal, 2010).

- **Development of Critical Thinking and Problem-Solving Skills**

Empowered educators use innovative and participatory teaching approaches that encourage critical thinking and problem-solving skills. They challenge students to analyze sustainability issues from multiple perspectives and propose creative solutions. This approach nurtures students' ability to think critically about complex sustainability challenges and empowers them to become informed decision-makers (Thomas, 2009).

- **Increased Environmental and Social Awareness**

Educators who are knowledgeable about sustainability issues can raise students' awareness of environmental and social concerns. Students gain insights into the impacts of human activities on the planet and society and become more conscious of their roles as global citizens. They can also guide discussions on ethical dilemmas related to sustainability, encouraging students to develop empathy and concern for social and environmental justice (Dale & Newman, 2005).

- **Inspiration and Leadership Development**

Empowered educators serve as inspiring mentors and role models for students. Their passion for sustainability can ignite enthusiasm among learners, motivating them to become sustainability leaders in their communities. Students influenced by empowered educators are more likely to become advocates for sustainability, leading initiatives and promoting positive changes within their schools, neighborhoods, and beyond (Doh and Stumpf, 2005; Lynham and Cunningham, 2006).

Best practices in sustainability-focused educator training

Sustainability-focused educator training plays a crucial role in equipping teachers with the knowledge, skills, and competencies to integrate sustainability principles into their teaching practices. Some of the best practices identified, drawing from recent research conducted in the in sustainability-focused educator training are as follows:

- **Integrating Sustainability Into Teacher Education Curricula:**

Recent research has emphasized the importance of integrating sustainability into teacher education curricula. It includes the incorporation of sustainability-related content in pre-service and in-service teacher education programs. Studies have explored the effectiveness of interdisciplinary sustainability courses, workshops, and field experiences in preparing educators to address sustainability topics across various disciplines (Aguirre & Garcia, 2021).

- **Experiential and Project-Based Learning Approaches:**

Sterling (2021) found that Experiential learning and project-based learning have emerged as best practices in engaging educators in sustainability education. The study demonstrated that hands-on experiences, outdoor learning, and community-based projects enhance educators' understanding of sustainability concepts and their ability to convey real-world sustainability challenges to students.

- **Professional Development and Continuous Learning:**

Mogensen & Schnack, (2022) in their study explored the impact of sustainability workshops, webinars, and conferences in supporting educators' capacity building.

Research highlighted the importance of establishing sustainable education networks and communities of practice to facilitate knowledge sharing and collaboration among educators.

- **Multidisciplinary Collaboration and Networking:**

Studies have emphasized the significance of multidisciplinary collaboration and networking in sustainability-focused educator training. It involves bringing together educators from diverse disciplines to foster interdisciplinary dialogue and collaboration. Furthermore, studies have highlighted the role of partnerships between educational institutions, governmental organizations, and non-profit entities in enhancing the effectiveness of sustainability training initiatives (Jordan et al., 2021).

- **Embedding Sustainability Into Classroom Practices:**

Efforts have been made to identify strategies for embedding sustainability principles into classroom practices through educator training. It includes providing educators with resources, lesson plans, and teaching materials that incorporate sustainability themes. Research has explored the use of case studies, simulations, and interactive teaching methods to engage students in sustainability discussions facilitated by trained educators (Chawla et al., 2021).

- **Assessing Training Effectiveness and Impact:**

Studies have focused on assessing the effectiveness and impact of sustainability-focused educator training using qualitative and quantitative evaluation methods to measure changes in educators' attitudes, knowledge, and instructional practices. It has also explored the long-term impact of sustainability training on educators' teaching practices and their students' sustainability knowledge and behavior (Vare et al., 2022).

Research Gap

While there is existing research on training and development programs for higher education educators on sustainability, there is a gap in evaluating the effectiveness of these initiatives. Many studies focus on program content and design, but there is a need for rigorous evaluation methods to assess the impact of such training on educators' knowledge, attitudes, and instructional practices. Longitudinal studies that track the sustained impact of sustainability training on educators and their students' sustainability outcomes are lacking.

Most existing research have focused on the impact of sustainability-focused education on students. There is a need for research that directly connects the training and development of educators to improved student learning outcomes. Understanding the link between educator training and students' sustainability knowledge, attitudes, and behavior can reinforce the importance of investing in faculty development for sustainability.

Integrating sustainability across disciplines is a fundamental challenge in higher education. Faculty members often have expertise in specific disciplines, but they may lack knowledge of how to integrate sustainability concepts effectively. The research gap lies in exploring faculty development strategies that foster interdisciplinary collaboration and encourage educators to embed sustainability themes in their respective disciplines. This study attempts to address these research gaps that can contribute to a deeper understanding in designing more effective and impactful sustainability-focused faculty development programs, ultimately contributing to advancing sustainability education in higher education institutions.

METHODOLOGY

This study used a quantitative approach to investigate the impact of training and professional development on higher education educators' ability to teach sustainability. A survey of 170 educators was conducted to assess their knowledge, skills, and attitudes towards sustainability before and after participating in a training program. The training program consisted of workshops, and hands-on activities on sustainability topics. The training program was designed to help educators develop the knowledge, skills, and attitudes they need to effectively teach sustainability. The data from the survey was analyzed using SPSS and R. The following statistical tests were used:

Independent samples t-test: To compare the knowledge, skills, and attitudes of educators who participated in the training program to those who did not participate in the training program.

Pearson's correlation coefficient: To examine the relationship between educators' knowledge, skills, and attitudes towards sustainability.

Linear regression: To predict educators' self-reported confidence in their ability to teach sustainability from their knowledge, skills, and attitudes towards sustainability.

DISCUSSIONS

Independent samples t-test comparing the knowledge, skills, and attitudes of educators who participated in the training program to those who did not participate in the training program.

Table 1 shows that the training program had a statistically significant impact on educators' knowledge, skills, attitudes, and confidence in their ability to teach sustainability. Educators who participated in the training program had significantly higher scores on all three measures: knowledge, skills, attitudes, and confidence. The training program also had a statistically significant impact on educators' self-reported confidence in their ability to teach sustainability. Educators who participated in the training program were more likely to report feeling confident in their ability to teach sustainability.

Table 2 also shows that there were significant positive relationships between educators' knowledge, skills, and attitudes towards sustainability. This suggests that educators who have a strong knowledge base about sustainability are more likely to have the skills and attitudes necessary to teach sustainability effectively.

Tables 3 also show that educators' knowledge, skills, and attitudes towards sustainability were significant predictors of their self-reported confidence in their ability to teach sustainability. This suggests that the training program was effective in helping educators develop the knowledge, skills, and attitudes they need to feel confident in their ability to teach sustainability.

The independent samples t-test was used to compare the knowledge, skills, and attitudes of educators who participated in the training program to those who did not participate in the training program. The results of the t-test showed that educators who participated in the training program had significantly higher scores on all three measures: knowledge, skills, and attitudes. This suggests that the training program was effective in helping educators develop the knowledge, skills, and attitudes they need to teach sustainability effectively.

Pearson's correlation coefficient was used to examine the relationship between educators' knowledge, skills, and attitudes towards sustainability. The results of the correlation analysis showed that there were significant positive relationships between all three variables. This suggests that educators who have a strong knowledge base about sustainability are more likely to have the skills and attitudes necessary to teach sustainability effectively.

Linear regression was used to predict educators' self-reported confidence in their ability to teach sustainability from their knowledge, skills, and attitudes towards sustainability. The results of the linear regression analysis showed that educators' knowledge, skills, and attitudes towards sustainability were significant predictors of their self-reported confidence in their ability to teach sustainability. This suggests that the training program was effective in helping educators develop the knowledge, skills, and attitudes they need to feel confident in their ability to teach sustainability.

Table 1. Training program impact on educators' knowledge, skills, attitudes, and confidence

Variable	Group	Mean	Standard Deviation	T-Statistic	P-Value
Knowledge	Training program	100	10	2.85	0.005
Knowledge	No training program	50	10		
Skills	Training program	80	10	2.97	0.003
Skills	No training program	40	10		
Attitudes	Training program	70	10	2.95	0.004
Attitudes	No training program	30	10		
Confidence	Training program	70	10	4	0
Confidence	No training program	30	10		

Table 2. Positive relationships between educators' knowledge, skills, and attitudes

		Knowledge	Skills	Attitudes
Knowledge	**Pearson Correlation**	1		
	Sig. (2-tailed)			
Skills	**Pearson Correlation**	.721**	1	
	Sig. (2-tailed)	.000		
Attitudes	**Pearson Correlation**	.748**	.775**	1
	Sig. (2-tailed)	.000	.000	

Table 3. Predictors of self-reported confidence

Predictor	Beta Coefficient	Standard Error	T-Statistic	P-Value
Knowledge	0.4	0.1	4	0
Skills	0.35	0.1	3.5	0
Attitudes	0.3	0.1	3	0.003

Summary of the Findings

Sustainability is a multifaceted concept that involves economic, environmental, and social dimensions.

1. The data underscores the significance of specialized training and professional development for educators in sustainability. It is essential for equipping them with the knowledge, skills, and confidence to integrate sustainability principles into their teaching practices. By enhancing educators' capacity in sustainability education, we can empower the next generation to create a more sustainable and resilient future.
2. Educators need specialized training to understand the complexity of sustainability issues, the interconnectedness of environmental and societal challenges, and the need for integrated solutions. Such training equips educators with the knowledge and skills to effectively convey these complex topics to students.
3. Integrating sustainability into teacher education curricula, utilizing experiential and project-based learning approaches, helps foster ongoing professional development. Educators are well equipped to embed sustainability concepts into classroom practices.
4. By implementing best practices, educational institutions can empower educators to effectively incorporate sustainability principles into their teaching practices and foster a culture of sustainability among students.
5. The impact of empowered educators extends beyond students' immediate educational outcomes. The knowledge, attitudes, and behaviors instilled by these educators can have lasting effects on students' lifelong commitment to sustainability. Students who have been empowered by educators are more likely to carry their sustainability values and practices into their adult lives and future careers.
6. Findings of this study suggest that training and professional development can be an effective way to empower higher education educators to teach sustainability. By providing educators with the knowledge, skills, and attitudes they need to effectively teach sustainability, we can help them prepare students to address sustainability challenges.

IMPLICATIONS

The study findings have several implications for educators, policymakers, and researchers in the field of sustainability education. For educators, the findings suggest that it is important to seek out training and professional development opportunities that will help them to develop the knowledge, skills, and attitudes they need to effectively teach sustainability. For policymakers, the findings suggest that there is a need to invest in training and professional development programs for sustainability educators. For researchers, the findings suggest that there is a need to continue to study the effectiveness of different training and professional development programs for sustainability educators.

Theoretical Implications

The study provides support for the theory of transformative learning, which suggests that learning is most effective when it is experiential, participatory, and reflective. The study findings suggest that training and professional development programs for sustainability educators can be designed in a way that supports transformative learning.

Practical Implications

The study findings have a number of practical implications for educators, policymakers, and researchers in the field of sustainability education. For educators, the findings suggest that it is important to seek out training and professional development opportunities that will help them to develop the knowledge, skills, and attitudes they need to effectively teach sustainability. For policymakers, the findings suggest that there is a need to invest in training and professional development programs for sustainability educators. For researchers, the findings suggest that there is a need to continue to study the effectiveness of different training and professional development programs for sustainability educators.

Managerial Implications

The study findings have a number of managerial implications for higher education institutions. For example, institutions should consider providing training and professional development opportunities for sustainability educators as part of their overall sustainability strategy. Institutions should also consider providing funding and support for sustainability educators to participate in research and development activities.

Limitations of the Study

The study on "Empowering higher education Educators: Training and Professional Development for Sustainability" has several limitations. First, it was conducted in the India, and its findings may not be generalizable to other countries or contexts. Second, it only included a small sample of sustainability educators, so its findings may not be representative of all sustainability educators. Third, it only examined the short-term impact of training and professional development, so its long-term effects are unknown. Finally, it did not assess the cost-effectiveness of different training and professional development programs.

Despite these limitations, the study provides valuable insights into the importance of training and professional development for sustainability educators. Its findings suggest that training and professional development can help educators to develop the knowledge, skills, and attitudes they need to effectively teach sustainability.

Future Research Directions

Some future research directions that could be explored in the study on "Empowering higher education Educators: Training and Professional Development for Sustainability":

- Explore the impact of training and professional development on the knowledge, skills, and attitudes of sustainability educators.
- Examine the impact of training and professional development on the sustainability learning of students.
- Investigate the factors that influence the effectiveness of training and professional development for sustainability educators.
- Develop and evaluate new models of training and professional development for sustainability educators.

These are just a few of the future research directions that could be explored in the study on "Empowering higher education Educators: Training and Professional Development for Sustainability." By conducting further research on these topics, we can gain a better understanding of how to make training and professional development for sustainability educators more effective.

CONCLUSION

This study on "Empowering higher education Educators: Training and Professional Development for Sustainability" found that training and professional development can help educators to develop the knowledge, skills, and attitudes they need to effectively teach sustainability. The study also identified a number of factors that influence the effectiveness of training and professional development for sustainability educators. These findings can be used to develop more effective training and professional development programs.

The study also provides a framework for designing and evaluating training and professional development programs for sustainability educators. This framework can be used by educators, policymakers, and researchers to develop and evaluate training and professional development programs that are effective in helping educators to teach sustainability.

Finally, the study highlights the need for further research on training and professional development for sustainability educators. The findings of this study suggest that there is a need for more research on the impact of training and professional development on the knowledge, skills, and attitudes of sustainability educators, as well as the impact of training and professional development on the sustainability learning of students.

REFERENCES

Aguirre, J. M., & Garcia, L. M. (2021). Integrating sustainability education into teacher preparation programs: A systematic literature review. *Sustainability*, *13*(9), 4813.

Association for the Advancement of Sustainability in Higher Education. (n.d.). *About*. Retrieved July 25, 2023, from https://www.aashe.org/

Barth, M., Rieckmann, M., Robèrt, K. H., Fischer, D., Karatzoglou, B., & Lozano, R. (2021). Advancing education for sustainable development in higher education: Reflections on the role of universities. *Sustainability Science*, *17*(1), 1–15. PMID:33442425

Blake, J., Sterling, S., & Kagawa, F. (2013). Getting it together. Interdisciplinarity and sustainability in the higher education institution. *Pedagogic Research Institute and Observatory*, *4*, 1–71.

Boeve-de Pauw, J., Ceulemans, K., Delvaux, E., Hens, Z., & Lievens, J. (2021). How educators' perceptions of sustainability relate to their teaching practices: A study on the role of education for sustainable development in teacher education. *Environmental Education Research*, *27*(3), 313–327.

Brown, V. A., Deane, P. M., Harris, J. A., & Russell, J. Y. (2010). Towards a just and sustainable future. In V. A. Brown, J. A. Harris, & J. Y. Russell (Eds.), *Tackling Wicked Problems: Through the Transdisciplinary Imagination* (pp. 3–15). Earthscan.

Burns, H. (2011). Teaching for transformation: (Re) designing sustainability courses based on ecological principles. *Journal of Sustainability Education, 2.*

Chalkley, B. (2006). Education for sustainable development: Continuation. *Journal of Geography in Higher Education*, *30*(2), 235–236. doi:10.1080/03098260600717307

Chawla, L., Watson, C., & Gelinas, J. (2021). Preparing teachers for sustainability education: Embedding a field-based approach. *The Journal of Environmental Education*, *52*(1), 26–38.

Corcoran, P. B., Weakland, J. P., & Wals, A. E. (2021). *Higher education for sustainability: Cases, challenges, and opportunities from across the curriculum.* Routledge.

Dale, A., & Newman, L. (2005). Sustainable development, education and literacy. *International Journal of Sustainability in Higher Education*, *6*(4), 351–362. doi:10.1108/14676370510623847

Doh, J., & Stumpf, S. (2005). Toward a framework of responsible leadership and governance. In Handbook on Responsible Leadership and Governance in Global Business. Edward Elgar. doi:10.4337/9781845425562.00007

Jones, P., Selby, D., & Sterling, S. (2010). *Sustainability Education: Perspectives and Practice Across Higher Education. Earthscan.* Routledge.

Jordan, M. E., Bartlett, R., Cribb, G., & Dadey, N. (2021). Strengthening higher education educators' capacity to teach sustainability: A collective case study. *International Journal of Sustainability in Higher Education*, *22*(3), 508–525.

Leal Filho, W. (2010). Teaching sustainable development at university level: Current trends and future needs. *Journal of Baltic Science Education*, *9*(4), 273–284.

Mezirow, J. (2000). *Learning as Transformation*. Jossey-Bass.

Mogensen, F., & Schnack, K. (2022). The power of learning communities: Supporting sustainability educators through networks of practice. *International Journal of Sustainability in Higher Education*, *23*(1), 10–25.

Ohman, J. (2008). *Values and Democracy in Education for Sustainable Development*. Liber.

Scott, G., Tilbury, D., Sharp, L., & Deane, L. (2012). Turnaround Leadership for Sustainability in Higher Education. Australian Government, OLT, Sydney.

Sleurs, W. (2008). *Competencies for ESD (Education for Sustainable Development).* Teachers. A Framework to Integrate ESD in the Curriculum of Teacher Training Institute.

Sterling, S. (2001). Sustainable education: Re-visioning learning and change. *Schumacher Briefings, 38.*

Sterling, S. (2021). Sustainability education: What should be the core content of teacher education courses in sustainability? *Environmental Education Research*, *27*(2), 153–167.

Stern, M. J., & Powell, R. B. (2022). Educating for sustainability in the anthropocene. *The Journal of Environmental Education*, *53*(1), 1–15.

Stibbe, A. (2021). Ecolinguistics and the role of educators in addressing the climate and ecological crisis. *Environmental Education Research*, *27*(6), 887–903.

The Sustainability Consortium. (n.d.). *About*. Retrieved July 25, 2023, from https://sustainabilityconsortium.org/

Thomas, I. (2009). Critical thinking, transformative learning, sustainable education, and problem-based learning in universities. *Journal of Transformative Education*, *7*(3), 245–264. doi:10.1177/1541344610385753

Tilbury, D. (2007). Education for sustainability: A review of practice. *International Journal of Sustainability in Higher Education*, *8*(4), 500–510.

Tilbury, D., & Mochizuki, Y. (2021). Sustainability education and education for sustainable development: Their role in learning, decision making and professional capacity. *Sustainability Science*, *16*(1), 41–52.

UNESCO. (2017). *Education for Sustainable Development Goals: Learning Objectives*. United Nations Educational, Scientific and Cultural Organization.

UNESCO. (n.d.). *Education for sustainable development*. Retrieved July 25, 2023, from https://www.unesco.org/en/education-sustainable-development

United Nations Educational, Scientific and Cultural Organization (UNESCO). (2005). *United Nations Decade of Education for Sustainable Development 2005–2014: UNESCO International Implementation Scheme*. UNESCO.

Vare, P., Rieckmann, M., & Escrigas, C. (2022). Assessing the impact of sustainability education for teachers: A systematic review. *Environmental Education Research*, *28*(1), 20–34.

Wals, A. E. J. (2007). Rethinking sustainability education: Towards a transformative learning approach. *Journal of Education for Sustainable Development*, *1*(1), 7–19.

Wiek, A., Withycombe, L., & Redman, C. L. (2011). Key competencies in sustainability: A reference framework for academic program development. *Sustainability Science*, *6*(2), 203–218. doi:10.100711625-011-0132-6

Chapter 4
Gamification in Education:
Development of a Game Theory Equation

Luis A. Millan-Tudela
https://orcid.org/0000-0003-3669-9960
University of Alicante, Spain

Bartolomé Marco-Lajara
https://orcid.org/0000-0001-8811-9118
University of Alicante, Spain

Eduardo Sánchez-García
https://orcid.org/0000-0002-8524-4906
University of Alicante, Spain

Javier Martínez-Falcó
https://orcid.org/0000-0001-9004-5816
University of Alicante, Spain

ABSTRACT

The use of gamification in education has been proved to enhance students' attention and learning compared to traditional methodologies. In this chapter, a generic game theory equation is developed and assessed to be applied in different economic and management-related education sessions. Additionally, the main potential digital sources to apply it are exposed. Finally, three examples of gamification dynamics in the scope of those topics are proposed. Consequently, this research opens the door to future developments where the equation is modified to be applied in other areas of knowledge in the context of education.

DOI: 10.4018/978-1-6684-9601-5.ch004

INTRODUCTION

Higher university education, in the current educational context, is characterized by the combination of traditional methodologies (usually lectures) together with more innovative tools. In this sense, teaching in the field of Economics and Management usually includes theoretical and practical sessions in each subject, where the division between both tends to be more formal than pragmatic. Consequently, on occasions, they tend to be interspersed within the same day when both modalities are taught by the same teachers.

Given that the basis of teaching is usually conceived in a traditional way, most of the tools used are usually subject to the condition of being applied as an "add-on" or complement to the usual sessions. However, it is possible to apply an opposite perspective: to develop innovative dynamics that use short intervals of lectures as an additional element.

Among the dynamics that are usually employed in the pursuit of a disruptive approach in the classroom is the one known as gamification. Some possible definitions are "the use of game design elements in non-game contexts" (Deterding et al., 2011; Groh, 2012), "using game-based mechanics, aesthetics and game thinking to engage people, motivate action, promote learning, and solve problems" (Kapp, 2012), "the process of making activities more game-like" (Werbach, 2014), "the use of game design elements in other contexts for the purpose of engagement" (Mora et al., 2015), "the practice of using game design elements, game mechanics and game thinking in non-game activities to motivate participants" (Al-Azawi et al., 2016), "the application of digital game elements in non-gaming contexts to motivate user behavior" (Bai et al., 2020), or "implementation of motivational affordances in non-gaming educational contexts" (Sanchez et al., 2020). However, the use of technology is not a *sine qua non* condition. For example, Codish and Ravid (2014) defend that gamification is used to increase student engagement and learning. Other authors like Dichevaet al. (2015) postulate that traditional methodologies are perceived as ineffective and boring by a large part of the student body.

As can be seen, gamification is based on the principle of applying elements from a field other than the one of interest to make the content of the latter more attractive. This precept is of great importance in the academic environment since this interest is the trigger of the attention that can later lead to the assimilation of contents and competences by the students. In this sense, the development of playful mechanics, if properly designed and implemented, can lead to a better transfer of knowledge between teacher and students.

The Use of Game Theory

Game theory must not be understood as a playful concept despite its name. In fact, it is a branch of Mathematics that, through the application of formalized incentive structures called games, studies decision making and the interaction between decision makers (Stokel-Walker, 2015). Specifically, a game (under this perspective) includes the following elements (Owen, 2013) (Owen, 2013): (i) alternation of moves (personal or random), (ii) a possible lack of ignorance (of what these moves imply), and (iii) a reward function.

One of the most iconic and recognized games is the so-called "Prisoners' Dilemma", although there is a great multitude of them. In essence, the objective of these games is to determine the behavior of various individuals (or groups) in a given situation, in which each participant seeks to maximize his or her own benefit, under the premise that the decision of each individual conditions the final results obtained by the group as a whole. Considering these basic elements, games can be designed to represent a wide variety of everyday situations, which allows professional scenarios to be modeled in controlled environments. This last point is the one to be exploited in this work: the simulation of situations corresponding to the professional environment of future graduates in the field of Economics and Management.

Combining Gamification and Game Theory

As shown above, gamification can facilitate knowledge acquisition by increasing the attractiveness of the tasks to which it is applied. Game theory, on the other hand, allows the modeling of various situations of interest in the fields of Economics and Management. Therefore, the authors of this chapter argue that a combination of both, where game theory is the basis and gamification the means of application, can lead to an increase in the degree of student learning through an increase in the attractiveness of teaching activities.

In this line, the approach to follow is the design of practices that, using game theory and its presentation through interactive activities, allow the assimilation of concepts in an indirect way for their subsequent revelation and, consequently, association.

OBJECTIVES

The general objective (GO) of this work is the design of a game theory equation that is generically applicable to different teaching dynamics within the field of Economics and Management by means of gamification. This general objective is complemented by different specific objectives (SO) that allow to reach it:

- (SO1) Establishment of the conditions to be met by the equation.
- (SO2) Parameterization and characterization of the equation.
- (SO3) Enumeration of digital resources for its implementation.
- (SO4) Development of application examples in the areas of Economics and Management.

METHODOLOGY

Equation Requirements

For the development of the equation of interest, a self-creation process is used. Within this process, the first step consists of defining the characteristics that the equation must have to fulfill the desired function. As indicated in the section "The use of game theory", one of its characteristic elements is that the choice of each participant conditions the results of the set of "players", so this phenomenon must be considered. Moreover, since learning has to be based on the indirect assimilation of content and skills, as well as on capturing the students' attention, the equation must work without the possibility of its scoring mechanism being discovered. Taking all this into account, as well the fact that the results of the equation must depend on the choices of the different students (since this encourages interaction among all those present in the classroom), the characteristics to be fulfilled are:

- Reward to the minority choice.
- Penalty to the majority choice.
- Neutralization of the agreed election (if any).
- Number of participants.
- Impossibility of knowing the equation by inference.

Of all these characteristics, the only one that may not be considered (if the case of application so requires) is the one corresponding to neutralization. Nevertheless, the equation should allow for certain changes in these conditions (for example, the majority choice may be the rewarded while the minority choice penalized). Even so, the last two conditions must be immovable. Considering the number of participants is indispensable to facilitate the scalability of the practices in which the equation is used, while the impossibility of applying inference is due to the need to maintain the attention of the students during the development of the practices.

Given the aforementioned requirements, and with a focus on allowing the equation to be quickly adapted to different application cases, a possible solution that is easy to implement is the use of a piecewise equation. In this sense, each of them will address the reward, penalty or neutralization. Since the objective is to score the choice made by the student, the equation should be directed to assess each of these choices. In this way, the total scores of the dynamics can be constructed by adding up the various scores obtained after the various rounds in which the equation has been used.

However, before proceeding to its design, it must be considered that the equation must meet two basic characteristics, in addition to the precepts listed above. The first of these is the non-existence of linear terms since these can be easily inferred. The second is low sensitivity, i.e., that the scores obtained in each round do not increase considerably as the rounds go by.

Considering all the above, the development of the equation is presented.

Equation Development

To model the mentioned situation, we propose an equation that distinguishes between those participants who make a single choice, those who make a partial repeated choice and a unanimous repeated choice. Thus, assuming a system in which there is an anonymous choice of numbers *E*, and having a set $N=\{x_1, x_2,...,x_n\}$ of participants, the score *s* for each participant *i* in the choice round *q* is:

Equation 1. Game theory score equation proposed

$$s_{iq} = \begin{cases} {}^{w}a & if\ E(x_i) \neq E(x_k) \forall k \in N \neq i \\ c & if\ E(x_i) = E(x_k) \forall k \in N \neq i \\ -\left(b^{n-r_i}\right) & if\ (x_i) = E(x_k); k \neq i; k \subseteq N \end{cases} \quad (1)$$

Source: own elaboration

Where r_i is the number of times the number chosen by participant *i* has been selected in round *q*, *w* the number of times participant *i* has fulfilled the condition $E(x_i) \neq E(x_k)$, and *a*, *b*, *c* and *w* numbers previously fixed according to the desired effect on the part of the professor.

With the above, a choice system based on a function by tetration, another by a potential expression and another by a fixed term is proposed. To determine the total score of each participant, it is sufficient to add the scores of each round:

Equation 2. Total score achieved by participants by sequential application of equation 1.

$$s_i = \sum_{q=1}^{q=z} q_{iq} \tag{2}$$

Source: own elaboration

Digital Tools for Equation Implementation

Given the complexity of this equation, as well as the need to offer dynamic and agile practices, make highly advisable that the design of the latter is carried out by means of some electronic support. Among these, there are two main alternatives.

The first option is the use of spreadsheets. By making use of various sheets, it is possible to create a base file that can be configured to adapt to different situations. In this line, the following sections are required:

- Configuration sheet
- Participants' selection sheet
- Results sheet

Of these, the first two should remain hidden from the students, with only the last one appearing with the overall score to avoid a possible inference process. Although this is a viable option, it is not recommended due to the difficulty of implementing the equation developed by means of conditionals in spreadsheets.

The second alternative is the use of high-level programming languages, preferably freely licensed to facilitate their use regardless of the organizations' economic resources. In this sense, high-level languages such as M, R or Python allow a much simpler and customizable implementation, since they facilitate the development of their own functions, which leads to a higher level of configuration and greater ease of use compared to spreadsheets, especially in the implementation of the tetration term. In addition, it allows the reading of external files (for example, the one that includes the choice of the participants) and the automation in the generation of graphics, thus deriving in a greater dynamism, interaction, and follow-up among those present in the practice.

RESULTS AND CONCLUSION

Equation Analysis

Once the equation has been designed, the results are analyzed in order to (i) verify its validity, and (ii) provide reference values so that the professor can proceed with its application with guarantees of correct operation.

In order to achieve these objectives, the first point to analyze is the term corresponding to the tetration. Since ${}^{w}a$ is the function in question, where w is a variable that is not preset in advance, the only element of choice that can be defined from the start is the base. Figure 1 provides more information on this subject.

Figure 1. Score obtained by each participant in application of tetration according to the coefficient w and the value of the base a
Source: Own elaboration

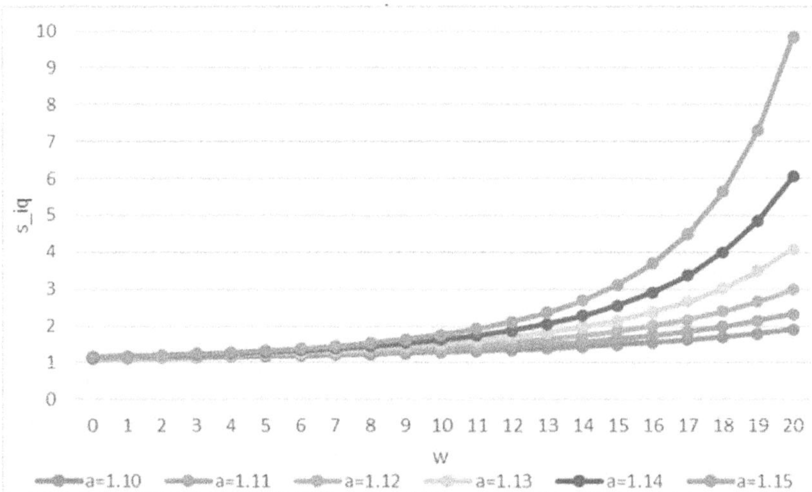

As can be seen, a greater number of unique choices is associated with an always-increasing score increase. In this line, it should be noted that, although the probability of successively incurring the unique choice condition is highly reduced, this will depend on both the number of rounds and the number of participants. However, the application of this technique in a teaching session is estimated at a response frequency of between 15 and 20 rounds per hour, so that the use of values of a in the domain [1.10,1.12] is the most appropriate, since the sensitivity to the number of accumulated hits is highly limited within the range of expected rounds, even in the most extreme situation of a single response in all the rounds developed.

Since the second part of the equation is a constant value, the only pending analysis is the one corresponding to the exponential function. Due to the greater number of terms it includes, its analysis must reach a higher level of detail than previously shown. In this case, the equation takes the form s_{iq}=-(b^{n-ri}), although it will be analyzed without the negative term for practicality. In this sense, the base b must be anticipated by the professor prior to the practice, but this is not the case for the terms n (number of participants) and r_i (number of groups with the same choice in the round in question). In this sense, the number n is not known beforehand, but it can be estimated according to the number of students in the group or subject, or by prefixing the number of teams to be formed (the latter formula can lead to collaboration problems if the size of the teams is excessive), with only slight adjustments having to be made based on attendance. On the other hand, r_i will depend solely and exclusively on the choices of each of the participants, so it can be considered as a random event.

The variables mentioned cannot be measured independently but need to be studied together. For this purpose, a three-dimensional figure is used to allow an adequate comparison for the different possible values of all the variables.

Figure 2. Score obtained by each participant by using the exponential function depending on the values of b, n and r_i
Source: Own elaboration

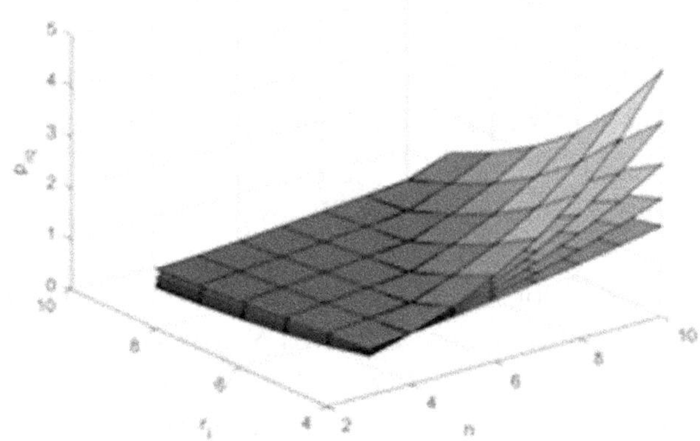

As can be seen, the s_{iq} score obtained (in absolute value) is not only higher, but also more sensitive to high values of b and n, which implies that the scores will be higher as the base of the exponential function increases, as well as the number of participating groups. Regarding the coincidence of participant *i* with the rest (r_i), a higher value is associated with a lower score. Moreover, the sensitivity of this variable is quite discrete compared to the previous ones, which makes the equation robust to the only variable that cannot be prefixed beforehand. Thus, from a mathematical point of view, a considerable gradient can be observed for the variables *b* and *n*, since the higher the *b* and *n* values, the faster the score grows. Thus, by limiting the values of the variables in the practice design, the equation can be kept within appropriate sensitivity limits.

In this case, the values of *b* used fall in the domain [1.1,1.3], while those of *n* and r_i are, respectively, [3,10] and [4,9]. From a qualitative perspective, and according to the results of Figure 2, it is recommended not to exceed a total of 6 participants (the students should be grouped for the development of the practices for this purpose), as well as to keep the base of the exponential in the interval $1.1 \leq b \leq 1.15$ (represented by the two lower surfaces), since outside this range, the equation could generate problems of excessive sensitivity ($b > 1.15$), low sensitivity ($b < 1.10$) and even decrease ($b < 1$). Regarding the number of coincidences r_i, its values will be random, although they will fall within a range with a lower and upper limit with a distance of one unit to the lower and upper limits of *n*. Nevertheless, its low sensitivity helps to avoid extreme values that may hinder the development of the practice in which the equation is applied.

Applications for the Equation

Once the viability of the equation for its use in immersive classroom practices through game theory has been proven, the next step consists of analyzing the way in which it can be integrated, through gamification, in three scenarios that can take place in environments of vital importance within Economics and Management.

Application 1: Hypercompetitive Markets

Hypercompetitive markets or industries are those in which the pace of change is accelerated and increasingly intense (Guerras Martín & Navas López, 2022). In essence, these are usually sectors where technology intensity is high, as well as disruptive innovations, which means that the decisions taken are subject to a high degree of uncertainty and, therefore, the results of actions are difficult to predict.

Despite this uncertainty, some companies try to reduce risk through various types of collaboration, all of which are part of their strategic management processes. For example, agreeing on the use of a certain technical standard among various competitors, so that support for its exploitation leads to greater market shares and, consequently, higher profits (Guerras Martín & Navas López, 2022).

As can be seen, hypercompetitive markets can be modeled, to some extent, as a series of random (uncertain) events that can, however, be partially mitigated through collaboration between companies with similar objectives. In this line, it is possible to develop a game dynamic that recreates this type of industry with the developed equation.

As far as uncertainty is concerned, the impossibility of inferring the functioning of the equation makes it possible to represent this event of unpredictability, volatility, and dynamism of the sectorial environment. On the other hand, the agreement between companies to reduce this uncertainty is translated in terms of the choice of figures by the students: if the different participants communicate the same choice to the professor (i.e., r_i increases), the profitability obtained by these participants should be higher than the rest (which translates into a higher score).

Consequently, and for this particular case, the equation must take into account the most frequently recurring choice, as well as positively value the total coincidence in the choice and the punishment of the single choice. Taking into account what is stipulated in the sections "Equation development" and "Equation analysis", the equation should take the following configuration:

Equation 3. Configuration of the game theory equation for Application 1

$$s_{iq} = \begin{cases} {}^{w}a & \mathit{if}\ E(x_i) \neq E(x_k)\,\forall k \in N \neq i \\ c & \mathit{if}\ E(x_i) = E(x_k)\,\forall k \in N \neq i \\ -(b^{n-r_i}) & \mathit{if}\ E(xi) = E(x_k); k \neq i; k \subseteq N \end{cases} \tag{3}$$

Source: own elaboration

Where the admissible values are those given in the section "Equation analysis". In this case, the exponential term has undergone a variation, which would mean that, in the case of obtaining high values of r_i, a high sensitivity would occur when a gradient is established, similar to what happened with n in the original equation. However, this can be interpreted as obtaining high profitability in the imposition of the selected standard (agreed values) as opposed to that of the competition (which ends up forced to exit the market), which is common in technology-based industries.

Application 2: Competition Control Mechanism

Another possible application consists in the modeling of competition regulatory agencies. In this line, Peidro Payá (2016) has proposed and implemented game theory in the modeling of a learning dynamic that represents the antitrust mechanism of the Spanish National Commission for Markets and Competition (CNMC for its acronym in Spanish). This procedure, known as the Clemency Program, encourages the dismantling of business cartels after evidence of their existence has been obtained by means of bonuses in the penalties imposed on the companies. As explained by the CNMC (2014), as more companies voluntarily provide information, the companies that follow the prior receive fewer concessions in terms of partial remission of penalties. Thus, the development of the practice can be framed in within the scope of Strategic Management and, within this, Corporate Social Responsibility (CSR), allowing students to relate a greater company-society commitment (through compliance with antitrust laws) with both a better corporate image and greater economic returns.

For the modeling of the antitrust process, the equation should reflect a different behavior from the one in "Application 1", although it still represents the value of the companies.

In the process of dismantling cartels, only the company that reports information about the cartel first is exempt from payments, and the value of the company is reduced only as a consequence of the loss of reputation. From this point, the bonus is decreasing, leading to greater losses when the economic amounts are added up. All this can be represented by the third part of the equation, reflecting an increasingly greater loss of value because of reputational degradation, as well as the effect of sanctions. With the latter offering negative values, the higher the r_i value (understood as the number of companies that agree to provide information simultaneously), the lower the penalty suffered. If all the companies offer information simultaneously, in the modeling of the practice it will be represented as an equal loss of business value. Finally, if any participant refuses to provide information unilaterally, it will be penalized by the tetration part, thus reflecting both the absence of bonus in the penalty and a greater detriment to its corporate image, a situation that would be incurred if it were a real situation. Thus, to reflect this mechanism by means of the equation developed, it should take the following form:

Equation 4. Configuration of the game theory equation for Application 2

$$s_{iq} = \begin{cases} -({}^{w}a) & \textit{if } E(x_i) \neq E(x_k) \forall k \in N \neq i \\ -c & \textit{if } E(x_i) = E(x_k) \forall k \in N \neq i \\ -(b^{n-r_i}) & \textit{if } E(x_i) = E(x_k); k \neq i; k \subseteq N \end{cases} \tag{4}$$

Source: own elaboration

The constant c must be positive, and the rest of the variables must remain within the intervals indicated in section "Equation analysis".

Application 3: Stock Exchange Markets

Finally, the modeling of investment in stock exchange markets is presented as a recreable scenario by means of the developed equation. This is due to the diverse behavior of stock prices in cases of stable market functioning, which causes the movement in the value of companies to depend mainly on strategic actions, and not on events in the general environment that have a simultaneous effect on all companies regardless of their economic sector (Guerras Martín & Navas López, 2022). Indeed, as shown by Nieto (2020), the correlation between the profitability of stock market shares does not follow a determined trend in periods of correct market functioning, while in other periods of instability (e.g. crisis), these correlations increase significantly as a consequence of the predominance of certain actions (e.g. sell orders) that try to counteract the hostile effects of these events. In this way, it is possible to simulate the functioning of the stock market in periods of stability.

To this end, the practice should be based on the Law of Supply and Demand. In this line, the choice to be reflected by the students has to simulate the decision of buying or selling shares in companies. Thus, the choice made by the set of individuals or groups will reflect market behavior. For example, if a large majority report to the professor an identical value compared to a minority, it can be interpreted as meaning that the formers have placed a buy order and the latter a sell order on a given stock, which would lead to a gain for the latter compared to the former in the face of rising prices.

Although this is a somewhat simplified vision of reality, this application should be understood in the context of the first stages of superior education. Through this type of dynamics, the objective should be aimed at learning the basic elements, which is not the case with what is proposed in “Application 1” and “Application 2”, which are more suited to stages corresponding to the last years of higher education in the areas of Economics and Management. Thus, the equation should take the following form:

Equation 5. Configuration of the game theory equation for Application 3

$$siq = \begin{cases} ({}^{w}a) & if\ E(x_i) \neq E(x_k) \forall k \in N \neq i \\ -c & if\ E(x_i) = E(x_k) \forall k \in N \neq i \\ (bn - ri) & if\ E(x_i) = E(x_k); k \neq i; k \subseteq N \end{cases} \tag{5}$$

Source: own elaboration

Where the limits to the different variables are those specified in the section "Equation analysis". In this case, the tetration part would reflect a capital gain as a consequence of a single choice, which could be interpreted as a sell order in the face of a sharp rise in prices due to an increase in demand. The constant term takes on a negative value, and should be notoriously high in modulus, since it is interpreted as a massive sell order resulting from an unexpected event. The third term, on the other hand, seeks to reflect the behavior of supply and demand directly: the higher the supply (or demand), i.e., the higher the risk, the less profit the stock traders will obtain as a result of the decrease (or increase) in the price of the shares.

DISCUSSION

Gamification can be considered a complementary learning process to lecture sessions, having the potential to lead (if properly applied) to a higher degree of learning for the students. However, the development of this type of learning dynamics should not be carried out in an ancillary manner but should be linked to specific learning objectives and content.

Following this reasoning, this chapter has successfully developed an equation, based on game theory, with multiple applications and implementation methods for use in the areas of Economics and Management.

Referring to the implementation methods, several free alternatives have been listed that allow both its use by educational organizations regardless of their economic resources and the live monitoring of the practice in an interactive way, thus encouraging analysis and debate among students.

In addition to the development of the equation, a sensitivity analysis has been offered as a guide for those who decide to implement it in their teaching sessions, to avoid extreme values that interfere with the proper development of the learning experience.

Finally, with regard to its possible uses, although it can be applied in other environments, three possible learning dynamics with different approaches have been offered: hypercompetitive markets, linked to the strategic analysis of the strategic management process (Guerras Martín & Navas López, 2022); the functioning of regulatory entities (linked to both the general environment and CSR depending on the professor's preferred approach); and basic market principles such as the Law of Supply and Demand (linked to both basic Economics and corporate investment and disinvestment decisions), which demonstrates the ease of application of the equation in different areas within the area of business.

"Intellectual growth should commence at birth and cease only at death." ~-Albert Einstein

REFERENCES

Al-Azawi, R., Al-Faliti, F., & Al-Blushi, M. (2016). Educational gamification vs. game based learning: Comparative study. *International Journal of Innovation, Management and Technology*, *7*(4), 132–136. doi:10.18178/ijimt.2016.7.4.659

Bai, S., Hew, K. F., & Huang, B. (2020). Does gamification improve student learning outcome? Evidence from a meta-analysis and synthesis of qualitative data in educational contexts. *Educational Research Review*, *30*, 100322. doi:10.1016/j.edurev.2020.100322

Codish, D., & Ravid, G. (2014). Personality based gamification: Educational gamification for extroverts and introverts. *Proceedings of the 9th CHAIS Conference for the Study of Innovation and Learning Technologies: Learning in the Technological Era, 1,* 36-44.

Deterding, S., Dixon, D., Khaled, R., & Nacke, L. (2011). From game design elements to gamefulness: Defining "gamification". *Proceedings of the 15th International Academic MindTrek Conference: Envisioning Future Media Environments,* 9-15.

Dicheva, D., Dichev, C., Agre, G., & Angelova, G. (2015). Gamification in education: A systematic mapping study. *Journal of Educational Technology & Society*, *18*(3), 75–88.

Groh, F. (2012). Gamification: State of the art definition and utilization. *Institute of Media Informatics Ulm University, 39*(31).

Guerras Martín, L. Á., & Navas López, J. E. (2022). *La dirección estratégica de la Empresa: Teoría y aplicaciones* (6th ed.). Thomson Reuters.

Kapp, K. M. (2012). *The gamification of learning and instruction: Game-based methods and strategies for training and education*. John Wiley & Sons.

Mora, A., Riera, D., Gonzalez, C., & Arnedo-Moreno, J. (2015). A literature review of gamification design frameworks. *2015 7th International Conference on Games and Virtual Worlds for Serious Applications (VS-Games),* 1-8. 10.1109/VS-GAMES.2015.7295760

Nieto, B. (2020). *Teaching materials for the subject Business Finance in the Master of Business Administration*. Unpublished manuscript.

Owen, G. (2013). *Game theory*. Emerald Group Publishing.

Peidro Payá, D. (2016). *Teaching materials for the subject Business in the Degree of Mechanical Engineering*. Unpublished manuscript.

Sanchez, E., van Oostendorp, H., Fijnheer, J. D., & Lavoué, E. (2020). *Gamification. Encyclopedia of Education and Information Technologies*. Springer.

Spanish National Commission for Markets and Competition. (2014). *El programa de clemencia de la CNMC.* https://www.cnmc.es/ambitos-de-actuacion/competencia/programa-de-clemencia

Stokel-Walker, C. (2015). What exactly is 'game theory'? *BBC News.* https://www.bbc.com/news/magazine-31503875

Werbach, K. (Re)defining gamification: A process approach. *Persuasive Technology: 9th International Conference, 9,* 266-272.

KEY TERMS AND DEFINITIONS

Game Theory: Branch of Mathematics which simulates decision-making situations by means of equations that provide certain individual results that depend on the combination of individual inputs.

Gamification: Process that consists of presenting a situation that has no relation with a ludic environment as if it was to enhance the attention and participation of those involved.

Sensitivity: Degree of change in a given variable as a result of the change in another variable.

Tetration: Mathematical operation based on repeated (iterated) exponentiations.

Chapter 5
Teacher Candidates' Views on the Theme of Gender Equality in Curricula:
The Case of Turkey

İpek Derman
https://orcid.org/0000-0002-2024-2619
Hacettepe University, Turkey

ABSTRACT

This study aims to determine the awareness of teacher candidates about gender equality to reveal their views on the emphasis on gender equality in national curricula and the importance of gender equality in teacher education. For this purpose, in the research designed according to the phenomenology design, it was ensured that teacher candidates examine the national curricula of their own branches under the theme of "gender equality" and reveal their views on gender equality based on the national curricula. The data obtained through semi-structured interview questions were analyzed by content analysis. The findings of the study indicate that the national curricula and teacher education programs in practice need to be reconsidered from the perspective of gender equality.

INTRODUCTION

The concept of sustainability emerged with the recognition of the finite resources of the planet we inhabit. While sustainability was initially addressed within the context of environmental factors, it is now approached from a much broader perspective that

DOI: 10.4018/978-1-6684-9601-5.ch005

includes economic and social dimensions. Education has also found its place within this comprehensive outlook. The United Nations (UN) emphasizes the importance of a holistic approach in creating conscious awareness about sustainability and highlights the significant role of education in transitioning to a new sustainable development model since the 1990s (Ferrer-Estevez & Chalmeta, 2021).

The concept of "Sustainable Development and Education" was first introduced in the comprehensive action plan formulated at the Earth Summit in Rio de Janeiro in 1992, which brought together 178 countries. Subsequently, significant concrete steps were taken through various institutional initiatives such as the United Nations Decade of Education for Sustainable Development (2005-2014) and the UN SDG Global Action Program (2014) (UN, 2023). Education takes center stage in the 2030 Agenda, launched in 2015, with Sustainable Development Goal (SDG) 4.7 defined as "ensuring that all learners acquire the knowledge and skills needed to promote sustainable development and sustainable lifestyles." Within this context, by 2030, education aims to promote sustainable development through the acquisition of knowledge and skills required for sustainable living, human rights, gender equality, the promotion of a culture of peace and non-violence, fostering global citizenship and appreciation of cultural diversity (UN, 2015). Education is seen as both a fundamental right for creating a more just, prosperous, and inclusive world and a force for economic, social, and political transformation (UNESCO, 2019). Consequently, many higher education institutions strive to incorporate the goal of "Qualified Education" into their action plans.

One of the significant aspects of quality education in higher education that needs to be reflected not only in action plans but also in implementation is gender equality. Gender equality in education refers to ensuring equal access to education and equal benefits from education for individuals of all genders, as well as avoiding discrimination or stereotypes based on gender within educational systems (Acar-Erdol & Gözütok, 2019; Pollock et al., 2021; Spencer et al., 2015). UNESCO (2023) calls for attention to gender equality in every component of the education system, from access to education, content, teaching practices, and learning outcomes, as well as life and employment opportunities. In many studies, the significance of gender equality initiatives in higher education is emphasized, weaknesses are identified, and improvement efforts are being sustained (CEPES, 2011; Enge, 2014; Okenwa-Emgwa & Straus, 2018; Yar et al, 2022).

Teachers play a critical role in creating educational environments where gender equality is ensured and awareness is fostered. To effectively fulfill this crucial role, it is deemed necessary to instill this awareness in the teacher training process. By providing a quality teacher education that promotes gender equality in education, more competent teachers can be trained, thereby supporting the success of SDGs. In this context, gender equality, which holds an important place among the sustainable development goals, should be integrated into the curriculum and presented to teacher candidates (Acar-Erdol & Gözütok, 2019; Pollock et al., 2021; Stake & Hoffmann, 2001).

While UNESCO (2015) recommends organizing training sessions on Gender Equality within the action plans developed by universities to support sustainable development goals, it is considered that integrating the theme of gender equality into existing curricula is a much more effective way to create a widespread impact rather than offering isolated courses disconnected from the context. Furthermore, it is believed that "Gender Equality" should not be confined to the scope of a single course but rather be integrated into the general awareness and understanding, becoming a way of life for individuals. Therefore, interventions aimed at promoting gender equality should take into account the characteristics of different stages of human development and educational contexts (Kollmayer et al., 2020).

Gender Equality in Education

Although significant progress has been made in gender equality, recognized as a global priority by UNESCO (2023), there still exist substantial gender disparities in many educational settings. The question of whether gender equality remains a problem in the early 2000s sadly continues to be relevant in some educational environments today. Even in democratic Western societies that uphold values of freedom, equality, and social justice, the patriarchal system continues to receive significant support and legitimize gender inequalities (Biemmi, 2015; Torre-Sierra & Guichot-Reina, 2022).

Globally, 118.5 million girls and 125.5 million boys are not attending school. Nearly two-thirds of illiterate adults are still women (UNESCO, 2023). The gains made in girls' education worldwide have not adequately reflected in the school completion rates of girls from impoverished rural areas, those from diverse ethnic backgrounds, or those with special needs. The disruptions experienced in education systems due to the pandemic have further worsened access for vulnerable groups consisting of girls and young women, deepening learning inequalities (UN Women, 2022).

According to the National Education Statistics of the Turkish Statistical Institute (TÜİK, 2022), the average years of education for the population aged 25 and over in Türkiye increased by 26% from 7.3 years in 2011 to 9.2 years in 2022. The average years of education were determined as 8.5 years for women and 10.0 years for men in 2022. When compared to the average school years in OECD countries, Türkiye has the lowest total education duration, being four years below the average for both women and men. In terms of quality in education, the Programme for International Student Assessment (PISA) results show that Türkiye ranks 34th in mathematical literacy and reading skills, with a score only higher than that of Mexico and Chile. In the field of scientific literacy, Türkiye ranked 32nd, surpassing the scores of Mexico,

Chile, Slovakia, and Greece (CEID, 2022). According to the PISA 2018 assessment in Türkiye, although the gender performance gap in mathematics has not significantly changed since 2009, male students have consistently outperformed female students (Ministry of Education, [MofEd], 2019). Considering the standardized nature of schools and their ability to reduce learning gaps between girls and boys (Downey et al., 2022), it is evident that there is still much to be done to achieve equality in school enrollment rates.

Gender inequalities in achievement, as indicated by the Programme for International Student Assessment (PISA), are a significant concern as they can have long-term consequences for the personal and professional futures of boys and girls in many countries (OECD, 2019). Socially constructed gender norms and stereotypes embedded in curricula, textbooks, and educational practices hinder girls' choices regarding what they will study at school and ultimately derail their careers and employment opportunities as adults (UN Women, 2022). Stereotypes related to gender roles for men and women can lead to biases in hiring and performance evaluations by perceiving women as unsuitable for tasks and professions typically associated with men. Conversely, the prescriptive aspect of gender stereotypes can legitimize the existing power structure favoring men and result in intentional discrimination against women who defy gender role prescriptions (Burgess & Borgida, 1999). If necessary measures are not taken, boys who lag behind in reading and lack basic competencies may face significant challenges in their further education, the labor market, and daily life. Similarly, the underrepresentation of girls among top-performing students in science and mathematics can perpetuate the gender gap in science, technology, engineering, and mathematics (STEM) careers, which often yield the highest income (OECD, 2019; UN Women, 2022). Consequently, while men are more represented in STEM fields and higher positions, women often find themselves working in lower-status jobs in health and social services (Biemmi, 2015; Kollmayer et al., 2020).

Considering the restrictive impact of socially constructed gender roles on both men and women, it is beneficial for educational guidance to aim at promoting diversity in the educational and career choices of both boys and girls. This objective is one of the key goals outlined in the European Parliament's decision on eliminating gender stereotypes in the European Union, dated March 12, 2013. In paragraph 26, the European Parliament emphasizes the need for policies that focus on breaking down gender stereotypes from an early age, providing awareness-raising education for teachers and students, and promoting and supporting career diversity for both young women and men (Biemmi, 2015). Education systems based on gender equality empower both boys and girls, fostering the development of life skills such as self-management, communication, negotiation, and critical thinking that young people need to succeed. They bridge skill gaps driven by economic factors, help children stay in school, and thus contribute to building well-being for all countries (UNICEF, 2023). Empowering the education of girls and young women will bring tangible returns in many areas, including faster poverty reduction, healthier mothers, reduced child mortality, HIV prevention and reduced violence (UN Women, 2022). An education free from negative gender norms is crucial not only for girls but also for boys. In many countries, norms associated with masculinity can fuel school dropout, child labor, gang violence, and involvement in armed groups. The need or desire to earn income also leads to boys leaving middle school because many believe that their schooling is unrelated to job opportunities (UNICEF, 2023).

Research indicates that male and female students receive different feedback from their teachers (Duffy et al., 2001). Female students are perceived by teachers as more motivated, well-behaved, and less demanding (Davies & Brember, 1995), whereas males are provided with more detailed explanations on how to improve their performance (Marshall & Smith, 1987). Teachers tend to attribute better mathematical skills to males and better language skills to females (Gajda et al., 2022). Traits such as emotional expression, willingness to collaborate, sensitivity, lack of self-confidence, and striving for perfection are seen as characteristic of girls, while boys are characterized by emotional restraint, toughness, a need for action, rationality, competitiveness, and greater behavioral problems (Aslan, 2015; Gajda et al., 2022). Students who do not conform to socially prescribed gender behaviors may become hesitant to express themselves confidently and exhibit their self-image, leading to self-repression. In this context, it would not be inaccurate to say that male and female students experience school differently (Frawley, 2005).

Considering the non-egalitarian discourses and environments present in educational processes, the importance of raising awareness about gender equality in teacher education becomes evident. This research, based on this premise, aims to determine the perspectives of prospective teachers in Türkiye regarding gender equality through their educational programs and to reveal the current situation.

Background

Theoretically, it can be observed that Türkiye recognizes the significance of creating an environment based on gender equality. In fact, there is a legal framework in place that holds all educational providers responsible for ensuring a gender-equitable environment. This is explicitly stated in the Constitution and the fundamental laws related to national education (Erden, 2009). Although the policy framework in this regard may not be well-defined, the implementation of these policies is crucial. The characteristics of individuals and society that an education system aims to foster shape the objectives of the curriculum. The curriculum, in turn, serves as the primary source and guide for teachers' classroom practices. Therefore, delivering gender equality goals to teachers through the curriculum is the first step to be taken. However, it would be appropriate to say that there are two parallel curricula in school settings that send conflicting messages. One is the official curriculum that offers equal opportunities for women and men explicitly, visibly, and on paper, while the other is the hidden curriculum that contains implicit messages contradicting the educational goals to be pursued (Biemmi, 2015). Children's perceptions of gender roles are influenced not only by explicit gender biases, such as being told whether they can perform a task based solely on their gender, but also by the "hidden curriculum" they encounter daily through teachers' behaviors, feedback, classroom differentiation, and instructional materials (Aslan, 2015; Frawley, 2005). Teachers are considered one of the most significant factors determining the quality of education due to their qualifications, subject expertise, and pedagogical competencies. In addition to their subject knowledge, teachers transmit their values, including explicit or implicit messages related to gender. Teachers' classroom practices are shaped by assumptions and stereotypes related to gender, which affect students' beliefs and learning (UNESCO, 2019). Moreover, teachers also have the power to stop the harm caused by the hidden messages conveyed through classroom biases (Acar-Erdol & Gözütok, 2017; Frawley, 2005). Therefore, teachers should reflect on their own classroom attitudes, critically analyze textbooks and curriculum, and question the alignment of textbooks and materials with the fundamental ideals and goals of the curriculum (Barton & Sakwa, 2012; Frawley, 2005; Kollmayer et al., 2020). Achieving gender equality and creating a fair educational environment requires teachers who possess unbiased gender beliefs and are aware of gender-related issues (Erden, 2009).

The significance of having well-educated teachers in promoting girls' enrollment and achievements has been emphasized in the existing literature. It is believed that providing gender equality education to teachers at all levels in the academic community can have a significant impact on challenging stereotypes and enhancing understanding of students and gender equality issues (Carinci, 2007; Gray & Leith, 2004). Furthermore, given that teachers work with hundreds of students throughout their careers, their education plays a crucial role in raising awareness and fostering consciousness about gender equality (Kollmayer et al., 2020).

Current teacher education programs and in-service training do not adequately prepare teachers to recognize and address gender biases that shape their classrooms (Barton & Sakwa, 2012; Biemmi, 2015; Erden, 2009; Gray & Leith, 2004; Sadker, 1999; Unterhalter et al., 2014). Teachers often fail to notice or overlook the gender-biased content in textbooks (Barton & Sakwa, 2012; Gajda et al., 2022; Kostas, 2023). Due to a lack of education on equitable approaches, educators are not ready to make the necessary changes in their assessment methods, curriculum choices, classroom interactions, or instructional strategies that would promote a gender-equitable learning environment (Carinci, 2007; Gaudet & Lapointe, 2013; Kollmayer et al.,

2020). The absence of gender equality education for teaching staff at all levels of education indicates the need to address this issue in university and adult education classrooms (Carinci, 2007; Unterhalter et al., 2014).

Gender bias can be mitigated through high-quality pre-service and in-service training programs, which provide teachers with opportunities to reassess their own gender assumptions, attitudes, and expectations, and learn ways to ensure equal treatment and participation in the classroom and the broader school community. To ensure the effectiveness and sustainability of gender equality awareness, teacher training institutions should develop programs that not only enhance teachers' general competencies but also explicitly engage with ideas related to gender equality (UNESCO, 2019; Unterhalter et al., 2014).

Research has shown that gender equality education provided to teacher candidates leads to positive changes in their attitudes (Erden, 2009; Gaudet & Lapointe, 2013; Kollmayer et al., 2020). Moreover, it is essential to provide teachers with critical tools to combat gender-biased stereotypes in their daily instructional practices (Biemmi, 2015). When teacher candidates are offered educational experiences and necessary tools that support the creation of an equitable learning environment, the implementation of teaching behaviors based on gender equality becomes inevitable in their classrooms. The crucial point is to continue applying and developing these skills throughout their entire careers (Tracy & Lane, 1999).

Purpose

Critical and democratic education plays an undeniable role in making the world more equitable and democratic by nurturing free-thinking and creative individuals. In order to achieve the intended goals of education, curriculum holds vital importance. The design and content determination of the curriculum reflects the understanding of individuals, society, and the world in educational policies (Aratemur-Çimen & Bayhan, 2018). A review conducted by UNESCO on over 110 national curricula in primary and secondary education across 78 countries revealed that many textbooks and learning materials contain gender biases, resulting in reduced classroom participation and limited expectations regarding education and life choices (UNESCO, 2019).

Education Goal 4.7, included in the Sustainable Development Goals, promotes education that eliminates gender-based discrimination or stereotypes as part of sustainable development.

Attaining this goal requires a teacher education that has internalized gender equality, which holds critical importance (Acar-Erdol & Gözütok, 2019; Kumlu & Çomoğlu, 2021; Kuruvilla, 2014; Pollock et al., 2021; Stake & Hoffmann, 2001). The education received by teachers at the university level should serve the internalization of gender equality, enabling them to comprehend their teaching programs effectively and transfer them into the instructional environment through the lens of gender equality.

Based on this context, this study focuses on the concept of "gender equality" through the national curriculum that will guide teachers when they start their profession. The implementation of gender equality education can only be realized by integrating it into national education policies, curricula, textbooks, and teacher training at all levels. No study was found in the literature that addresses the curriculum from this perspective. Thus, this study holds original value by allowing teacher candidates to examine their own subject area's curriculum under the theme of "Gender Equality" and integrate gender equality awareness into their curriculum literacy skills. In this regard, the aim of the study is to examine teacher candidates' own subject area curricula under the theme of "gender equality" and articulate their views on gender equality based on the curriculum. The study seeks to answer the following questions:

1. What is the gender equality awareness of teacher candidates?
2. How do teacher candidates evaluate the national curricula they examine in terms of emphasizing gender equality?
3. What are the views of teacher candidates on the integration of the gender equality theme into teacher education programs?

METHOD

Design

This study aims to determine the views of teacher candidates who take the course "Curriculum Development in Education" regarding gender equality within the context of their own subject area curricula. Teacher candidates were asked to examine their own subject-related curricula in terms of creating awareness about gender equality within the scope of the "Curriculum Development in Education" course and their opinions were obtained accordingly. The research design employed in this study is the phenomenology design, which provides rich data by determining the experiences, perceptions, and meanings attributed by teacher candidates to the researched phenomenon. The specific phenomenon focused on in this study is the "theme of gender equality in the national curriculum" (Patton, 2014; Yıldırım & Şimşek, 2018).

Study Group

This research was conducted with teacher candidates who were enrolled in higher education. The study group consisted of 54 teacher candidates who took the course

Table 1. Information about the teacher candidates in the study group

		Teacher Candidates' Branches					
Variables		**Science**	**Mathematic**	**Turkish**	**Foreign Language**	**Preschool**	**Sum**
Gender	Female	9	7	7	4	9	**36**
	Male	1	6	2	4	5	**18**
Class Level	2nd class	2	3	5	1	4	**15**
	3rd class	4	8	1	4	7	**24**
	4th class	4	2	3	3	3	**15**
Status of Receiving Training on Gender Equality	Yes	-	-	-	-	1	**1**
	No	10	13	9	8	13	**53**
SUM		**10**	**13**	**9**	**9**	**14**	**54**
Teacher candidates' codes		**S1-10**	**M1-13**	**T1-9**	**FL1-9**	**P1-14**	

"Curriculum Development in Education" during the spring semester of the 2021-2022 academic year at the Faculty of Education. The information about the teacher candidates comprising the study group is presented in Table 1.

Based on Table 1, it can be observed that the participating students are pursuing their education in the fields of Science (f=10), Mathematics (f=13), Turkish (f=9), Foreign Language (f=8), and Pre-school (f=14). The teacher candidates are enrolled in their 2nd year (f=15), 3rd year (f=24), and 4th year (f=15) of study. Out of the 54 participating teacher candidates, only one has indicated that they have previously attended any training related to "Gender Equality". Each teacher candidate was coded with the first letter of his/her branch and a number to indicate the statements noted during the interview. For example, the first student studying science education was given the code "S1".

Data Collection

The research utilized qualitative data collection methods. The views and suggestions of teacher candidates who took the elective course "Curriculum Development in Education" regarding the theme of gender equality in the curriculum were sought through structured interview questions. The questions included in the research are as follows:

1. In what contexts have you encountered gender equality?
2. What does gender equality mean to you?

3. How did you find the national curriculum you examined in terms of "Gender Equality"? Could you elaborate?
4. Is "Gender Equality Awareness" important for a teacher? Why is it important?
5. How can you incorporate "Gender Equality Awareness" into your future teaching practices?
6. How did the inclusion of the concept of gender equality in the curriculum development course affect your views?

Data Analysis

The data obtained from the interviews conducted with teacher candidates were analyzed using content analysis method. The content analysis was conducted in five stages. In the first stage, the interviews with teacher candidates were transcribed. In the second stage, the responses of teacher candidates were analyzed in detail. In the third stage, the data were analyzed by the researchers at different time intervals to create the codes for the study. In the fourth stage, the codes were grouped under categories and themes. Expert opinions were sought from professionals in the field regarding the codes, categories, and themes created by the researchers. In the final stage, the data, which were given their final form with the input of expert opinions, were organized and presented in detail in the findings section for the readers (Creswell, 2014).

Validity and Reliability Studies

The validity and reliability of the research have been thoroughly reported by the researcher for each piece of data. Additionally, expert opinions were sought for the data, and efforts were made to establish the validity and reliability of the research through expert evaluations. During this process, the similarities and differences between expert and researcher codings were examined. The discrepancies among expert codings were discussed and deliberated until 100% consensus was reached between the experts and the researcher (Yıldırım & Şimşek, 2016).

Findings

The aim of this research was to determine the perspectives of teacher candidates who took the course "Curriculum Development in Education" regarding gender equality within the context of their own subject area curriculum. In this section of the study, the awareness of teacher candidates regarding gender equality, their views on the emphasis on gender equality in the national curriculum they examined, and their suggestions will be presented.

Teacher Candidates' Awareness of Gender Equality

The awareness of teacher candidates regarding gender equality was examined in two stages. Firstly, an attempt was made to determine in which environments the teacher candidates heard about the concept of gender equality in their surroundings. Secondly, the statements of the teacher candidates regarding the meaning of gender equality were examined. Table 2 presents the frequencies of the environments in which the teacher candidates encountered the expression of gender equality.

Table 2. The environments teacher candidates where they acquire the concept of gender equality

	Teacher Candidates' Branches					
Environments	**Science**	**Mathematic**	**Turkish**	**Foreign Language**	**Preschool**	**Sum**
Social Media	7 *(S1, S4, S5, S6, S7, S8, S9)*	6 *(M4, M5, M6, M8, M11, M13)*	8 *(T1, T2, T3, T4, T5, T6, T7, T9)*	4 *(FL2, FL4, FL5, FL7, FL8)*	6 *(P1, P3, P9, P11, P12, P14)*	31
Course	3 *(S2, S3, S7)*	5 *(M1, M2, M6, M9, M12)*	2 *(T4, T8)*	-	7 *(P3, P6, P7, P8, P9, P11, P13)*	17
Television	1 *(S10)*	2 *(M6, M10)*	-	2 *(FL3, FL8)*	2 *(P3, P11)*	7
Friends	2 *(S2, S10)*	-	-	1 *(FL6)*	2 *(P10, P11)*	5
Family	1 *(S10)*	-	-	1 *(FL1)*	1 *(P11)*	3
Conference	-	1 *(M15)*	-	-	1 *(P5)*	2
Training	-	-	-	-	1 *(P4)*	1
Never met	-	1 *(M3)*	-	-	-	1

*There are teacher candidates who indicated more than one environment.

When Table 2 is examined, it is observed that teacher candidates mostly encounter the concept of gender equality on Social Media (f=31). 17 of the teacher candidates emphasized that the concept of gender equality is discussed in their courses. Television (f=7), Friends (f=5), Family (f=3), and Conference (f=2) were expressed as other environments where the concept of gender equality is encountered. Additionally, 1 teacher candidate mentioned participating in a certificate program aimed at raising awareness of gender equality. 1 teacher candidate stated that they have never encountered this concept before.

In order to determine the understanding of gender equality among teacher candidates, they were asked to define this concept. The codes and frequencies regarding how teacher candidates conceptualize "gender equality" are presented in Table 3.

When examining Table 3, it is observed that the teacher candidates' definitions of gender equality are mostly categorized under the code "Equal rights and responsibilities for women and men" (f=31). Under the code "Equality of all individuals in all aspects of social life regardless of their sexual identity," 8

Table 3. Teacher candidates' understanding of gender equality

	Teacher Candidates' Branches					
Codes for Definitions	**Science**	**Mathematic**	**Turkish**	**Foreign Language**	**Preschool**	**Sum**
Equal rights and responsibilities for women and men	3 *(S1, S3, S6)*	7 *(M2, M5, M6, M7, M8, M9, M10, M13)*	5 *(T4, T5, T7, T8, T9)*	5 *(FL1, FL3, FL5, FL7)*	11 *(P1, P2, P3, P5, P6, P7, P8, P9, P10, P12, P14)*	**30**
Equality of all individuals regardless of sexual identity in all areas of social life.	2 *(S7, S9)*	2 *(M3, M12)*		2 *(FL2, FL8)*	3 *(P4, P13)*	**8**
Equal rights and responsibilities for women and men to the extent permitted by biological sex.	2 *(S2, S10)*	1 *(M4)*	4 *(T1, T2, T3, T6)*			**7**
Women's struggle for equality.	1 *(S5)*	1 *(M11)*				**2**
Prevention of discrimination against women and girls.	1 *(S4)*					**1**
Thoughts and attitudes towards the sexes are influenced by culture.		1 *(M1)*				**1**
Equality in occupation.				1 *(FL4)*		**1**
Non-classification of women and men according to gender.					1 *(P11)*	**1**
Granting women the same rights as men.				1 *(FL6)*		**1**
Gender should not be an advantage or disadvantage.	1 *(S8)*					**1**
Discrimination.		1 *(M7)*				**1**

teacher candidates provided definitions, while under the code "Equal rights and responsibilities for women and men within the limits allowed by biological sex," 7 teacher candidates' definitions were found. It can be seen that the teacher candidates include definitions that take a female perspective on gender equality under the codes "Women's struggle for equality" (f=2), "Prevention of discrimination against women and girls" (f=1), and "Extending rights given to men to women as well" (f=1). Additionally, there are definitions expressed through the codes "Thoughts and attitudes towards genders shaped by culture" (f=1), "Equality in the profession" (f=1), "Non-classification of individuals based on gender" (f=1), "Gender not being an advantage or disadvantage" (f=1), and "Discrimination" (f=1). Here are some examples of the definitions provided by the teacher candidates:

Equal rights and responsibilities for women and men:

Gender equality is the equal representation of women and men in society, without depriving women of their abilities or restricting their freedom based on their gender. It is about ensuring that individuals can live freely and pursue their aspirations without being hindered by circumstances beyond their control. (T7)

Equality of all individuals in all aspects of social life regardless of their sexual identity:

In the society we live in, gender equality means that individuals of all genders, including those with different sexual orientations, should not experience any form of gender discrimination. In other words, it entails that individuals should not be judged or subjected to discriminatory practices or discourses solely based on their gender in social life. (S9)

Equal rights and responsibilities for women and men within the limits allowed by biological sex:

We acknowledge that men and women are not equal in terms of their biological makeup, and they have physical differences. For instance, it is recognized that man generally possess greater physical strength. In my perspective, gender equality entails that men and women should have equal rights within the legal framework, considering these physical disparities, and that they should be equally esteemed and valued in society. (M4)

Women's struggle for equality:

This concept invokes the struggle that women have been engaged in for equality and the challenges they have encountered throughout their lives. (S5)

The Emphasis on Gender Equality in Curriculum

The purpose of this study was to investigate the perspectives of teacher candidates regarding the emphasis on gender equality in their own subject area national curriculum (in Türkiye). The codes and frequencies of teacher candidates' views on the presence of gender equality emphasis in the national curriculum are presented in Table 4.

Table 4. Views of teacher candidates on the emphasis on gender equality in the examined curriculum

	Teacher Candidates' Branches					
Codes	**Science**	**Mathematic**	**Turkish**	**Foreign Language**	**Preschool**	**Sum**
Inadequate	9 *(S1, S2, S3, S4, S5, S6, S7, S8, S10)*	11 *(M3, M4, M5, M6, M7, M8, M9, M10, M11, M12, M13)*	9 *(T1, T2, T3, T4, T5, T6, T7, T8, T9)*	5 *(FL1, FL2, FL4, FL5, FL6)*	10 *(P2, P3, P4, P5, P6, P8, P9, P10, P11, P14)*	**44**
Partially Adequate	1 *(S9)*	1 *(M2)*		1 *(FL8)*	3 *(P1, P12, P13)*	**6**
Adequate		1 *(M1)*		2 *(FL3, FL7)*	1 *(P7)*	**4**

When examining Table 4, it is observed that 44 teacher candidates *(30 female, 14 male)* found the emphasis on gender equality in the examined curriculums to be inadequate. Six teacher candidates *(3 female, 3 male)* considered the curricula partially adequate, while four teacher candidates *(3 female, 1 male)* stated that the examined curricula adequately emphasized gender equality. Some teacher candidates who believed that the curriculum was inadequate in terms of gender equality also pointed out the presence of sexist discourses and visuals in the curriculum. For example:

Unfortunately, I noticed the presence of discriminatory distribution based on gender in visuals, stories, and similar content that perpetuate social roles in the curriculum. (P14)

I can say that it is not sufficiently addressed or explicitly included as a specific achievement in the curriculum. Additionally, it is possible to state that gender-biased expressions are present in the text selection, and we need to be careful in this regard. (T3)

Furthermore, some teacher candidates who believed that the curriculum was inadequate in terms of gender equality also mentioned that although the subject areas were highly suitable for establishing these connections, the necessary connections were not made in the curriculum. For example:

We did not come across any achievement related to the concept of gender equality in the middle school mathematics curriculum that we examined. Although it is very challenging to establish these connections, I realized that some topics are suitable for adding achievements related to gender equality. I believe that the curriculum is highly deficient in this regard and should be updated. (M4)

I examined the Turkish Language Curriculum, and I found it inadequate. Since it is a thematically designed program, every topic can be easily used in the curriculum, including gender equality. However, it has not been utilized. (T2)

Some of the teacher candidates who believed that the curriculum they examined had a "partially adequate" emphasis on gender equality expressed their views as follows:

Our curriculum does not delve deeply into gender equality issues, but it does touch on gender equality, justice, and task distribution between genders in many of its achievements. (P1)

Since we examined the Mathematics curriculum, we could not find much detail about this topic. Generally, there were visuals in problem questions related to the roles assigned to men and women in society. For example, the father going to the grocery store and the mother preparing the meal. (M12)

Some teacher candidates considered the emphasis on gender equality in the examined curriculum to be "adequate". Here are some sample statements from those who believed that the curriculum adequately emphasized gender equality:

I found the examined curriculum to be generally sufficient in terms of the concept of gender equality. The messages hidden in the subtext and the activities integrated into the lessons attempted to explain what gender equality is in a simple manner. Since they were prepared in accordance with the achievements, I can say that they achieved their objectives. (FL3)

By examining the curriculum, I grasped the importance of gender equality in early childhood. I saw that gender equality was not overlooked in the curriculum and that they emphasized its significance. (P7)

The Importance of Gender Equality in Teacher Education

Within the scope of the study, teacher candidates were asked whether they considered having awareness of gender equality important for a teacher. All of the teacher candidates (f=54) stated that having awareness of gender equality is crucial for teachers. The codes and frequencies of their views on how they would reflect this awareness in the classroom are presented in Table 5.

Table 5. Teacher candidates' views on how they will reflect gender equality awareness in their classrooms

	Teacher Candidates' Branches					
Codes	**Science**	**Mathematic**	**Turkish**	**Foreign Language**	**Preschool**	**Sum**
Utilizing supportive teaching activities and materials.	2 *(S6, S9)*	3 *(M5,M7, M9)*	2 *(T2, T6)*	2 *(FL2, FL7)*	6 *(P1, P3, P4, P8, P9, P13)*	**16**
By treating students in the classroom equally.	4 (S1, S2, S8, S10)	5 *(M3, M6, M8, M10, M13)*	3 *(T1, T3, T9)*		1 *(P2)*	**13**
By being a role model.	1 *(S7)*	2 *(M2, M4)*	1 *(T7)*		4 *(P5, P6, P11, P12)*	**8**
By explaining the concepts of gender equality to students.			1 *(T4)*	4 *(FL1, FL3, FL5, FL6)*	1 *(P14)*	**6**
I don't know how to do it.	2 *(S3, S5)*		1 *(T5)*	1 *(FL8)*	1 *(P10)*	**5**
I have confidence in myself.	1 *(S4)*	1 *(M12)*				**2**
By making use of the hidden curriculum.		1 *(M1)*	1 *(T8)*			**2**
By raising awareness among female students.		1 *(M11)*				**1**
By raising awareness about professions.				1 *(FL4)*		**1**

When examining Table 5, it can be observed that teacher candidates predominantly intend to foster gender equality awareness in their classrooms through the utilization of "supportive teaching activities and materials" (f=16) and "by treating students equally" (f=13). Furthermore, some teacher candidates expressed their belief that they can promote gender equality awareness in their classrooms by serving as "role models" (f=8), "explaining gender equality concepts to students" (f=6), and utilizing "implicit curriculum" (f=2). One candidate mentioned raising "awareness among female students" (f=1), while another mentioned creating "awareness about professions" (f=1). Two candidates expressed their confidence solely in themselves.

Although five teacher candidates acknowledged the significance of possessing gender equality awareness as a teacher, they expressed uncertainty about how to effectively implement it in their classrooms. Below are some examples of statements made by teacher candidates:

Utilizing supportive teaching activities and materials:

As a teacher, I consider it important to have awareness of 'Gender Equality' when I start my profession. I believe that I need to approach my classroom with an egalitarian mindset, considering that I will have students of all genders and orientations, in order to create a democratic environment. Over time, I am confident that my students will also acquire a solid understanding of this topic. To reflect this awareness in my classroom, I would implement various activities such as discussions involving dilemma-based scenarios and employing techniques like circle time for dialogue. (S9)

By treating students equally:

I firmly believe that teachers should possess this awareness. In my teaching career, I plan to treat all my students equally, without any gender distinctions, and I intend to instill this awareness in my students as well. (M3)

As a role model:

We know that teachers are the individuals who serve as examples to future generations. From this perspective, teachers are crucial in creating awareness, whether it is in the field of gender equality or any other area. Therefore, if we want to foster gender equality in society, we must educate effectively in this regard as well. If we desire our students to have awareness in this matter, it is essential for us to be good role models of gender equality for them. Even our own awareness in this aspect would be a positive step towards reflecting it in the classroom. (S7)

By explaining gender equality concepts to students:

First and foremost, I would like to teach students the importance of freedom and equality, and how these concepts are relevant in our lives. I believe that students who thoroughly understand these concepts will develop awareness in this area. (E4)

I don't know how to do it:

As an early childhood teacher, I consider it important, and my approach will be the same for all children. Since children are very young at this age, I don't think much will be visualized in their minds when I mention gender equality. Unfortunately, I don't know how to reflect this concept in my classroom. (P10)

Finally, as part of this study, teacher candidates were asked to explain how the inclusion of 'Gender Equality' in the context of the 'Curriculum Development in Education' course influenced their views on the subject. The obtained codes and frequencies related to the opinions of the teacher candidates are presented in Table 6.

Table 6. The impact of addressing the topic of gender equality in the "curriculum development in education" course on the opinions of teacher candidates

	Teacher Candidates' Branches					
Codes	**Science**	**Mathematic**	**Turkish**	**Foreign Language**	**Preschool**	**Sum**
Increased general awareness of gender equality.	2 (S3, S7, S8)	6 *(M1,M2, M5, M8, M11, M12, M13)*	1 *(T8)*	5 *(FL1, FL2, FL5, FL6, FL7)*	6 *(P1, P4, P6, P11, P12, P13)*	**22**
Learning how to ensure gender equality in the classroom.	4 *(S2, S4,S6, S9)*	4 *(M3,M6, M9, M10)*	4 *(T2, T4, T5, T7)*	2 *(FL3, FL4)*	6 *(P2, P3, P7, P8, P9, P14)*	**20**
Recognizing their own gender stereotypes.	3 *(S1, S5, S10)*	1 *(M4)*	2 *(T1, T6))*		2 *(P5, P10)*	**8**
Learning that gender equality is also important for men.		1 *(M7)*	1 (T9)			**2**
Keeping their views unchanged.			1 *(T3)*	1 *(FL8)*		**2**

When examining Table 6, it can be observed that teacher candidates, within the scope of the "Course on Curriculum Development in Education," primarily emphasized the aspects related to "increased general awareness of gender equality" (f=22) and "learning how to promote gender equality in the classroom" (f=20) when analyzing their own subject area national curriculum with a focus on gender equality. The following are examples of statements made by teacher candidates regarding these codes:

Increased general awareness of gender equality:

I have come to realize the deficiencies in terms of gender equality in all subject programs, whether they are included or not, not only in my own field of Turkish teaching but also in other disciplines, which has enhanced my overall awareness of gender equality. (T8)

Learning how to promote gender equality in the classroom:

At first, I didn't find it particularly significant to address gender equality within the scope of the Curriculum Development course. However, through the assignments and activities we have undertaken, I realized that the foundations of this topic can indeed be laid within the curriculum. This has led me to develop the perspective that the subject, which is aimed to be instilled in students, should first find its place in the curriculum and subsequently come to life through the practices of future teachers. (T4)

Additionally, the opinions of eight teacher candidates are grouped under the code "Recognizing their own gender stereotypes.".

I have realized that gender equality is actually a much broader concept. I used to approach it from a narrow perspective, focusing only on linguistic aspects, such as using 'scientist' instead of 'man of science.' However, I have come to recognize that our perceptions play a significant role in this matter. I am now striving to eliminate gender-biased expressions from my thinking and cleanse my mind of baseless associations. (T1)

Two teacher candidates expressed their awareness under the code "Learning the importance of gender equality from the perspective of men.".

I have come to see that both women and men face different issues. Prior to this course, I predominantly believed that women were more subjected to gender inequality. However, I now believe that men also experience various forms of gender inequality in many aspects. (T9)

Two other teacher candidates stated that their initial awareness of gender equality remained unchanged.

DISCUSSION

In this study, the aim was to determine the views of teacher candidates who have taken the course on curriculum development in education regarding gender equality within the context of their own subject area national curriculum. Within the scope of this aim, the awareness of teacher candidates about gender equality, their opinions on the inclusion of gender equality in the national curriculum, and their views on the importance of gender equality education in teacher education were attempted to be identified.

Firstly, it was revealed that teacher candidates encounter information about gender equality mostly through social media. Following this, teacher candidates stated that gender equality is also discussed in school lessons, on television, and among their peers. While one teacher candidate expressed that they had never come across the concept of gender equality until now, another candidate mentioned attending a conference on this topic, and only one candidate stated that they received education within the scope of a certificate program on gender equality. At this point, it is noteworthy that the knowledge of teacher candidates about gender equality mostly relies on social media. Numerous search engines and recommendation websites on the internet contain stimuli that reinforce gender stereotypes (Raj et al., 2021). In this sense, while social media can be an effective communication channel, it can also serve as an environment that allows the rapid dissemination of unreliable information. From the statements of teacher candidates, it can be understood that although gender equality is mentioned in the courses they take, there is a lack of courses in the teacher education programs that aim to raise awareness about gender equality. This finding supports the results of other studies in the field (Barton & Sakwa, 2012; Biemmi, 2015; Erden, 2009; Gray & Leith, 2004; Sadker, 1999; Unterhalter et al., 2014). Indeed, it is believed that discussing gender equality in courses that do not intentionally focus on gender equality will be insufficient in creating awareness at the desired level.

Secondly, the teacher candidates were asked to define gender equality. It is observed that the majority of teacher candidates express their understanding of gender equality as "equal rights and responsibilities for women and men." Koyuncu Şahin et al. (2018) also draw attention to similar definitions provided by teacher candidates in their study. In addition, some teacher candidates define gender equality as "equal opportunities for all individuals in all spheres of social life, regardless of their sexual identity" and "equal rights and responsibilities for women and men to the extent permitted by biological sex." It is worth noting that a few teacher candidates associate gender equality with definitions such as "the struggle for equality for women," "preventing discrimination against women and girls," and "granting women the same rights as men," indicating a perception of gender equality as a process focused solely on women. Gray & Leith (2004) similarly expressed concerns among teachers that equality issues tend to be more focused on girls than boys.

When examining the opinions of teacher candidates regarding the emphasis on gender equality in their own subject area's national curriculum, it is seen that the majority of teacher candidates did not come across sufficient evidence of creating awareness of gender equality in the curriculum they examined. This finding is consistent with the results of studies in the literature (Bayhan, 2020; Bayhan & Aratemur-Çimen, 2019). Additionally, the opinions of teacher candidates regarding the inclusion of sexist discourse and visuals in the national curriculum can be considered as an indication of the need to eliminate sexist language from the curriculum. In their research on early childhood curricula implemented in Türkiye, Koyuncu-Şahin et al. (2018) similarly stated that curricula pose problems in achieving gender equality and contain content that perpetuates inequality.

Lastly, the opinions of teacher candidates regarding the importance of gender equality in teacher education were examined. The findings indicate that all teachers are aware of the importance of gender equality in teacher education. This view of teachers is consistent with the findings of other research in the field (Duffy et al., 2001; Koyuncu-Şahin et al., 2018; Öngören, 2019). The fact that teacher candidates express their desire to utilize supportive teaching activities to create an egalitarian classroom environment and to serve as role models for their students through their behaviors can be considered an indication of their awareness of this issue and their willingness to translate this awareness into practice. Teachers being the individuals who spend the most time with students after their families (Aslan, 2015) further highlights the importance of teacher candidates' awareness. Furthermore, some teacher candidates stating that they are aware of gender equality but unsure of how to reflect it in the classroom points to the need for measures to be taken in teacher education (Carinci, 2007; Gaudet & Lapointe, 2013; Gray & Leith, 2004; Kollmayer et al., 2020). Similarly, the fact that teacher candidates express their desire to explain the concept of gender equality to their students, raise awareness among girls, and utilize the hidden curriculum to carry out this process suggests the need to direct teacher candidates toward reliable sources of information and effective educational processes instead of relying solely on social media. In fact, research findings indicate that while teacher candidates aim to create an egalitarian teaching environment, they unintentionally reinforce inequality (Biemmi, 2015; Gajda et al., 2022).

Teacher candidates stated that they found the process of examining their own subject areas' national curriculum from the perspective of gender equality within the scope of the "Curriculum Development in Education" course to be highly beneficial. They mentioned that this process increased their overall awareness, enabled them to learn how to reflect gender equality in their future classrooms, and most importantly, made them aware of their own biases. Other studies in the literature also indicate that teachers hold gender-biased stereotypes (Duffy et al., 2001; Öngören, 2019). The findings support the existing studies and reveal the inadequacy of integrating a gender equality perspective into teacher education (Barton & Sakwa, 2012; Biemmi, 2015; Carinci, 2007; Erden, 2009; Gray & Leith, 2004; Sadker, 1999; Unterhalter et al., 2014).

CONCLUSION AND RECOMMENDATIONS

In general, it is evident that teacher candidates have limited exposure to the concept of "Gender Equality" beyond social media, classroom discussions, and conversations with friends. They have not undergone a target educational process on this topic. While teacher candidates can define the concept of gender equality as equal rights and responsibilities for women and men, it is observed that they lack sufficient depth of understanding. When examining the national curriculum, teacher candidates could not find sufficient evidence of gender equality. They discovered indications of avoidance of expressions that would reflect gender equality and the presence of sexist discourse in the curriculum. They express the need for the curriculum to be redesigned from a gender equality perspective and emphasize the significant role of teachers in implementing the curriculum. Teacher candidates are aware of the importance of gender equality in teacher education and express their willingness to receive education on this topic and reflect on its outcomes in their future classrooms.

This research reveals that gender equality does not have a significant place in the current national curriculum and teacher education practices in Türkiye. Based on this point, it is deemed necessary to integrate the perspective of gender equality into teacher education programs as outlined in universities' action plans. It is believed that developing educational programs and organizing conferences in universities to raise awareness of gender equality and encourage students to adopt it as a way of life would be effective in meeting the needs in this area. Additionally, the existing national curriculum for preschool, primary, and secondary education levels in Türkiye should be reviewed from a gender equality perspective and rid of sexist discourse (Acar-Erdol & Gözütok, 2017; Bayhan, 2020). Curricula, which are a set of training situations for teachers, should be made to serve the creation of teaching environments that support gender equality with their objectives, content, teaching-learning process and assessment and evaluation elements (ERG, 2023; Kostas, 2023). It is believed that teachers who have undergone effective teacher education processes and have awareness of gender equality will contribute to the cultivation of individuals with critical thinking skills by incorporating national curriculum and other instructional materials prepared with this awareness into their teaching environments. This, in turn, will foster the development of individuals with advanced thinking skills in the 21st century. The improvements made in both teacher education and the curriculum are expected to contribute significantly to achieving the goal of "Quality Education" as part of the sustainable development plan.

LIMITATIONS AND DIRECTIONS FOR FUTURE STUDIES

This research was conducted by examining teacher candidates on the preschool, mathematics, science, and foreign language curriculums prepared by the Board of Education of the Ministry of National Education of the Republic of Türkiye. Obtaining findings through the analysis of another curriculum may lead to more comprehensive results regarding the curriculum. Furthermore, the findings obtained regarding the curriculum are limited to the perspectives of teacher candidates. Of the total 44 teacher candidates in the study group, 36 were female, and 18 were male. The difference in the number of people in terms of gender variable is due to the fact that there are more female students in the faculty of education. In this sense, the obtained results are influenced by the teacher candidates', gender, cognitive levels, grade levels, and cultural backgrounds. A research study conducted with the analysis by experts in the field may yield different results. It is believed that for identifying gender equality in educational environments, more detailed data can be obtained through the analysis of textbooks and other materials, observation of teaching environments, and the opinions of practicing teachers. Such data would contribute to the existing literature in the field.

REFERENCES

Acar-Erdol, T., & Gözütok, F. D. (2017). Ortaöğretim öğrencileri için toplumsal cinsiyet eşitliği öğretim programı ihtiyaç analizi:Bir Anadolu lisesi örneği [Needs assessment for gender equality curriculum among high school students: An example of an Anatolian High School]. *Eğitim ve Bilim, 42*(190), 39–65. doi:10.15390/EB.2017.6429

Acar-Erdol, T., & Gözütok, F. D. (2019). Toplumsal cinsiyet eşitliği eğitim programının değerlendirilmesi [Evaluation of gender equality curriculum]. *İlköğretim Onlne, 18*(4), 1497-1519.

Aratemur-Çimen, C., & Bayhan, S. (2018). *Değişen ders kitaplarında sekülerizm ve toplumsal cinsiyet eşitliği araştırması*. Karşılaştırmalı Eğitim Derneği.

Aslan, G. (2015). Öğretmen adaylarının toplumsal cinsiyet algılarına ilişkin metaforik bir çözümleme [A metaphoric analysis regarding gender perceptions of preservice]. *Eğitim ve Bilim, 40*, 363–384. doi:10.15390/EB.2015.2930

Barton, A., & Sakwa, L. N. (2012). The representation of gender in English textbooks in Uganda. *Pedagogy, Culture & Society, 20*(2), 173–190. doi:10.1080/14681366.2012.669394

Bayhan, S. (2020). *2005'ten 2020'ye öğretim programları toplumsal cinsiyet eşitliği ve sekülerizm araştırması sonuç raporu*. Karşılaştırmalı Eğitim Derneği.

Bayhan, S., & Aratemur-Çimen, C. (2019). 2017 müfredat reformu sonrası ders kitaplarında toplumsal cinsiyet eşitliği [Gender equality in the textbooks of 2017 curricula]. *Eğitim Bilim Toplum Dergisi, 17*(67), 62-88.

Biemmi, I. (2015). Gender in schools and culture: Taking stock of education in Italy. *Gender and Education, 27*(7), 812–827. doi:10.1080/09540253.2015.1103841

Burgess, D. J., & Borgida, E. (1999). Who women are, who women should be: Descriptive and prescriptive gender stereotyping in sex discrimination. *Psychology, Public Policy, and Law, 5*(3), 665–692. doi:10.1037/1076-8971.5.3.665

Carinci, S. (2007). Inclusion of gender in curriculum and instructional teaching practices and its impact on the adult learner. *Western Region Research Conference on the Education of Adults (WRRCEA).*

CEİD. (2022). *Türkiye'de Toplumsal Cinsiyet Eşitliğini İzleme Raporu 2021-2022*. Cinsiyet Eşitliği İzleme Derneği.

CEPES. (2011). *From Gender Studies to Gender IN Studies: Case Studies on Gender-Inclusive Curriculum in Higher Education*. UNESCO.

Davies, J., & Brember, I. (1995). Attitudes to school and the curriculum in year 2, year 4 and year 6: Changes over four years. *European Conference on Educational Research.*

Downey, D. B., Kuhfeld, M., & van Hek, M. (2022). Schools as a relatively standardizing institution: The case of gender gaps in cognitive skills. *Sociology of Education, 95*(2), 89–109. doi:10.1177/00380407211070319

Duffy, J., Warren, K., & Walsh, M. (2001). Classroom interactions: Gender of teacher, gender of student, and classroom subject. *Sex Roles, 45*(9/10), 579–593. doi:10.1023/A:1014892408105

Enge, M. R. (2014). Gender equality in public higher education institutions of Ethiopia: The case of science, technology, engineering, and mathematics. *Discourse and Communication for Sustainable Education, 5*(1), 3–21. doi:10.2478/dcse-2014-0001

Erden, F. T. (2009). A course on gender equity in education: Does it affect gender role attitudes of preservice teachers? *Teaching and Teacher Education, 25*(3), 409–414. doi:10.1016/j.tate.2008.11.001

ERG. (2023). *Türkiye'de Kapsayıcı ve Nitelikli Eğitim için Öncelikler: 14 Mayıs 2023 Seçimlerinde Göreve Gelecek Hükümete ve Meclise Öneriler*. Eğitim Reformu Girişimi.

Ferrer-Estévez, M., & Chalmeta, R. (2021). Integrating sustainable development goals in educational institutions. *International Journal of Management Education*, *19*(2), 1–19. doi:10.1016/j.ijme.2021.100494

Frawley, T. (2005). Gender bias in the classroom: Current controversies and implications for teachers. *Childhood Education*, *81*(4), 221–227. doi:10.1080/00094056.2005.10522277

Gajda, A., Bojko, A., & Stoecker, E. (2022). The vicious circle of stereotypes: Teachers' awareness of and responses to students' gender-stereotypical behaviour. *PLoS One*, *17*(6), 1–24. doi:10.1371/journal.pone.0269007 PMID:35704605

Gaudet, J., & Lapointe, C. (2013). What could an equitable instructional design achieve? An experiment with trainers. *New Horizons in Adult Education and Human Resource Development*, *16*(3), 12–26. doi:10.1002/nha3.10149

Gray, C., & Leith, H. (2004). Perpetuating gender stereotypes in the classroom: A teacher perspective. *Educational Studies*, *30*(1), 3–17. doi:10.1080/0305569032000159705

Kollmayer, M., Schultes, M.-T., Lüftenegger, M., Finsterwald, M., Spiel, C., & Schober, B. (2020). REFLECT – A teacher training program to promote gender equality in schools. *Frontiers in Education*, *5*(136), 1–8. doi:10.3389/feduc.2020.00136

Kostas, M. (2023). Textbooks, students and teachers talk around gender: A new materialist approach to children's agency. *Teaching and Teacher Education*, *125*, 104052. Advance online publication. doi:10.1016/j.tate.2023.104052

Koyuncu Şahin, M., Esen Çoban, A., & Korkmaz, A. (2018). Toplumsal cinsiyet eşitliği ve Türk eğitim sistemindeki yeri: Okul öncesi öğretmen adaylarının gözünden. *Uluslararası Bilimsel Araştırmalar Dergisi*, *3*(2), 735–752. doi:10.21733/ibad.457232

Kumlu, E., & Çomoğlu, H. İ. (2021). Gender issues in english language teacher education: Cindrella's awakening through comparative literature. *Educational Policy Analysis and Strategic Research*, *16*(1), 99–102. doi:10.29329/epasr.2020.334.6

Kuruvilla, M. (2014). Gender studies course at UG/PG levels and gender awareness training to teachers. *Higher Education for the Future*, *1*(1), 29–38. doi:10.1177/2347631113518393

Marshall, S., & Smith, J. (1987). Sex differences in learning mathematics: A longitudinal study with item and error analyses. *Journal of Educational Psychology, 79*(4), 372–383. doi:10.1037/0022-0663.79.4.372

Ministry of National Education (MEB). (2019). *PISA 2018 Türkiye ön raporu.* Türkiye Cumhuriyeti Milli Eğitim Bakanlığı.

OECD. (2019). *PISA 2018 Results* (Vol. 2). Where All Students Can Succeed.

Okenwa-Emgwa, L., & Straus, E. (2018). Higher education as a platform for capacity building to address violence against women and promote gender equality: The Swedish example. *Public Health Reviews, 39*(31), 1–6. doi:10.118640985-018-0108-5 PMID:30473904

Öngören, S. (2019). Öğretmen adaylarının toplumsal cinsiyet algılarına göre okul öncesi öğretmenliğine ilişkin görüşleri [Teacher candidates' views on preschool teaching according to social gender perceptions]. *OPUS International Journal of Society Researches, 11*(18), 1773–1796. doi:10.26466/opus.516102

Patton, M. Q. (2014). *Qualitative research & evaluation methods: Integrating theory and practice.* Sage publications.

Pollock, E. R., Young, M. D., Lubans, D. R., Coffey, J. E., Hansen, V., & Morgan, P. J. (2021). Understanding the impact of a teacher education course on attitudes towards gender equity in physical activity and sport: An exploratory mixed methods evaluation. *Teaching and Teacher Education, 105*, 1–15. doi:10.1016/j.tate.2021.103421

Raj, A., Milton, A., & Ekstrand, M. D. (2021). Pink for princesses, blue for superheroes: The need to examine gender stereotypes in kids' products in search and recommendations. arXiv preprint arXiv:2105.09296.

Sadker, D. (1999). Gender Equity: Still Knocking at the Classroom Door. *Educational Leadership, 80-83.* Advance online publication. doi:10.1080/1066568000330112

Spencer, R., Rehman, S., & Kirk, S. (2015). Understanding gender norms, nutrition, and physical activity in adolescent girls: A scoping review. *The International Journal of Behavioral Nutrition and Physical Activity, 12*(6), 1–10. doi:10.118612966-015-0166-8 PMID:25616739

Stake, J., & Hoffmann, F. (2001). Changes in student social attitudes, activism, and personal confidence in higher education: The role of women's studies. *American Educational Research Journal, 38*(2), 411–436. doi:10.3102/00028312038002411

Torre-Sierra, A. D., & Guichot-Reina, V. (2022). The influence of school textbooks on the configuration of gender identity: A study on the unequal representation of women and men in the school discourse during the Spanish democracy. *Teaching and Teacher Education, 117*, 1–13. doi:10.1016/j.tate.2022.103810

Tracy, D. M., & Lane, M. B. (1999). Gender-equitable teaching behaviors: Preservice teachers' awareness and implementation. *Equity & Excellence in Education, 32*(3), 93–104. doi:10.1080/1066568990320311

TÜİK. (2022). *Türkiye Aile Yapısı Araştırması 2021*. Türkiye İstatistik Kurumu.

UN. (2015). General Assembly. *Transforming our world: the 2030 Agenda for Sustainable Development*. https://sdgs.un.org/2030agenda

UN. (2023). https://sdgs.un.org/goalsadresinden alındı

UNESCO. (2015). *A guide for gender equality in teacher education policy and practices*. https://unesdoc.unesco.org/ark:/48223/pf0000231646

UNESCO. (2019). *From access to empowerment: UNESCO strategy for gender equality in and through education 2019-2025*. United Nationas Educational, Scientific and Cultural Organization., doi:10.54675/IJGQ3826

UNESCO. (2023). *Gender equality and education*. UNESCO Priority Gender Equality. https://www.unesco.org/en/gender-equality/education

UNICEF. (2023). *Gender equality in education*. UNICEF. https://www.unicef.org/education/girls-education

Unterhalter, E., North, A., Arnot, M., Lloyd, C., Moletsane, L., Murphy-Graham, E., . . . Saito, M. (2014). Interventions to enhance girls' education and gender equality. Education Rigorous Literature Review. Department for International Development.

Women, U. N. (2022). *Progress on the sustainable development goals: The gender snapshot 2022*. https://www.unwomen.org/en/digital-library/publications/2022/09/progress-on-the-sustainable-development-goals-the-gender-snapshot-2022

Yar, D., Ulaş, S., & Vefikuluçay-Yılmaz, D. (2022). Üniversite öğrencilerinin cinsiyet rollerinin benlik saygısı ve yaşam doyumu üzerindeki etkisi. *IBAD Sosyal Bilimler Dergisi*, (13), 324–343. doi:10.21733/ibad.1143175

Yıldırım, A., & Şimşek, H. (2018). *Sosyal Bilimlerde Nitel Araştırma Yöntemleri* (11 b). Seçkin Yayıncılık.

ADDITIONAL READING

Aikman, S., Halai, A., & Rubagiza, J. (2011). Conceptualising gender equality in research on education quality. *Comparative Education*, *47*(1), 45–60. doi:10.1080/03050068.2011.541675

Auwarter, A. E., & Aruguete, M. S. (2008). Effects of student gender and socioeconomic status on teacher perceptions. *The Journal of Educational Research*, *101*(4), 242–246. doi:10.3200/JOER.101.4.243-246

Avery, P. G., & Walker, C. (1993). Prospective teachers' perceptions of ethnic and gender differences in academic achievement. *Journal of Teacher Education*, *44*(1), 27–37. doi:10.1177/0022487193044001005

Blumberg, R. L. (2008). The invisible obstacle to educational equality: Gender bias in textbooks. *Prospects*, *38*(3), 345–361. doi:10.100711125-009-9086-1

DeJaeghere, J., Parkes, J., & Unterhalter, E. (2013). Gender justice and education: Linking theory, policy and practice. *International Journal of Educational Development*, *33*(6), 539–545. doi:10.1016/j.ijedudev.2013.05.006

Ehrenberg, R. G., Goldhaber, D. D., & Brewer, D. J. (1995). Do teachers' race, gender, and ethnicity matter? Evidence from the National Educational Longitudinal Study of 1988. *Industrial & Labor Relations Review*, *48*(3), 547–561. doi:10.1177/001979399504800312

Gråstén, A., Kokkonen, J., & Kokkonen, M. (2022). Gender bias and gender equality beliefs in teaching efficacy profiles of Finnish physical education teachers. *Teachers and Teaching*, *28*(2), 246–262. doi:10.1080/13540602.2022.2062714

Jones, S. M., & Dindia, K. (2004). A meta-analytic perspective on sex equity in the classroom. *Review of Educational Research*, *74*(4), 443–471. doi:10.3102/00346543074004443

Sadker, D., Sadker, M., & Bauchner, J. (1984). Teacher reactions to classroom responses of male and female students. Washington, DC: National Institute of Education. (ERIC Document Reproduction Service No. ED245839)

Wilkinson, L. C., & Marrett, C. B. (Eds.). (2013). *Gender influences in classroom interaction.* Academic Press.

KEY TERMS AND DEFINITIONS

Curriculum: It is an education plan that explains how subjects in the relevant disciplines will be taught and the order in which topics will be introduced.

Gender Equality: Gender equality is the equality of individuals with different sexual identities in all areas of life.

Quality Education: Ensuring inclusive and equitable educational opportunities for all to serve the sustainable development goals.

Sustainable Development: It is a development model that can meet the needs of present generations without prejudice to the ability to meet the needs of future generations.

Teacher: It denotes the person who carries out teaching activities in education environments in a planned and organized manner.

Teacher Candidates: University students who continue their education in teacher training departments of higher education institutions but have not yet received their diplomas.

Teacher Candidates Branches: Various teaching departments according to subject areas under faculties of education. For example; science, math, etc.

Teacher Education: It is a bachelor's degree consisting of knowledge of the subject matter, knowledge of teaching, and knowledge of learners and their development in social contexts, which takes place under the roof of the faculties of education of higher education institutions.

Chapter 6
Accidents in Mass Transit Systems:
The Case of Trolleybus in Mexico City

Jaime Santos-Reyes
https://orcid.org/0000-0002-3758-9862
SEPI-ESIME, ZAC, Instituto Politecnico Nacional, Mexico

Rosario Moran-Ramirez
Instituto Politécnico Nacional, Mexico

ABSTRACT

The chapter presents the results of a spatiotemporal analysis of accident occurrence of the trolleybus mass transit system in Mexico City. Chi-square analysis for goodness of fit method was employed to assess the relationship between the temporal variables and the accident occurrence. The ArcGis software was used for the analysis of the spatial visualization. The employed data were for 2017. The main conclusion were a) accidents do not have a uniform distribution weekly and hourly (All lines and Line A), on weekdays and the seasons of the year for Lines B and C, and crashes occurring hourly for the case of lines D-G; b) there was not a relationship between weekdays and hourly accident occurrence when considering the seven lines altogether and Line A; c) the most-affected boroughs were those related to Cuauhtémoc (27%), Coyocan (21.6%), and GAM (20.5%); they accounted for 69.1% of the accidents; and d) the peaks of trolleybus accident occurrence depend on the lines, that is, Line F (one peak of hours), Lines B-E (two), and Lines A and G (three).

DOI: 10.4018/978-1-6684-9601-5.ch006

INTRODUCTION

The UN agenda for Sustainable Development Goals (SDGs) aim at building an inclusive, sustainable, and resilient future for people and planet (UN General Assembly, 2015). To achieve these aims, 17 Goals were adopted in 2015; in relation to road safety, two sets of goals were established: first, "by 2030, to halve global deaths and injuries from road traffic accidents" (Goal 3), and second, "to provide access to safe, affordable, and sustainable transport systems for all, improving road safety, notably by expanding public transport, with special attention to the needs of those in vulnerable situation, women, children, persons with disabilities" (Goal 11) (UN General Assembly, 2015).

In order to achieve these goals, it becomes of paramount importance to gain a better understanding of factors that contribute to traffic accidents, and the negative impacts of such road crashes. There has been a vast amount of research on these issues worldwide (Liao, et al., 2023; Lu, et al., 2021; Mohammed, et al., 2023; Fort, et al., 2011; Geng, et al., 2023; Karacasu, et al., 2011). For example, one of the negative impacts of traffic crashes is that related to congestion and therefore contributors to CO2 emissions (Bharadwaj et al., 2017; Liao, et al., 2023; Deng, et al., 2023; Xia, et al., 2020). Moreover, it may be argued that congestion becomes even worst if traffic crashes occur at rush hours (Geng, et al., 2023). In a way, the increase of CO2 emissions constitutes a lateral effect of traffic accidents in urban environments, which has also been one of the targets of Goal 13 of the SDGs, i.e.: "Integrate climate change measures into national policies, strategies and planning" (UN General Assembly, 2015). In particular, it has been argued that energy-related CO2 emissions rose by 6% for 2021, reaching their highest level ever (UN-SDG, 2023).

Road accidents are the result of several factors such as road conditions, environmental, mechanical, behavioral. Moreover, traffic accidents vary spatially and temporally (Mohammed, et al., 2023; Ziakopoulos and Yannis, 2020). A number of approaches to spatial modelling have been developed over the years to investigate spatial traffic crashes, spatial correlation, spatial autocorrelation that contribute traffic accidents, among others (Mansour, et al., 2022; Mollalo, et al., 2020; Mohammed, et al., 2023). In relation to the temporality, it has been reported that meteorological conditions in winter and summer time influences negatively on drivers' performance and increasing the risk of road accidents; in a similar study, it has been reported that traffic accidents occurred mostly between April and October (Farrel, et al., 1996). More recently, it has been found that accidents occurred mostly between 18:00-19:00 p.m. (Karacusu, et al., 2011).

Research on traffic crashes of the mass transit trolleybus system is lacking in the context of Mexico. The chapter addresses traffic accidents by considering the temporal and spatial aspects of these. The aim of the study is to shed some light on accident occurrence in the trolleybus system in Mexico City. The employed data on accidents is for 2017 only, given the lack of data (more details are given in the Limitations of the study section). Moreover, the availability of data corresponds only for seven lines (i.e., trolleybus lines A-G; currently, the system comprises nine lines). The aims of the chapter are:

1. To investigate the accident occurrence patterns on seasonal, monthly, daily, and hourly basis.
2. To investigate the relationship between the hourly accidents and weekdays.
3. To shed some light on the spatial visualization of the accidents in the city.

The chapter is organized as follows: first, a very brief description of the mass trolleybus transit system in Mexico City. Second, a description of the employed materials and methods are briefly described; third, the most relevant results are presented. Fourth, a discussion of the most relevant findings is presented in the section; also, some limitations and future directions are given within the discussion section. Finally, some relevant conclusions are reported.

THE TROLLEYBUS MASS TRANSPORT SYSTEM

One of the mass public transit systems that constitute a network of different modes of transport in Mexico City is the trolleybus; its given name is because it operates by means of a catenary of two superior cables that take electrical energy through two trolleys. It is believed that the system has been operating in the metropolis for 71 years, beginning operations with its first line on March 9, 1951, with a fleet of ten units purchased by the company "Servicio de Transportes Eléctricos del D.F.", with a capacity of 40 passengers and a cost of 20 cents a ticket (El universal, 2019).

Currently, the trolleybus service operates a network of 10 lines and one of them started its operations in 2022; the current system comprises a total length of 203.64 kilometres in lanes with confined and semi-confined sections. It is believed that the vehicle fleet is of 293 electric trolleybuses with capacities that range from between 85 and 142 passengers (STECDMX, 2023). Figure 1 and Table 1 highlights some of the trolleybus characteristics (Table 1 shows some characteristics of lines it shows A-G only).

Figure 1. The trolleybus network system
Source: Trolebús (2023)

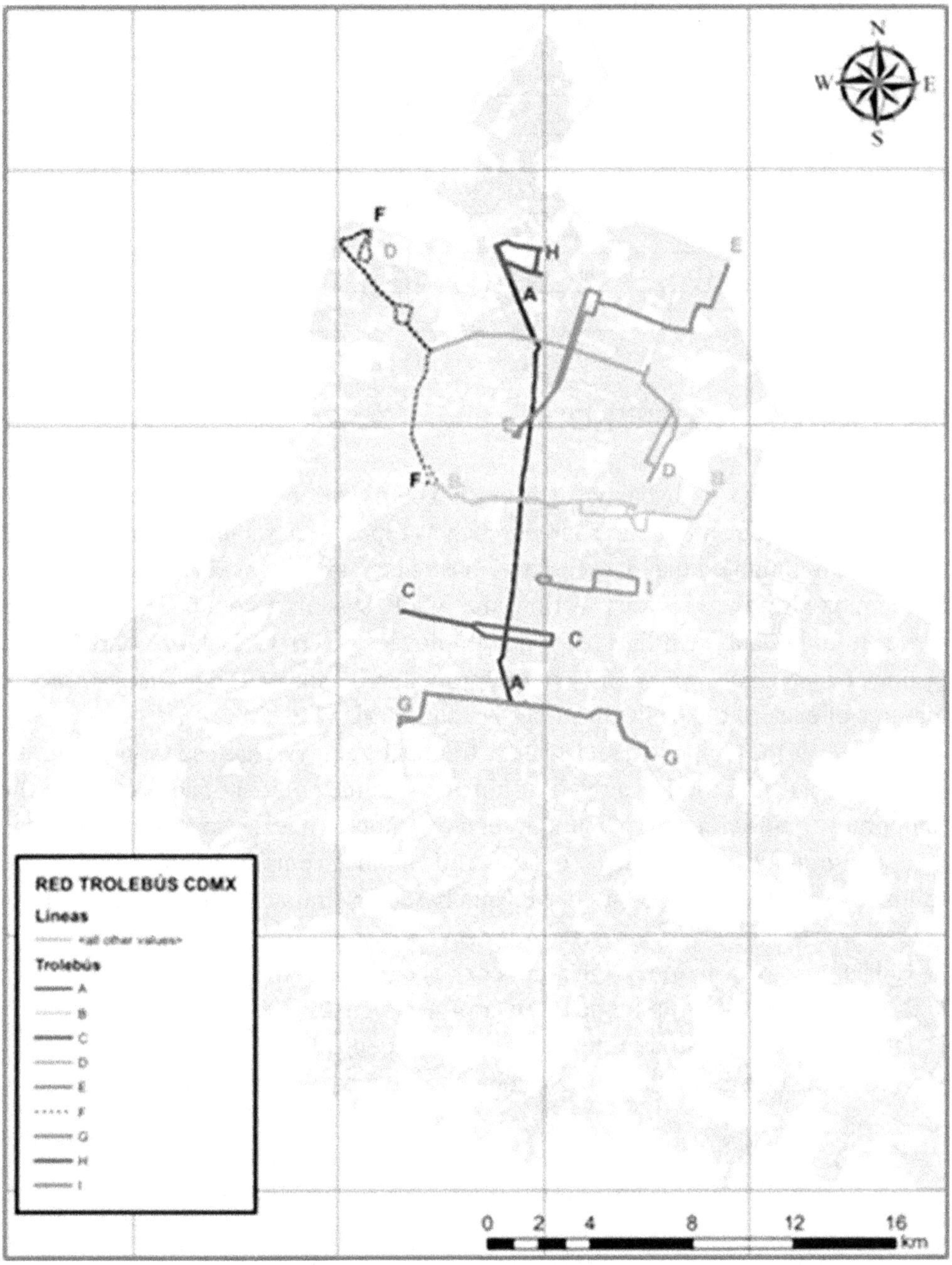

Table 1. Some characteristics of the trolleybus transit system (Trolebús, 2023)

Trolleybus Lines	Starting Operations Date	Total Length (km)	Number Stations
Line A	June 1979	36.6	52
Line B	June 1979/December 2010	27.5	40
Line C	November 2012	12.3	27
Line D	June 1979	44.9	68
Line E	March 2001	26.14	56
Line F	June 1979	30.2	55
Line G	June 1979	50.5	48
Total		164.04	346

MATERIALS AND METHODS

In the present work, trolleybus accident crashes data were obtained by the organization running the system, and other sources, such as reported news in the mass media. It should be highlighted that data for 2017 have been used in the analysis; moreover, as mentioned in the previous section, the whole system accounts for eight lines, however, only data from lines A-G are employed, given the lack of accident data of line H-J (see the Limitations section for further details on this). The operational features of each of the trolleybus lines are summarized in Table 1.

To analyze the temporal accident data, frequency analysis has been conducted. To investigate whether there is a relationship between the number of accidents and the temporal variables related to hours, weekdays, months and seasons of the year, the Chi-square analysis for goodness of fit method has been employed. In particular, the following hypotheses were established and based on similar studies (Yalcin, 2013):

$H_0$1: There is no variation in the number of accidents by the hours of the day.
$H_0$2: There is no variation in the number of accidents by the days in a week.
$H_0$3: There is no variation in the number of accidents by months in 2017.
$H_0$4: There is no variation in the number of accidents by the seasons.

Finally, the following hypothesis was established in order to investigate whether there is a relationship between the hours of accident occurrence and the weekdays, i.e.:

$H_0$5: The days on which accidents occur and the hours of accidents are independent from each other.

Regarding the spatial accident analysis, the ArcGis 10.8 software was used; the tool is a geographic information system (GIS) that enables data-driven mapping and analysis tools. It combines different types of data layers using spatial locations, including images, features, and base maps linked to tables and graphs. It incorporates a geostatistical framework, developed by the National Institute of Statistic and Geography (INEGI), which provides a geostatistical distribution data of the country, including Mexico City, at different levels, e.g., geographic data, census data, census statistics and institutional surveys and government units integrated into the system (INEGI, 2023).

RESULTS

Overall, a total of 630 accidents occurred in 2017 and the accident distribution per line were the following: line A (56.7%, 357/630), line B (5.9%, 37/630), line C (4.8%, 30/630), line D (9.5%, 60/630), line E (7.5%, 47/630), line F (4.1%, 26/630), and line G (11.6%, 73/630). The frequency data highlights that the most critical line of accident occurrence is the line A (the line that crosses the city center from the South-North-South (see a later section for further details on this)). This was followed by lines G and D; line F, on the other hand, resulted the least critical of the seven lines considered in the present study.

Temporal Accident Distribution

In order to have a better understanding of trolleybus accident patterns, the seasons of the year were considered, and on monthly, weekly, and hourly basis. This may help decision-makers running the system to better plan to reduce accident occurrence of this type of mass transit system.

Figure 2 shows the results when considering the seasons of the year, and by considering the seven lines, it can be observed that winter registered the highest number of accident occurrence (26.8%); when considering each of the lines individually, it was found that lines C-E follow this trend, and line C registered the highest percentage of accidents (50%). However, the frequency of accidents in line G decreased from summer to winter with 31.5% and 21.6%, respectively.

Figure 3 highlights the accident distribution monthly; when considering the seven lines altogether, the lowest frequency of accidents occurs in December (7.6%, 48/630) and the highest in March (9.5%, 60/630), and with a Mean=6.40, Median=6.00, and Std. Deviation=3.479. The figure also shows that line C registered the highest number of accidents in March with 26.7%.

The accident distribution of the seven lines weekly (Mean=3.71, Median=4.00, and Std. Deviation=1.819) highlights that Thursday registered the highest number

Figure 2. Accidents according to the season of the year

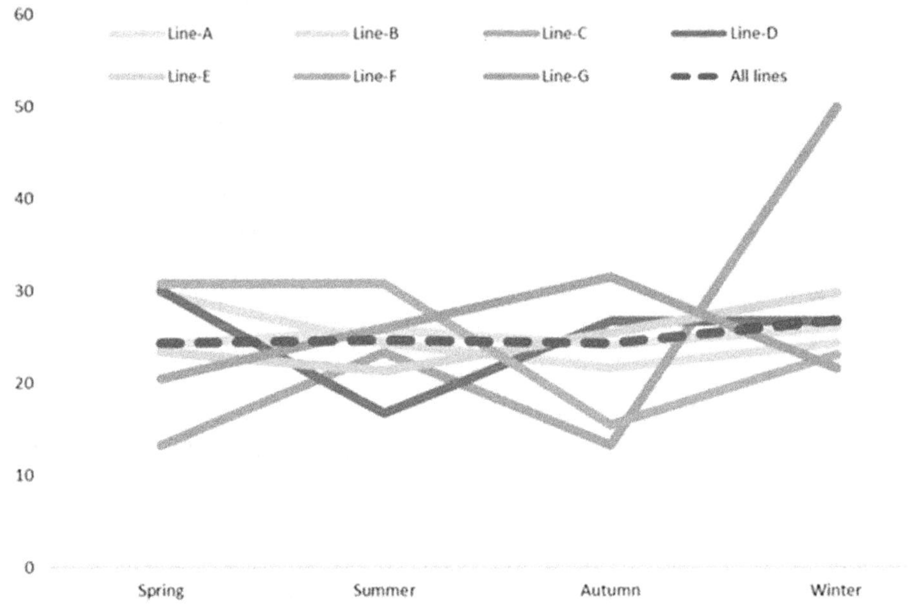

Figure 3. Accident distribution monthly

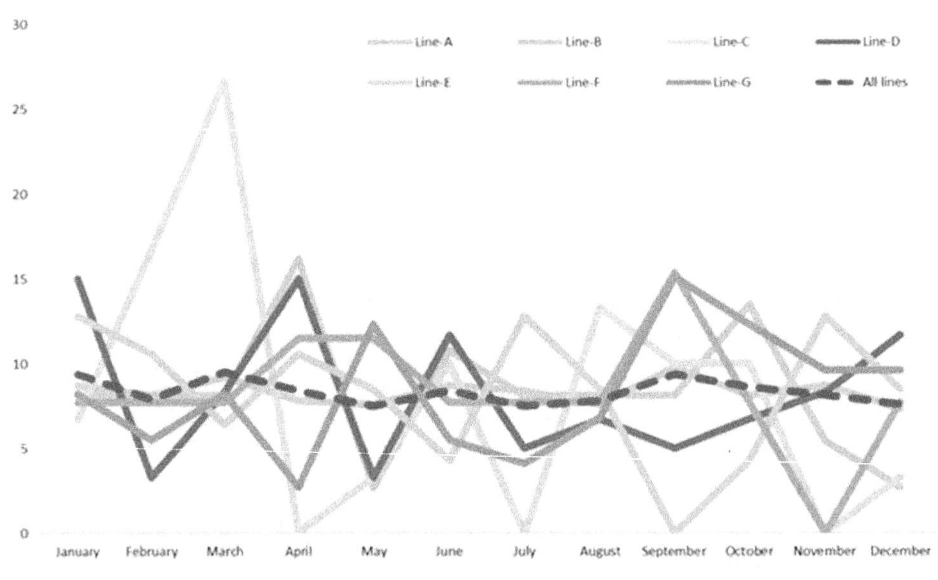

of accidents (19.0%, 120/630); further, Wednesday to Friday accounted for 52.2% (329/630) accidents (Figure 4). As expected, the weekends, Saturday, and Sunday, registered the lowest accident occurrence, with 10.3% (65/630) and 8.1% (51/630), respectively.

Lines B,C,D,F & G followed the above pattern of accident occurrence from Wednesday to Friday; for example, crashes in line B peaked on Thursday (32.4%), line D on Wednesday (25.0%), lines C,F&G on Friday (23.3%, 23.1% & 20.5%, respectively).

Figure 5 shows accident distribution hourly; overall, and when considering the seven lines, the pattern of accident occurrence exhibits a 'M' shape, i.e., there are two peaks: 08:00-11:59 a.m. (25.4%, 160/630) and 16:00-19:59 p.m. (27.6%, 174/630). A similar pattern was found for the cases of trolleybus lines B (08:00-11:59 a.m. (27.0%), 16:00-19:59 p.m. (27.0%)), D (08:00-11:59 a.m. (31.7%), 16:00-19:59 p.m. (28.4%), and E (08:00-11:59 a.m. (38.3%), 16:00-19:59 p.m. (27.7%).

However, the pattern of accident occurrence for lines A,C,F&G shows a different picture. In particular, lines A (08:00-11:59 a.m. (22.4%), 12:00-15:59 p.m. (23.0%) & 16:00-19:59 p.m. (28.6%)) and G (08:00-11:59 a.m. (21.9%), 12:00-15:59 p.m. (27.4%) & 16:00-19:59 p.m. (28.8%)) appear to have three ascending peak hours.

Line C appears to have two peak hours, i.e., 08:00-11:59 a.m. (26.7%), and 20:00-23:59 p.m. (33.3%). Finally, line F shows only one peak range of hours (08:00-11:59 a.m.) with 34.6%.

Table 2 shows the results of the relationship between the number of accidents and the temporal variables by employing the Chi-square analysis for goodness of fit method. Overall, the results highlighted that accidents do not have a uniform distribution weekly and hourly, when considering the seven trolleybus lines altogether; similar findings have been found for the case of line A.

In relation to the other trolleybus lines, it has been also found that accidents do not have a uniform distribution for the case of weekdays and the seasons of the year in lines B and C, respectively, and crashes occurring hourly for the case of lines D-G.

In the related literature, authors such as Karacasu et al. (2011) have assessed the relationship between the hours and the weekdays of accident occurrence, i.e., see the hypothesis $H_0 5$. The frequency analysis is shown in Figures 6&7, and for the cases of the seven lines and line A, respectively. The Chi-square analysis highlighted that there is not significant relationship between these variables, i.e., seven lines altogether, χ^2 (24, n=630) = 35.173, p= .066, Cramer's V= .118, and line 1, χ^2 (24, n=357) = 34.607, p= .074, Cramer's V= .156. These findings are consistent with those reported by Karacasu et al. (2011).

Figure 4. Accident distribution weekly

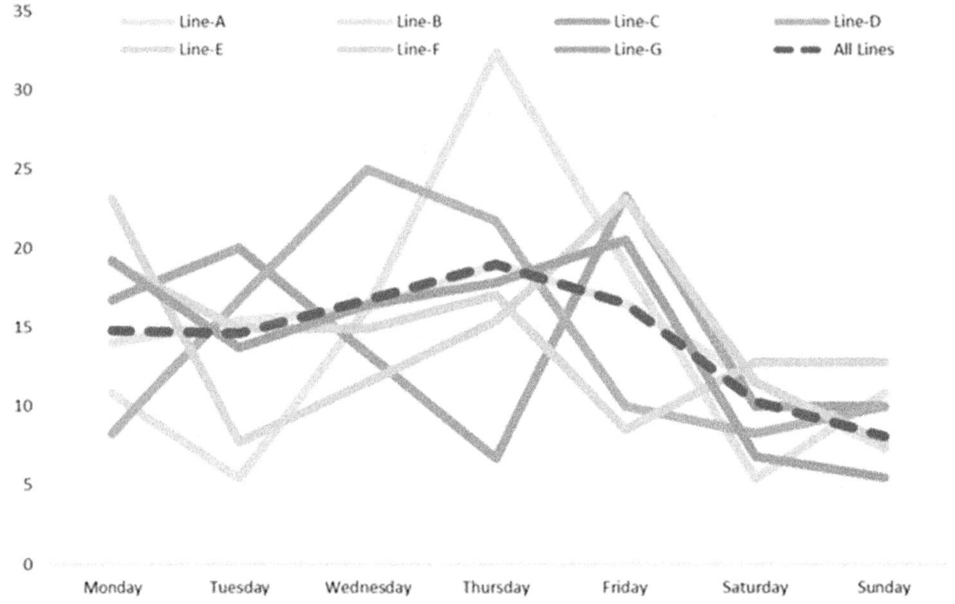

Figure 5. Accident distribution hourly

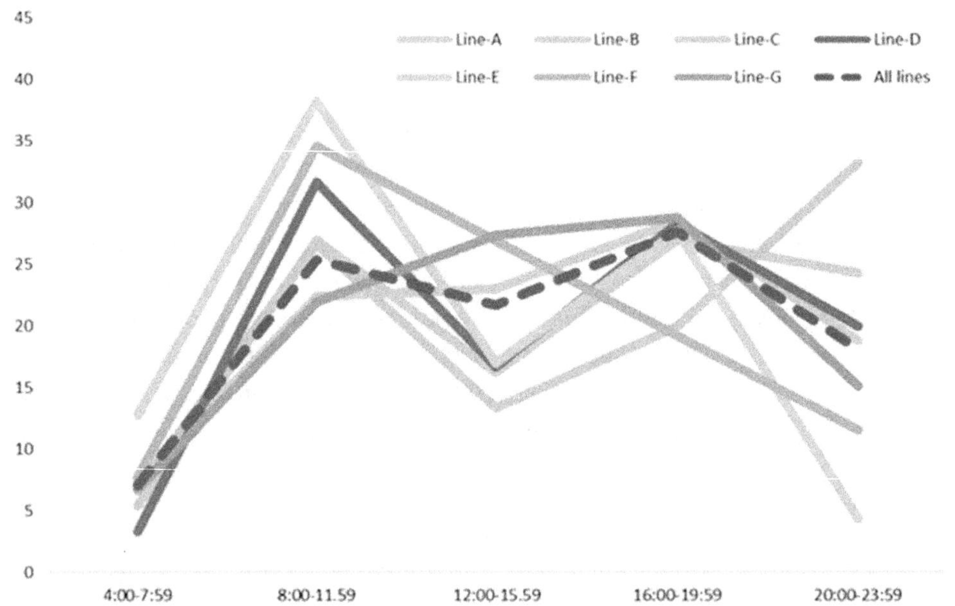

Table 2. Chi-square results

			χ^2_{-Test}	df	p*			$\chi^2_{-Critical}$	χ^2_{-Test}	df	p*
All lines						*Line D*					
	Seasons	*7.81*	*1.137*	*3*	*.768*		*Seasons*	*7.81*	*2.40*	*3*	*.494*
	Months	*19.70*	*4.667*	*11*	*.946*		*Months*	*19.70*	*13.60*	*11*	*.256*
	Weekdays	*12.60*	*38.70*	*6*	*.000*		*Weekdays*	*12.60*	*11.870*	*6*	*.065*
	Hours	*9.50*	*81.63*	*4*	*.000*		*Hours*	*9.50*	*14.83*	*4*	*.005*
Line A						*Line E*					
	Seasons	*7.81*	*.480*	*3*	*.923*		*Seasons*	*7.81*	*.745*	*3*	*.863*
	Months	*19.70*	*2.563*	*11*	*.995*		*Months*	-	-	-	-
	Weekdays	*12.60*	*22.43*	*6*	*.001*		*Weekdays*	*12.60*	*2.30*	*6*	*.890*
	Hours	*9.50*	*44.86*	*4*	*.000*		*Hours*	*9.50*	*16.51*	*4*	*.002*
Line B						*Line F*					
	Seasons	*7.81*	*.514*	*3*	*.916*		*Seasons*	*7.81*	*1.692*	*3*	*.639*
	Months	-	-	-	-		*Months*	-	-	-	-
	Weekdays	*12.60*	*13.89*	*6*	*.031*		*Weekdays*	-	-	-	-
	Hours	*9.50*	*6.40*	*4*	*.173*		*Hours*	*9.50*	*6.31*	*4*	*.177*
Line C						*Line G*					
	Seasons	*7.81*	*10.80*	*3*	*.013*		*Seasons*	*7.81*	*2.123*	*3*	*.547*
	Months	-	-	-	-		*Months*	*19.70*	*12.98*	*11*	*.295*
	Weekdays	-	-	-	-		*Weekdays*	*12.60*	*10.90*	*6*	*.091*
	Hours	*9.50*	*6.70*	*4*	*.155*		*Hours*	*9.50*	*12.13*	*4*	*.016*

*Assumed significance level, α=0.05.

If $\chi^2_{-Test} > \chi^2_{-Critical}$, then H_o is rejected.

Spatial Accident Distribution

As mentioned in the introduction section, spatial analysis of accidents may provide a valuable information to identify the location where a crash (or crashes) occurs; moreover, it may be useful to see and assess the patterns of accident occurrence in the trolleybus lines. Hence, the section presents some of the main findings in this regard.

Figure 8a-g shows the location of the crashes of the seven trolleybus lines considered in the present study. For example, Figure 8b shows the locations of intersections where the accidents occurred. As already shown in the previous section, line A runs from the south-north-south of the megacity. The fact that runs through the city center makes it the most critical in terms of accident occurrence. (See the Discussion & limitations sections for more this.)

Figure 9 shows the distribution of the crashes in eight boroughs of the megacity; in particular, it has been found that the most affected boroughs by accident occurrence

Figure 6. Number of accidents occurring hourly for all the lines

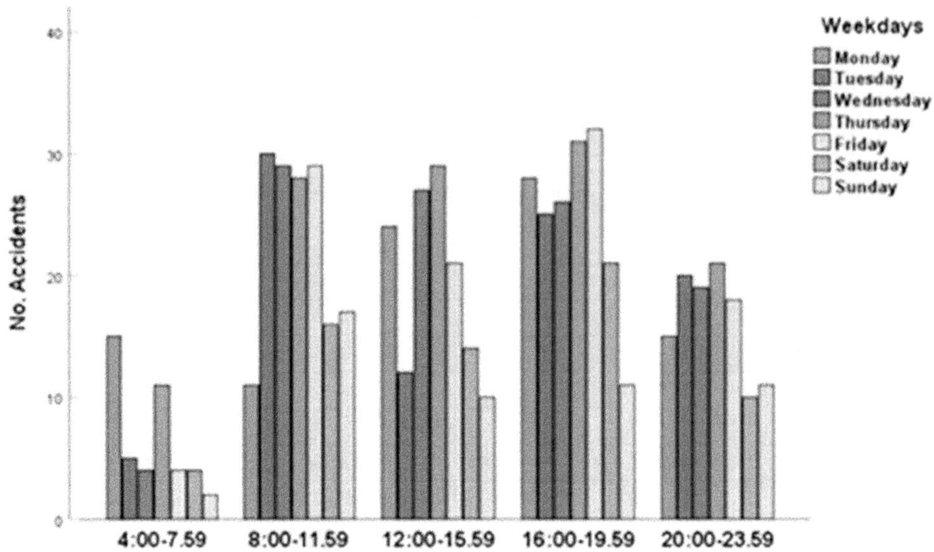

Figure 7. Number of accidents occurring hourly for the case of line A

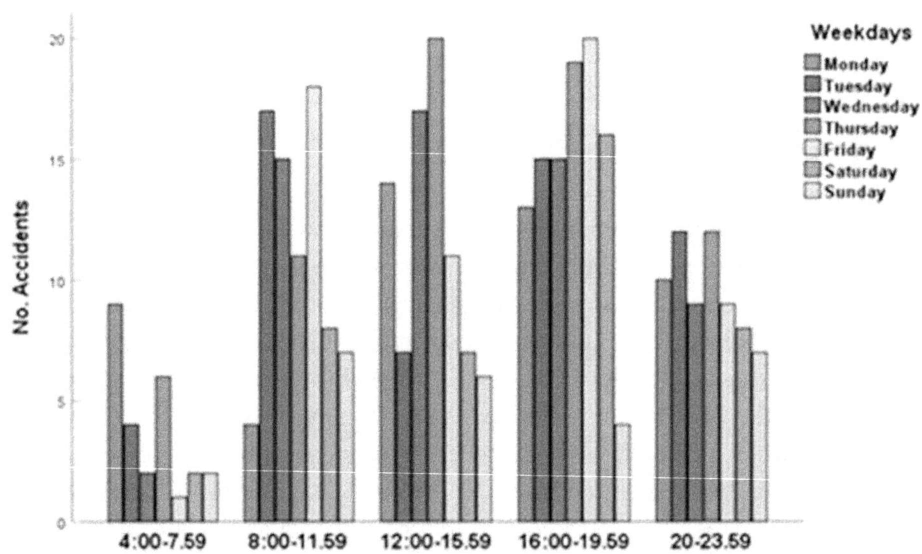

Figure 8. Spatial distribution of accident occurrence

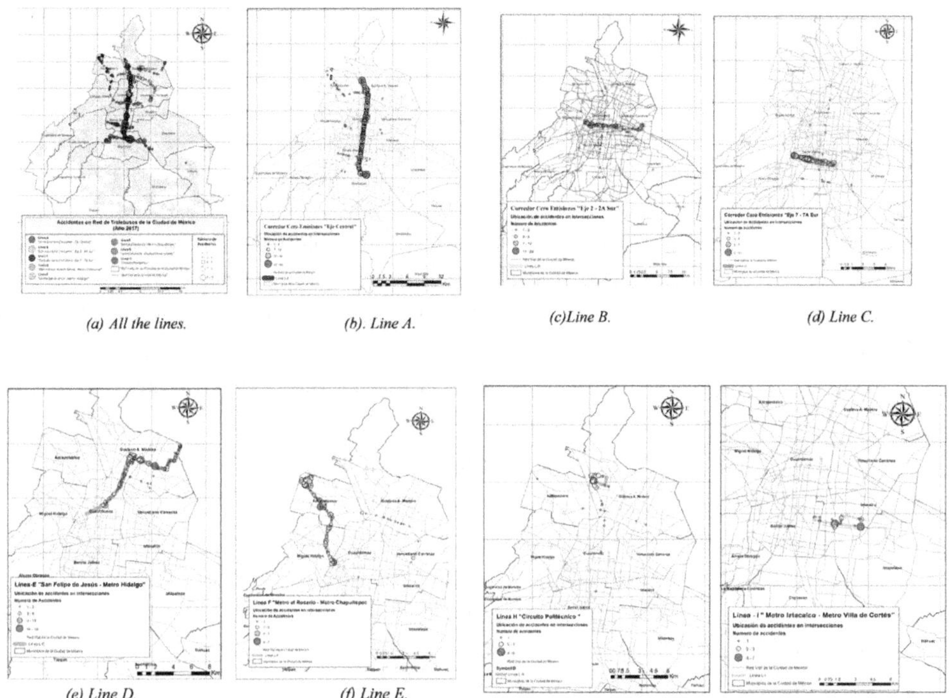

(a) All the lines. (b). Line A. (c)Line B. (d) Line C.

(e) Line D. (f) Line E. (g) Line F. (h) Line G.

Figure 9. Accidents and the city's boroughs

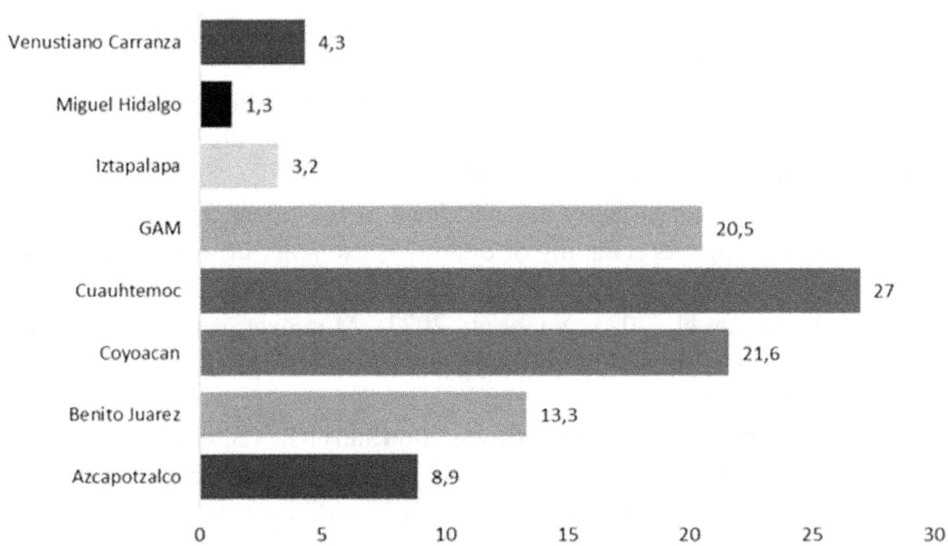

are Cuauhtémoc (27%, 170/630), Coyocan (21.6%, 136/630), and GAM (20.5%, 129/630). Overall, they account for 69.1% of the accidents; again line 1 runs through these boroughs (Figure 8b).

DISCUSSION

In general, road safety and traffic accidents have severely affected the economy and social development of cities (Mohammed, et al., 2023), negative health impacts (Moller-Jensen, 2021; Krecl et al., 2019; WHO, 2020), to the environment (Wang et al., 2020; Lu et al., 2021; Abbasi et al., 2020), among other. These negative impacts have led the UN agenda for Sustainable Development Goals to established two goals; first, "by 2030, to halve global deaths and injuries from road traffic accidents" (Goal 3), and second, "to provide access to safe, affordable, and sustainable transport systems for all…" (Goal 11) (UN General Assembly, 2015).

In recent years, a great deal of effort has been put to the process of shifting from fossil-fueled vehicles to electric driven transport systems by governments running cities worldwide (Lin & Du, 2017; Grigorieva & Nikulshin, 2022). Moreover, the population growth and urbanization in the developing world, megacities such as Mexico City, have adopted electric public mass transit systems, such as the trolleybus (Chan, 2023; Wolek, et al., 2021).

The book chapter presented the spatial and temporal pattern of trolleybus mass transit system crashes for the case of Mexico City; in particular, the employed data were for 2017. Overall, the results have highlighted the spatial and temporal patterns of accident occurrence. For example, it has been lines A & G appear to have three ascending peak hours; lines B-E, on the other hand, exhibited two peaks of accident occurrence, whereas line F only one peak range of hours.

The aforementioned raises the following questions: does this pattern of accident occurrence affect the urban mobility? Do crashes occur during rush hours? What about the congestion and CO2 emissions?

Regarding the urban mobility and accidents, it is evident that urban mobility is disrupted affecting the economy of cities, for example, in work time loss and productivity (Schrank, et al., 2019; Forth, et al., 2011; Chantith, et al., 2021; Geng, et al., 2023; Raza, et al., 2023; Xu, et al., 2023). Moreover, it has been reported that in the US alone, around 8.8 billion hours were lost due to traffic accidents (Schrank, et al., 2019).

Traffic accidents also contribute to congestion and an increase of CO2 emissions (Liao, et al., 2023; Deng, et al., 2023; Xia, et al., 2020). This is clearly the case of the present study where trolleybus accidents occur within the recognized three rush hours in the city: 07:30-09:30 a.m.; 14:00-15:00 p.m.; and 18:00-19:00 p.m. (Molina,

2023). In fact, it was not possible to analyze the users demand for the trolleybus lines due to the lack of data. However, in a similar study conducted on Bus Rapid Transit (BRT) systems in Mexico City, three peak hours were identified when considering the data on the users demand in one of the BRT lines (i.e., 06:00–08:00 a.m.; 13:00– 15:00 p.m.; and 17:00–20:00 p.m.) where it was found that 48.28% of accidents occurred during these ranges of peak hours.

Back to the discussion of the rush hours criteria in the city, the results highlighted that there are no fixed rush hours for each of the mass transit systems. According to a study on Origin-Destination (OD) (INEGI, 2017), between 6:00-8:59 a.m., is when the highest number of trips to work and school were observed, and between 6:00-6:59 p.m., the highest number of trips returning home were reported. Moreover, it has been reported that the metro transit system exhibited rush hours occurring in the morning (05.00 hrs-10.00 a.m.) and from 17.00 - 21.00 p.m. (Santos-Reyes & Padilla-Pérez, 2015, p.1430). More recently, and in relation to the trolleybus extension of line E, it was reported that:

"Its hours of operation will be from 6:00 a.m. to 10:00 a.m. and from 6:00 p.m. to 10:00 p.m., that is, during peak hours when thousands of workers go to work and then to their homes…" (El Financiero, 2022).

In short, there is not well-established criteria of rush hours, it is very confusing. More research is needed on this.

Regarding the spatial analysis of trolleybus accidents, it has highlighted the locations of the accidents through the map visualization (the details of the intersections where accidents occur were not presented herein). As expected, line A was the most critical given that it crosses the city center where there are plenty of traffic, crowded due to small business along the main avenue, and pedestrians walking around. In many cases, pedestrians and car drivers do not follow the traffic lights or invade the trolleybus lane.

In general, it may be argued that mass transit related accidents contribute significantly to traffic congestion and therefore to the CO2 emission, and in particular traffic congestion is augmented if accidents occur during the rush hours (i.e, normal traffic fluctuations, bottlenecks, etc.). According to Schrank, et al., (2019) around 3.3 billion gallons of fuel were wasted due to traffic congestion in the US. Moreover, it has been reported that congestion associated with traffic accidents is the most frustrating to travelers due to their unexpected and undesirable consequences (e.g., late delivery, missed flights, delayed meetings) (Liao, et al., 2023); furthermore, the presence of accidents may also lead to higher risks of secondary crashes (Dingus et al. 2016). It should also be noted accidents in any of the lines of the trolleybus system not only can affect other lines within the system (Figure 1), but also other mass transit systems (e.g., metro, BRT), given that they are highly interconnected (Santos-Reyes, et al., 2019).

Limitations of the Study and Future Work

As with any study, the presented analysis has several limitations: first, the lack of quality and reliability of trolleybus accident data; one of the reasons why only data from 2017 were employed; therefore, the presented results should be taken with caution (Smith, 2008). Further research is needed by considering data from 2018-to date. Secondly, the lack of data on key variables, such as the road conditions when the crashes occurred, the gender and age of those involved in the crashes, the consequences of the accidents such as the fatalities/injuries, among other, it was not possible to estimate, for example, the fatality rate (FR) in the mode of transport considered in the study. Thirdly, in relation to the spatial analysis, because of the lack of data, it was not possible to conduct a more robust analysis, for example, regarding the intersections size, geometry, traffic parameters, crash rates, injury severity rates (Ghofrani et al., 2022; Gilardi et al., 2020; Mohammed, et al., 2023), among other. Finally, more research is needed not only on trolleybus system, but also other mass transit systems such as BRT, metro, in the capital city.

CONCLUSION

In general, traffic crashes have negative consequences in terms of fatalities/injuries, congestion, mobility disruption, economic and environmental impacts, among other (e.g., Mohammed, et al., 2023; Moller-Jensen, 2021; Krecl et al., 2019; WHO, 2020; Wang et al., 2020; Lu et al., 2021; Abbasi et al., 2020). Given the impact of road accidents, the United Nations' Sustainable Development Goals, for example, has targeted to halve global deaths and injuries from traffic accidents by 2030, among other (Mohammed, et al., 2023).

The chapter has presented the results of a spatiotemporal analysis of the trolleybus mass transit system in Mexico City; in particular, the employed data were from 2017 only. Frequency analysis and the Chi-square analysis for goodness of fit method was employed to assess the relationship between the temporal variables and the accident occurrence. The ArcGis 10.8 software, on the other hand, was used to the analysis of the spatial visualization of accidents. The main conclusions were the following:

a) Overall, the results highlighted that accidents do not have a uniform distribution weekly and hourly, when considering the seven trolleybus lines altogether, similar findings have been found for the case of line A; it has also been found that accidents do not have a uniform distribution on weekdays and the seasons of the year in lines B and C, and crashes occurring hourly for the case of lines D-G. (See the hypothesis $H_0 1$- $H_0 4$).

b) There was not relationship between weekdays and hourly accident occurrence when considering the seven lines altogether and line A. (See hypothesis $H_0 5$).
c) The most affected boroughs by accident occurrence were those related to Cuauhtémoc (27%, 170/630), Coyocan (21.6%, 136/630), and GAM (20.5%, 129/630). Overall, they accounted for 69.1% of the accidents; again line 1 runs through these boroughs (Figure 8b).
d) The peaks of trolleybus accident occurrence depend on the lines, for example, line F (one peak range of hours), lines B-E (two), and lines A&G (three).

More generally, it may be argued that, in principle, an electric trolleybus system may be a sustainable mode of transport, but when accidents occur, there are negative effects to the environment, people, economy. Therefore, it is of paramount importance to consider not only a greener vehicle but also to see the totality of the system. That is, to consider all the variables that are in place and to assess the whole system rather than the individual parts. In short, there is a need to see the transport system from a systemic point of view.

ACKNOWLEDGMENT

This project was funded under the following grant: SIP-IPN: No-20230179.

REFERENCES

Abbasi, M. H., Hosseinlou, M. H., & Fadaki, S. M. J. (2020). An investigation of Bus Rapid Transit system (BRT) based on economic and air pollution analysis. *Case Studies on Transport Policy*, *8*(2), 553–563. doi:10.1016/j.cstp.2019.11.008

Bharadwaj, S., Ballare, S., Rohit, & Chandel, M. K. (2017). Impact of congestion on greenhouse gas emissions for road transport in Mumbai metropolitan region. *Transportation Research Procedia*, *25*, 3542–3555. doi:10.1016/j.trpro.2017.05.282

Chan, L. S. (2023). Transition from fossil fuel propelled transport to electrified mass transit railway system-Experience from Hong Kong. *Energy Policy*, *173*, 113372. doi:10.1016/j.enpol.2022.113372

Deng, X., Chen, W., Zhou, Q., Zheng, Y., Li, H., Liao, S., & Biljecki, F. (2023). Exploring spatiotemporal pattern and agglomeration of road CO2 emissions in Guangdong, China. *The Science of the Total Environment*, *871*, 162134. doi:10.1016/j.scitotenv.2023.162134 PMID:36775171

Dingus, T. A., Guo, F., Lee, S., Antin, J. F., Perez, M., Buchanan-King, M., & Hankey, J. (2016). Driver crash risk factors and prevalence evaluation using naturalistic driving data. *Proceedings of the National Academy of Sciences of the United States of America, 113*(10), 2636–2641. doi:10.1073/pnas.1513271113 PMID:26903657

El Financiero. (2022). *Nueva ruta del Trolebús Línea 5 en Reforma: horario, costo y estaciones*. Available at: https://www.elfinanciero.com.mx/cdmx/2022/12/27/nueva-ruta-del-trolebus-linea-5-en-reforma-horario-costo-y-estaciones/

Farrel, T. M., Sutton, J. E., Clark, D. E., Horner, W. R., Morris, K. I., Finison, K. S., Menchen, G. E., & Cohen, K. H. (1996). Moose-Motor Vehicle collisions. *Archives of Surgery, 131*(4), 377–381. doi:10.1001/archsurg.1996.01430160035005 PMID:8615722

Fort, E., Bouffard, E., Charnay, P., Bernard, M., Boisson, D., Laumon, B., Hours, M. (2011). Return to work following road accidents: Factors associated with late work resumption. *Journal of Rehabilitation Medicine, 43*(4), 283-291. doi:. doi:113372

Geng, K., Brands, D. K., Verhoef, E. T., & Wang, Y. (2023). The impact of tradable rush hours permits on peak demand: Evidence from an on-campus field experiment. *Transportation Research Part D, Transport and Environment, 120*, 103775. doi:10.1016/j.trd.2023.103775

Ghofrani, A., Zaidan, E., & Abulibdeh, A. (2022). Simulation and impact analysis of behavioral and socioeconomic dimensions of energy consumption. *Energy, 240*, 122502. doi:10.1016/j.energy.2021.122502

Gilardi, A., Mateu, J., Borgoni, R., & Lovelace, R. (2020). Multivariate hierarchical analysis of car crashes data considering a spatial network lattice. *Journal of the Royal Statistical Society. Series A, (Statistics in Society)*. Advance online publication. doi:10.48550/arxiv.2011.12595

Grigorieva, O., & Nikulshin, A. H. (2022). Electric buses on the streets of Moscow: experience, problems, prospects. *X International Scientific Siberian Transport Forum, 63*, 670-675. 10.1016/j.trpro.2022.06.061

INEGI. (2017). *Encuesta origen-destino en hogares de la Zona Metropolitan del Valle de Mexico* [Origin-Destination survey in households of the metropolitan Zone of the Valley of Mexico]. INEGI.

INEGI. (2023). *Marco geoestadístico*. Available at: https://www.inegi.org.mx/temas/mg/#Descargas

Karacasu, M., Er, A., Bilgic, S., & Barut, H. B. (2011). Variations in traffic accidents on seasonal, monthly, daily and hourly basis: Eskisehir Case. *Procedia: Social and Behavioral Sciences, 20*, 167–775. doi:10.1016/j.sbspro.2011.08.085

Khare, R., Villuri, V. G. K., & Chaurasia, D. (2021). Urban sustainability assessment: The evaluation of coordinated relationship between BRTS and land use in transit-oriented development mode using DEA model. *Ain Shams Engineering Journal*, *12*(1), 107–117. doi:10.1016/j.asej.2020.08.012

Liao, X., Wu, G., Yang, L., & Barth, M. J. (2023). A real-world data-driven approach for estimating environmental impacts of traffic accidents. *Transportation Research Part D, Transport and Environment*, *117*, 103664. doi:10.1016/j.trd.2023.103664

Lu, J., Li, B., Li, H., & Al-Barakani, A. (2021). Expansion of city scale, traffic modes, traffic congestion, and air pollution. *Cities (London, England)*, *108*, 1–15. doi:10.1016/j.cities.2020.102974

Mansour, S., Alahmadi, M., & Abulibdeh, A. (2021, August). 2022. Spatial assessment of audience accessibility to historical monuments and museums in Qatar during the 2022 FIFA World Cup. *Transport Policy*, *127*, 116–129. doi:10.1016/j.tranpol.2022.08.022

Mohammed, S., Alkhereibi, A. H., Abulibdeh, A., Jawarneh, R. N., & Balakrishnan, P. (2023). GIS-based spatiotemporal analysis for road traffic crashes; in support of sustainable transportation planning. *Transportation Research Interdisciplinary Perspectives*, *20*, 100836. doi:10.1016/j.trip.2023.100836

Molina, M. (2023). *¿Notas la diferencia? Detectan nueva hora pico un la Ciudad de México*. Available at: https://www.debate.com.mx/cdmx/Notas-la-diferencia-Detectan-nueva-hora-pico-en-la-Ciudad-de-Mexico-20230227-0199.html

Moller-Jensen, M. (2021). Frictions of everyday mobility: Traffic, transport and gendered confrontations on the roads of Accra. *Mobilities*, *16*(4), 461–475. doi:10.1080/17450101.2021.1917969

Raza, A., Zhong, M., Akuh, R., & Safdar, M. (2023). Public transport equity with the concept of time-dependent accessibility using geostatistical methods, Lorenz curves, and Gini coefficients. *Case Studies on Transport Policy*, *11*, 100956. doi:10.1016/j.cstp.2023.100956

Santos-Reyes, J., Avalos-Bravo, V., & Padilla-Pérez, D. (2017). Bus Rapid Transit Systems Road Safety: A Case Study of Mexico City. In H. Yaghoubi (Ed.), *Urban Transport Systems* (pp. 125–144). INTECH Publishers Press. doi:10.5772/65583

Santos-Reyes, J., Padilla-Pérez, D., & Beard, A. N. (2015). Modeling Critical Infrastructure Interdependency: The Case of the Mexico City Metro Transport System. *Human and Ecological Risk Assessment*, *21*(5), 1428–1444. doi:10.1080/10807039.2014.957956

Santos-Reyes, J., Padilla-Pérez, D., & Beard, A. N. (2019). Transport Infrastructure interdependency: Metro's failure propagation in the road transport system in Mexico City. *Sustainability (Basel)*, *11*(17), 4757. doi:10.3390u11174757

Schrank, D., Eisele, B., & Lomax, T. (2019). *Urban mobility report 2019*. Available at: https://static.tti.tamu.edu/tti.tamu.edu/documents/umr/archive/mobility-report-2019.pdf

Smith, E. (2008). Pitfalls and Promises: The Use of Secondary Data Analysis in Educational Research. *British Journal of Educational Studies*, *56*(3), 323–339. doi:10.1111/j.1467-8527.2008.00405.x

Trolebús. (2023). *Trolebus – CDMX*. Available at: https://www.ste.cdmx.gob.mx/red-de-servicio/lineas-de-trolebus

UN General Assembly. (2015). *Transforming our world: the 2030 agenda for sustainable development, 21 October 2015*. A/RES/70/1.

UN-SDG (UN-Sustainable Development Goals). (2023). *Goals 13: Take urgent action to combat climate change and its impacts*. Available at: https://www.un.org/sustainabledevelopment/climate-change/

Wang, Y., Li, Y., Liu, C., & Li, Z. (2020). Is traffic accident related to air pollution? A case report from an island of Taihu Lake, China. *Atmospheric Pollution Research*, *11*(5), 1028–1033. doi:10.1016/j.apr.2020.02.018

WHO. (2020). *Road Traffic Injuries*. https://www.who.int/news-room/fact-sheets/detail/road-traffic-injuries

Wolek, M., Wolanski, M., Bartłomiejczyk, M., Wyszomirski, O., Grzelec, K., & Hebel, K. (2021). Ensuring sustainable development of urban public transport: A case study of the trolleybus system in Gdynia and Sopot (Poland). *Journal of Cleaner Production*, *279*, 123807. doi:10.1016/j.jclepro.2020.123807

Xu, M., Shuai, B., Wang, X., Liu, H., & Zhou, H. (2023). Analysis of the accessibility of connecting transport at high-speed rail stations from the perspective of departing passengers. *Transportation Research Part A, Policy and Practice*, *173*, 103714. doi:10.1016/j.tra.2023.103714

Yalcin, G. (2013). Non-spatial analysis for the road traffic accidents. *Procedia: Social and Behavioral Sciences*, *92*, 1033–2038. doi:10.1016/j.sbspro.2013.08.795

Ziakopoulos, A., & Yannis, G. (2020, February). A review of spatial approaches in road safety. *Accident; Analysis and Prevention*, *135*, 105323. doi:10.1016/j.aap.2019.105323 PMID:31648775

Chapter 7
Developing an Undergraduate Module on Climate Change, Social Justice, and Sustainability:
Learning from the Designing Process

M. Mahruf C. Shohel
https://orcid.org/0000-0002-4048-0577
The Bedford College Group, UK

Charles Edward Whewell
The Bedford College Group, UK

Alex Mortby
The Bedford College Group, UK

Farzana Khan
The Bedford College Group, UK

ABSTRACT

This chapter is based on researchers' experience of developing an undergraduate module (Level 4) in collaboration with students and colleagues supported by a curriculum investigation grant from the British Educational Research Association (BERA). It presents a research-led module development initiative that was designed to engage and empower students by making them the co-creators of the curriculum which promotes practical examples of education towards sustainable development. This project utilised participatory research to promote teacher's agency and student's voices, empowering them as partners in co-creating the curriculum. The project aimed to foster teaching and learning that enables institutions and individuals to make informed choices that impact on individuals and communities to develop practice that promotes sustainability. This chapter reflects on the process of developing the module and learning from the process, outlining future directions for embedding climate change mitigation and adaptation issues across different disciplines.

DOI: 10.4018/978-1-6684-9601-5.ch007

INTRODUCTION

Climate change has become one of the highest-priority global challenges threatening mankind, with a level of threat that could be argued as existential (IPCC, 2022; MIT, 2021). The issues of social justice and sustainability have never been more profound than at the current time, and each of these concerns, when woven together, will have a significant impact on the future of the planet (Greenpeace, 2023). The changing climate has become one of the leading issues for young people across the world as they fight to raise awareness of the growing impact it has on everyday lives, both now and increasingly in the future (UNFCCC, 2013). It is crucial to empower the next generation of learners, the leaders of tomorrow, with a comprehensive understanding of the implications of the current trajectory of changes in climate on the planet, along with the potential solutions and mitigations for addressing these growing challenges. In this context, educational institutions can play a pivotal role in nurturing informed, compassionate, and proactive individuals who can drive positive change (FCDO, 2022). This project was an opportunity to stimulate and maintain the momentum developed in this area across The Bedford College Group (TBCG). By embracing sustainability and incorporating it into the curriculum, educational institutions can play a crucial role in creating a more sustainable and responsible future (Holst, 2022).

CONTEXT OF THE PROJECT

The Bedford College Group and University Centre is a prestigious educational institution and a prominent leader in the UK's further education (FE) landscape. Renowned for its excellence and its wide educational scope, it stands as one of the largest further education colleges in the country, serving as a catalyst for learning and personal development. It is also recognised regionally and nationally for its work on sustainable futures (TBCG, 2021).

Encompassing a diverse array of specialised colleges, including those dedicated to Agriculture, Animal Sciences, and Motor Sport, The Bedford College Group offers a comprehensive and dynamic educational experience. The Group's Higher Education (HE) offer is supported by University Centre status, with the institution being the largest college-based provider of HE in the South Midlands. The College aims to empower students to pursue their academic aspirations and reach new heights in their chosen fields whilst working in an organisation that has sustainability as a central pillar of its approach. The Group is driven by a passion for developing knowledge, innovation, and skills and its range of specialist campuses employ educators who are dedicated to nurturing students' talents and preparing them for successful futures. With state-of-the-art facilities, buildings and infrastructure that utilise and demonstrate cutting-edge sustainable technologies, the college provides a vibrant learning environment that encourages creativity, research, critical thinking, and practical application of skills and technologies.

The Bedford College Group and Its Sustainability Strategy 2022-2027

The Bedford College Group (TBCG) is deeply committed to sustainable development policy and practice, holding the top rank in the SE Midlands Local Enterprise Partnership (SEMLEP) for the highest number of students entering green jobs and acquiring green skills. The college has launched a comprehensive strategy for addressing its sustainability and is leading the further education sector to embrace the changes needed to ensure a sustainable future. This strategy highlights the college's comprehensive approach towards achieving its sustainability goals. To further this commitment, TBCG has devised a 5-year plan known as the Sustainability Strategy 2022-2027, aiming to make substantial progress in becoming a carbon net-zero organisation by 2050 and a leader in sustainability practices. To spearhead sustainability and climate action efforts, a dedicated sustainability committee has been established. The Research and Scholarship Team within the Group, alongside curriculum teams, is working diligently to integrate climate change and sustainability content into all courses offered across the institution. As an education provider, the Group is actively developing qualifications aligned with the government's green skills agenda, capitalising on strong opportunities for collaboration between colleges and the community to address climate change issues. This collaborative approach will not only benefit individuals but also contribute positively to local communities. In line with this vision, the Group is dedicated to creating qualifications that empower individuals in their journey towards a greener future. Through these initiatives, the Bedford College Group seeks to play a crucial role in driving sustainability and making a significant impact on environmental challenges through the engagement of young people in the process. Thus, TBCG has set forth an ambitious set of sustainability targets that the Group takes great pride in. Not only do these targets serve as a source of inspiration for students and staff, but they also position the college as a role model for other colleges and institutions to follow.

Research Project and the Module Development Process

The Research Team won a BCF Curriculum Investigation Grant 2022-2023 from the British Educational Research Association (BERA) in conjunction with the British Curriculum Forum (BCF) on the theme of 'Developing a Curriculum for

Climate and Sustainability Education'. The title of the project is 'Students as the Co-creators of the Sustainability Curriculum: Developing an Introductory Course on Climate Change, Social Justice and Sustainability'. As local authorities announce climate emergencies and the UK government sets 2050 net-zero carbon targets, and the Department for Education set out 'green strategies', climate change mitigation and adaptation have become key priorities for educational institutions to strengthen sustainability policy and practice (DfE, 2022; Staufenberg, 2022). The Bedford College Group have identified sustainability as a core value, receiving recognition by achieving The Association for Decentralised Energy's National Award in 2022, setting a standard for others to follow both across the education landscape and corporate organisations (TBCG, 2022). The redevelopment of the Brundtland Building at the Cauldwell Street campus in Bedford, UK, used cutting-edge technologies to reduce its carbon footprint and showcase financial and sustainability benefits by utilising power generation from wind, sun and river-based water-source heat pumps. The college sector has the responsibility and opportunity for taking action to empower communities to engage in education and training, promoting sustainable living and consumption (UNFCCC, 1992). The College ethos is to engage communities in local actions, thus changing behaviours to address global problems. This curriculum investigation project employed participatory approaches to develop student awareness and influence sustainable practices through the creation of an undergraduate level 4 module.

AIM AND OBJECTIVES OF THE CURRICULUM INVESTIGATION PROJECT

The aim of the research project was to develop materials for an introductory module on climate change, social justice and sustainability to offer to students at TBCG as well as share with other colleges across the country. The objectives of this participatory study were:

- To engage colleagues and students for participating in the curriculum development process;
- To foster student ownership in their learning design and delivery;
- To utilise teacher's agency and students' voices;
- To empower students as partners in co-creating curriculum for teaching and learning;
- To assist institutions and individuals in making informed choices that impact on sustainability at individual and community levels.

This project reflects a pioneering approach to college-based higher education; one that embraces student involvement in shaping the very fabric of their learning experiences. This chapter delves into the core ideas, principles, and outcomes of the project, exploring the design and development process of the undergraduate Level 4 module on climate change, social justice and sustainability.

RATIONALES FOR DEVELOPING A MODULE ON CLIMATE CHANGE, SOCIAL JUSTICE, AND SUSTAINABILITY

Literature suggests those that contribute least to global warming will be disproportionately affected, including and not least through negative health outcomes caused by climate change (Nicholas and Breakey, 2017). For college-based higher education programmes, the development of a core undergraduate Level 4 module on climate change, social justice, and sustainability is a response to the pressing need for action. The rationale behind creating such a transformative module was to address these global challenges, identifying the unprecedented threats that they pose to the planet whilst acknowledging that this is exacerbated by the increasing interconnected and interdependent nature of modern and future society. The research team felt it was vital to equip students and young people with a comprehensive understanding of the complexity and interconnected nature of many of the issues, providing facts that highlight the seriousness of the challenge we face. The team were also conscious of highlighting solutions that can be used to address the global issues at local, regional and national levels. By introducing an interdisciplinary module that delved into the multifaceted nature of these challenges, the module aimed to empower students to become informed global citizens who can engage in and contribute to climate change adaptation and mitigation. It was also felt important to foster critical thinking and problem-solving skills related to real-life scenarios. The module was designed to emphasise the significance of climate change, social justice and sustainability by encouraging students to engage with everyday real-world issues, applying their knowledge to propose innovative and pragmatic solutions. The pedagogical approach was to nurture problem-solving skills and a sense of agency among students by exploring higher order thinking and developing critical analytical skills in identifying the issues and proposing suitable solutions.

The module development project aimed to empower agency by involving students in co-creating the curriculum, with the project granting them a sense of ownership over the development of material which could benefit not just themselves but their wider peer group and beyond. This approach not only fostered a deeper commitment to the subject matter but also aimed to bolster student motivation, engagement, and ownership of their future learning and interests. As climate change, social justice, and sustainability are intrinsically interdisciplinary topics that require holistic and interconnected approaches, this module will promote interdisciplinary learning by drawing from disciplines such as environmental science, sociology, economics, ethics, and policy studies. It aimed to facilitate a comprehensive understanding and exploration of the issues discussed. It was felt that this interdisciplinary lens would help to nurture a well-rounded and adaptable skill set in students and develop understanding of different approaches that is crucial when addressing complex global challenges.

This module also aimed to instil ethical awareness and social responsibility in individual students, who as future leaders would need to grapple with the ethical dimensions of their decisions and actions. It prompted discussions on ethical frameworks and social responsibility, aiming to support them in becoming conscientious leaders capable of making decisions that prioritise environmental integrity, human rights, and social equity. By introducing a new module that places emphasis on climate change, social justice, and sustainability, the institution demonstrates a commitment to creating socially aware graduates who can make a positive impact on the world. However, climate change and social justice are complex issues with both local and global implications. Designing a Level 4 module that explores these challenges allows students to understand how these issues manifest within their communities while appreciating the global interconnectedness of these problems. This context-awareness was emphasised to empower students to become active participants in shaping both local and global solutions.

The approach also aimed to support current employability strategy by preparing students for the workforce. The current job market increasingly values sustainability-minded employees who can navigate complex environmental and social challenges. By exposing students to the latest infrastructure developments, it was felt that they were more likely to promote and recommend methods for improving sustainable practices in the workplace. The development of the undergraduate module therefore aimed to equip students with the competencies sought by employers whilst preparing them to become leaders in a rapidly changing world.

PLANNING OF THE WORKSHOPS

Through a pre-planned workshop series, this initiative delved into the intricacies of designing a forward-thinking undergraduate module, one that transcended conventional teaching methods and redefined the traditional teacher-student dynamic. At the heart of this innovative endeavour lay a fundamental belief in students as co-creators of knowledge, actively engaging with complex issues and contributing their perspectives, creativity, and insights (Ashwin & McVitty, 2015; Bovill et al.,

2011; Lubicz-Nawrocka & Owen, 2022). The workshops were developed to ensure that there were opportunities for deep discussion between students and staff. The planning of these workshops included developing our approach with students and ensuring that we were able to gather qualitative data that impacted the design of the curriculum. As part of the planning, pre-designed workshops included different styles of delivery and data-gathering techniques such as student-centred activities, digital tools (i.e. Mentimeter, Padlet), live documents and physical resources (printed forms, documents and sticky notes). These methods enhanced our workshops and student engagement and allowed us to maximise student participation in our research.

MODULE DESIGN PROCESS

Throughout the development process, this project promoted ESG (environmental, social and governance) credentials of the Group facilities to the student body and education sector, providing case studies for others to follow. By demonstrating effective policy implementation, this research-led curriculum development initiative will promote practical examples of sustainable development implementation (Shohel & Howes, 2011). Participatory research was utilised to promote teacher's agency (Priestley et al., 2012) and students' voices (Bland & Atweh, 2007) and empower students as partners in co-creating curriculum (Bovill et al., 2016) and teaching and learning (Lubicz-Nawrocka, 2018) in a bid to evidence that it can assist institutions and individuals to make informed choices that impact on individual and community sustainability.

Throughout the development process, a unique research framework was employed. Collaborative partnerships between educators and students from diverse disciplines across the college group were developed, which in turn fostered a rich and dynamic learning and development environment. Such an approach enabled students and staff to explore their ideas; to establish the interconnections between climate change, social justice, and sustainability and to boost their understanding of the intersections and complexities that drive these global challenges.

During student induction week in the academic year 2022-23, an induction event was held. Speakers with sustainability credentials from the college were invited, who presented the college sustainability strategy and provided examples of how the college puts into practice its ambitious sustainability strategy. We explored the TBCG sustainability policy, exploring the sustainable development initiatives identified above, including water efficiency and waste reduction policies, evidenced by the recent investment in water efficiency in the college heating system from the local river. It was important that, as part of the design process, we reflected upon the improvements in sustainability at the college and that these were at the forefront of our minds. Allowing students to review their own behaviour and how this contributes to impact on the climate alongside how they can make their place of study more sustainable through their practices such as sustainable travel choices was also important. A clear focus on and commitment to highlighting personal actions and informed choices in relation to packaging and waste reduction, recycling, food, transportation and 'fast fashion' were identified, to build agency and empowerment by thinking globally and acting locally (Steel, 1996). This intention was for the module design process to act as a call to action for students and the wider community.

METHODOLOGY FOR RESEARCH-INFORMED DESIGN AND DEVELOPMENT PROCESS

The project is a staff and student-led, mixed-methods interpretative study to generate module content and empirical data about the module development process. Criterion-based sampling (Patton, 2002) was employed to select motivated participants to create resources on making informed choices on personal, organisational and societal sustainability issues. Following the initial presentation to staff and students, participants were invited to attend pre-planned workshops (n=5) to review current sustainability issues and to consider additional actions that can be taken to further promote the notions of social justice and sustainability in relation to climate change and how it impacts on individuals, organisations and local communities. In the next stage of the project, questionnaires, interviews and focus group discussions were used as data collection tools to understand the process, individuals' perspective and organisational culture to support future improvements. It also adapted the Grounded Theory approach for analysing data using the thematic analysis technique. Ethical concerns were dealt with by following the BERA Ethical Guidelines (2018) and BPS Code of Ethics and Conducts (2018), especially in relation to informed consent, safeguarding participants and ensuring that the right to withdraw was promoted throughout the research process. Research data was anonymised, stored on secure college-based systems and destroyed after the project was completed.

WORKSHOPS WITH STAFF AND STUDENTS

In total five workshops were run within the group (Figure 1). All workshops were recorded on Microsoft Teams for the project's future use and for adapting activities for students. Though the participants were encouraged to attend sessions face-to-face (F2F), most colleagues attended virtually, utilising the hybrid nature of the delivery. This was important in order to build engagement from each of our college campuses and enabling colleagues to feel engaged in the process even when unable to travel between sites to attend in-person. The intention to operate the sessions in a sustainable manner was also promoted to all participants throughout the research process. The workshops were delivered primarily from the Cauldwell street campus, with one workshop being delivered in parallel, face-to-face from the Tresham College, Kettering campus. The use of the varied methods of attendance enabled a broader representation across the group both in terms of geography but also demography. There was clearly a range of experience and understanding in both the student and staff attendees, and this enriched the workshop experience and enabled the team to gather important insights into the topics and level of content that was required to meet future participants at a level suitable to their ability. This enabled deeper, focused activities on building awareness of the issues of climate change and we were able to demonstrate how the module would develop student understanding and their potential impact and influence on these local, national and international social issues.

The *introductory workshop* was delivered on 30th November 2022 to introduce the curriculum investigation project to the staff and students from the Bedford College Group. It also created awareness about climate change and some associated issues so that young people saw the relevance of participating in the project. This raised awareness about sustainability and inspired staff and students to participate in the research, fostering a sense of ownership and responsibility for developing the module. The participants received a comprehensive briefing on their roles and responsibilities within the project. They were informed about how they could actively contribute to the development process of the proposed module for the college. As a point of reflection, we were surprised that participants were well-versed in the issues of climate change but conscious that they were unable to understand how to act upon these to make a difference. Upon reflection, it was important that this development of social action and how to make a change was built into the module.

The *follow-up four workshops* were delivered from November 2022 to May 2023 to develop the curriculum for this Level 4 module for the TBCG students (Table 1). The first workshop was carried out with staff, and it focussed on exploring the importance of a module on climate change, social justice and sustainability, possible key topics, their contents, teaching and learning approaches, and assessment techniques. The second workshop was the same as the first one, but with students. Before the third workshop, all the derived contents were put together and used during the third workshop to develop further, but mostly given emphasis to finalise key topics, their relevant contents, teaching and learning strategies and assessment techniques. Then through the fourth workshop, key topics were finalised and then during the final workshop, the selected key topics were revisited, and the structure of the module was finalised.

Figure 1. Dates of the arranged workshops to develop module content and the module handbook

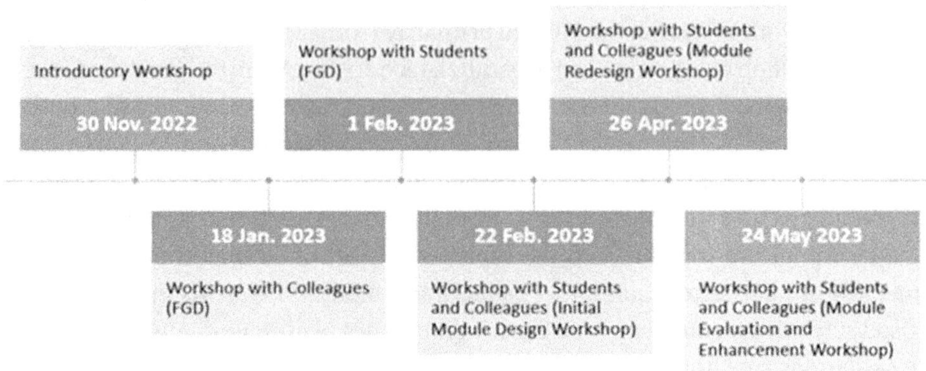

Table 1. Arranged workshops, their focus, and participants

SL	Induction and Workshop	Participants	Focus
1	Induction workshop	Both staff and students	Briefing about the project, participants' roles and responsibilities
2	First workshop	Staff	Exploring initial ideas of module content with the participants including key topics and their relevant examples and case studies
3	Second workshop	Students	Exploring initial ideas of module content with the participants including key topics and their relevant examples and case studies
4	Third workshop	Both staff and students	Developing initial module design including elaborating key issues for each topic and teaching and learning strategies
5	Fourth workshop	Both staff and students	Redesigning the module structure including elaborating teaching and learning approaches and assessment strategies
6	Fifth workshop	Both staff and students	Evaluating and enhancing the draft module handbook and getting final feedback on chosen content of the module

The aim of the first two sessions after the introductory workshop, organised separately with staff and students, was to encourage participants to think and reflect on sustainability issues that impacted them either locally or globally, followed by a focus on the design of the module. These sessions were started by asking the participants to think about the importance of introducing a course on climate change, social justice and sustainability. This sparked lively discussions among the lecturers and students as they shared their experiences and perspectives. Participants recorded their thinking and thoughts electronically using the workshop materials. Throughout these sessions, students and colleagues volunteered their ideas about the importance of such a course, what core topics should be included and what these key topics should cover. Participants outlined the module topics and associated concepts, examples and case studies for each topic. They also suggested different assessment approaches and what could be done in the next stage of the course development. One of the interesting discussion highlights of the voice of students was to have alternative assessments and to make assessments more life-oriented for their submission. In the follow-up workshop, participants (both staff and students) assessed and finalised the proposed topics and the content of the module and they extended their thoughts about the teaching and learning strategies as well as the assessment for this module. In the final workshop, participants were given the opportunity to evaluate the draft handbook of the module. Participants reviewed the documentation and provided their feedback on different aspects of the designed module outline.

REFLECTION AND LEARNING FROM THE PROCESS

Participants, both staff and students, were involved in the development process and contributed to the planning, designing and development of the module content. During the discussions, different assessment strategies were proposed. One of the proposed assessment strategies was to give students the opportunity to submit their work as a multimedia production using participatory photography and filming (Prosser, 1998) to explore and understand the issues around them related to climate change impacts and social justice for changing their behaviour to develop a sustainable society. The intent was to enable the co-design of multi-media sustainability projects on personal and community-based sustainability solutions. Students will be encouraged to develop projects related to their own interpretations of the three pillars of sustainable development.

During the first and second workshops with staff and students, the topic of preliminary discussion was about climate change and its significance for individuals. It was clearly described to the participants that they were to be actively involved in the development process of the proposed new module. As both the students and staff were to collaborate in creating these resources, best effort was made to ensure that students could connect with the material. Empowering students to take the lead in driving change and co-creating the content was one of the major objectives. It became evident that student perspectives differed from those of staff. As some of us have more experience and diverse viewpoints on what approaches might be practical, as well as distinct interpretations of sustainability and young people's engagement in decision-making, a holistic approach was used to harness and accommodate these differences.

Sustainable approaches can take various forms, including individual, organisational, and community-based efforts. The effectiveness of these approaches depends on the context and the people involved. For some individuals, it may involve adopting sustainable practices in their personal lives. For organisations, it might entail a concerted effort to improve their sustainability practices collectively. Additionally, community involvement can be crucial for others, as they work together towards positive change. It is important to note that the intention behind developing these modules extends beyond their use within Bedford College Group courses. They are meant to be shared and utilised in various educational settings. By doing so, the module aims to raise awareness about climate change and provide simple, actionable steps individuals can take to make a difference. Ultimately, the module's goal is to engage with communities, empowering individuals to become more aware and motivated to drive change. Whether it be individuals, communities, or organisations, the modules seek to inspire positive transformations in their approach to sustainability.

Finally, ten topics were chosen from the fourteen topics identified in the workshops, merging several together.. The finalised module structure is presented in Table 2.

Table 2. Finalised sessions derived from the workshops

Session	Topic
Session 1	Climate Change, Social Justice and Sustainability: Key Concepts and their Interconnectedness or Inter-relationships
Session 2	Climate Change: Global Warming, Planetary Health, Adaptation and Mitigation
Session 3	Social Justice: Importance and Relevance with Climate Change Activism
Session 4	Sustainability: Sustainable Development Goals, Individual and Collective Responsibilities for Sustainable Future
Session 5	Individual Projects: Introduction and Planning
Session 6	Climate Change Solutions 1: Disciplinary Understanding
Session 7	Climate Change Solutions 2: Disciplinary Approaches
Session 8	Climate Change Solutions 2: Evaluating Sustainability Policy and Practice
Session 9	Individual Projects: Presentation
Session 10	The Way Forward: Link between Society, Culture, Economy and the Environment

IMPLICATIONS

It was intended that the impact of this new module extends far beyond the classroom, touching on institutional practices and educational paradigms. By challenging conventional module development processes, the initiative should prompt broader discussions about the role of students in shaping their own learning experiences and the co-production of knowledge. It should inspire educators to reimagine their roles as facilitators and collaborators, working hand in hand with students to craft meaningful and impactful learning designs for enjoyable learning experiences and journeys.

The module development process has assisted teachers' (practitioners) and students' (service users) understanding of available technologies and approaches from cross-disciplinary and transdisciplinary perspectives to promote sustainability as well as giving them ownership of developing the curriculum resources, enabling their voices as agents of change within their institutional settings. Findings from this study enhanced the understanding of student participation and their engagement in curriculum development and implementation within the college environment. It enabled them to contribute towards changing practices that help to mitigate climate change, to ensure social justice and to progress towards the sustainable practices required to ensure a positive future.

Research as a continuous process has informed the policy and practice within the college via its sustainability working groups that were established in the 2022-23 academic year. Based on the findings and learning from this study, global issues of climate change and sustainable development will be embedded into different programmes and courses TBCG will be offering from cross-disciplinary and transdisciplinary approaches.

CHALLENGES AND LIMITATIONS

The research team aims to learn from the initiative and understand the limitations of this study, whilst also identifying the potential for further studies in this area.

Technological challenges: The Bedford College Group has multiple campuses. The researchers endeavoured to conduct workshops in a hybrid mode, combining in-person and virtual attendance. However, this approach posed challenges, particularly in facilitating seamless interactions for both researchers and participants. Integrating the experiences of those attending virtually with those present in person proved to be a complex task.

Timing and attendance issues: The way the college's timetable works made it challenging to bring people together at a particular time. However, after considering the college context and discussing with colleagues, the workshop was designed and delivered during lunch. Thus, considering the time constraint attendance was an issue and that approximately one-third of the participants attended more than one workshop.

Challenges of incorporating different perspectives for decision-making: As participants were from diverse backgrounds, their views were multifaceted and it was challenging to accommodate everyone's view about structuring the module and diversity of topics and contents as well as teaching, learning and assessment strategies.

FUTURE DIRECTIONS

Through this kind of future-oriented educational initiative, students and staff will have the opportunities to build awareness where employability skills and agency could be embedded in their programmes of study. This module development experience will be used in future curriculum development and implementation activities within Bedford College Group. Teaching staff will be encouraged to link climate justice and sustainability in their module delivery and assessments to ensure that students develop their understanding and are able to take appropriate action to raise concerns and address the issues. In the future, as part of the college curriculum design process, we will increase our focus on areas of climate change, social justice and sustainability. The Group will provide further training to its academic developers to ensure that during their course design stage of the curriculum, units are chosen that complement the areas of climate change, social justice and sustainability to maximise opportunities for students. This will involve ensuring that courses provide students with this content as part of their wider skill development and are treated the same as academic skills. The development in the future will also focus on how we can ensure climate change, social justice and sustainability are embedded into teaching and assessment and how practices of assessment can be more sustainable. In the future, there might be further support from the British Curriculum Forum (BCF) to adapt the newly developed module to different levels and disciplines. The Bedford College Group is aiming to deliver its own HE programmes and courses in future as its University Centre status develops. However, the impact of this module development project and the module itself on student recruitment, retention and academic outcomes needs to be explored so that lessons can be learnt to consider future developments in this key area.

CONCLUSION

The rationale behind developing an undergraduate Level 4 module on climate change, social justice, and sustainability for college-based higher education lies in its transformative potential and addressing the Sustainable Development Goals (SDGs) in education. By empowering students with interdisciplinary knowledge, critical thinking skills, ethical awareness, and a sense of social responsibility, this innovative module aimed to cultivate a generation of change-makers who can lead the world towards a more sustainable future and a just society by highlighting the significance of equipping students with the knowledge, skills, and ethical framework to navigate the complex challenges of the 21st century. This curriculum investigation study provided the opportunity to engage students and staff to understand and embed the practices to address the issues caused by climate change, social injustice and, thus, to promote sustainability within their choices as local actors in addressing global problems.

ACKNOWLEDGMENT

The authors express their gratitude to the British Educational Research Association (BERA) for funding the project and Prof Dr Gerry Czerniawski, University of East London for his support as the mentor of the BERA BCF Curriculum Investigation Grant 2022. They are also grateful to Dr Andrew Howes, from Manchester Institute of Education, School of Environment, Education and Development, the University of Manchester and Dr Faisal Tariq, from James Watt School of Engineering, the University of Glasgow as the critical friends for research integrity and quality enhancement of the project. The authors would also like to show gratitude for support from the Bedford College Group through their support of the authors and support from college services. The authors would also like to thank those speakers who attended the workshops and shared their work on climate change, social justice and sustainability within the College. The authors would finally like to show their gratitude towards their colleagues and students who participated in module development activities, who's names have been noted in the module handbook as the co-creators of this new module.

REFERENCES

Ashwin, P. & McVitty, D. (2015). The Meanings of Student Engagement: Implications for Policies and Practices. In Curaj, A., Matei, L., Pricopie, R., Salmi, J., Scott, P. (eds) The European Higher Education Area. Springer, Cham. https://doi.org/10.1007/978-3-319-20877-0_23

Bland, D., & Atweh, B. (2007). Students as researchers: Engaging students' voices in PAR. *Educational Action Research, 15*(3), 337–349. doi:10.1080/09650790701514259

Bovill, C., Cook-Sather, A., & Felten, P. (2011) Students as co-creators of teaching approaches, course design and curricula: implications for academic developers. International Journal for Academic Development, 16 (2). pp. 133-145. ISSN 1360-144X

Bovill, C., Cook-Sather, A., Felten, P., Millard, L., & Moore-Cherry, N. (2016). Addressing potential challenges in co-creating learning and teaching: Overcoming resistance, navigating institutional norms and ensuring inclusivity in student–staff partnerships. *Higher Education, 71*(2), 195–208. doi:10.100710734-015-9896-4

British Educational Research Association (BERA). (2018). *Ethical Guidelines for Educational Research* (4th ed.). British Educational Research Association.

British Psychological Association (BPS). (2018). *Code of Ethics and Conduct.* British Psychological Association.

Department for Education (DfE). (2022). *Sustainability and climate change: a strategy for the education and children's services systems.* Policy paper published on 21 April 2022. Retrieved from Sustainability and climate change: a strategy for the education and children's services systems. www.gov.uk

Foreign, Commonwealth, and Development Office (FCDO). (2022). *Addressing the climate, environment, and biodiversity crises in and through girls' education.* FCDO Position Paper. Available at: https://assets.publishing.service.gov.uk/government/uploads/system/uploads/attachment_data/file/1122368/Addressing_the_climate__environment__and_biodiversity_crises_in_and_through_girls__education.pdf

Green Peace. (2023). *Climate Justice and Social Justice: Two Sides of the Same Coin.* Available at: https://www.greenpeace.org/international/story/58334/climate-justice-and-social-justice-two-sides-of-the-same-coin/

Holst, J. (2023). Towards coherence on sustainability in education: A systematic review of Whole Institution Approaches. *Sustainability Science, 18*(2), 1015–1030. doi:10.100711625-022-01226-8

Intergovernmental Panel on Climate Change (IPCC). (2022). *Climate Change: A Threat to Human Wellbeing and Health of the Planet. Taking Action Now Can Save Our Future*. Available at: https://www.ipcc.ch/2022/02/28/pr-wgii-ar6/

Lubicz-Nawrocka, T. (2018). Students as partners in learning and teaching: The benefits of co-creation of the curriculum. *International Journal for Students as Partners*, *2*(1), 47–63. Advance online publication. doi:10.15173/ijsap.v2i1.3207

Lubicz-Nawrocka, T., & Owen, J. (2022). Curriculum Co-creation in a Postdigital World: Advancing Networked Learning and Engagement. Postdigit Sci Educ 4, 793–813. https://doi.org/10.1007/s42438-022-00304-5

MIT. (2021). *Why do people call climate change an "existential treat"?* Available at: https://climate.mit.edu/ask-mit/why-do-some-people-call-climate-change-existential-threat

Nicholas, P. K., & Breakey, S. (2017). Climate Change, Climate Justice, and Environmental Health: Implications for the Nursing Profession. *Journal of Nursing Scholarship*, *49*(6), 606–616. doi:10.1111/jnu.12326 PMID:28749596

Patton, M. Q. (2002). *Qualitative research and evaluation methods* (3rd ed.). Sage Publisher.

Priestley, M., Edwards, R., Priestley, A., & Miller, K. (2012). Teacher agency in curriculum making: Agents of change and spaces for maneuver. *Curriculum Inquiry*, *43*(2), 191–214. doi:10.1111/j.1467-873X.2012.00588.x

Prosser, J. (Ed.). (1998). *Image based research: a sourcebook for qualitative researchers*. Falmer Press.

Randstad. (2021). Sustainability at Work 2021: Toward 2030 Together. Available at: https://www.randstad.com/s3fs-media/rscom/public/2021-09/sustainability-at-work-2021.pdf

Shohel, M. M. C., & Howes, A. J. (2011). Models of Education for Sustainable Development and Nonformal Primary Education in Bangladesh. *Journal of Education for Sustainable Development*, *5*(1), 129–139. doi:10.1177/097340821000500115

Staufenberg, J. (2022). Focus feature: decarbonising college estates. *FE Week*. feweek.co.uk

Steel, B. S. (1996). Thinking globally and acting locally?: Environmental attitudes, behavior and activism. *Journal of Environmental Management*, *47*(1), 27–36. doi:10.1006/jema.1996.0033

The Bedford College Group (TBCG). (2021). *Bedford College Sustainability Policy*. Available at: bedfordcollegegroup.ac.uk

The Bedford College Group (TBCG). (2022). *Skills for a sustainable future: strategic plan 2022-2027*. Author.

United Nations Framework Convention on Climate Change (UNFCCC). (1992). *What is action for climate empowerment?* UNFCCC.

United Nations Framework Convention on Climate Change (UNFCCC). (2013). *Youth in action on climate change: Inspirations from around the world.* Available at: https://unfccc.int/resource/docs/publications/publication_youth_2013.pdf

Chapter 8
Empowering Sustainability: The Role of Artificial Intelligence in Renewable Energy

Prerna Tundwal
College of Technology and Engineering, Udaipur, India

ABSTRACT

This chapter explores the significant role that artificial intelligence (AI) plays in advancing renewable energy technologies and promoting sustainability. It discusses how AI can address the challenges and complexities associated with renewable energy systems, improve their efficiency, and enable their seamless integration into existing power grids. The chapter also explores various AI applications in renewable energy generation, forecasting, grid optimization, energy management, and demand response. Additionally, it highlights the potential benefits of AI-driven solutions in accelerating the global transition to a sustainable energy future.

INTRODUCTION

Sustainable Development

According to the theory of sustainable development, human civilizations must continue to exist and meet their own needs without jeopardising the ability of subsequent generations to do the same. The original definition of sustainable development was presented in the Brundtland Report, which was published in 1987. The 2030 agenda for sustainable development, which was approved by all United Nations Member States in 2015, offers a united road map for peace and prosperity for people and the

DOI: 10.4018/978-1-6684-9601-5.ch008

planet, both now and in the future. At its core are the 17 Sustainable Development Goals, an urgent call to action for both developed and developing countries to work together in a global partnership (Hák et al., 2016).

A framework for attaining sustainable development by 2030 is provided by the Sustainable Development Goals (SDGs) of the United Nations. The 17 objectives and 169 targets of the SDGs are intended to address the sustainability's economic, social, and environmental facets. Among other objectives, the SDGs call for the eradication of poverty, the reduction of inequality, access to education, the promotion of sustainable energy, the preservation of biodiversity, and the advancement of sustainable cities and communities (Henderson & Loreau, 2023).

The ability of future generations to meet their own needs is not compromised when current demands are met through sustainable development. It is a comprehensive and integrated strategy to development that aims to strike a balance between social progress, environmental preservation, and economic prosperity. A future that is economically, socially, and environmentally sustainable and that ensures a good standard of living for all people, both now and in the future, is what sustainable development entails (Nasir et al., 2023).

The foundation of sustainable development is the idea that economic growth must be long-term and shouldn't harm the environment or people's quality of life. This entails conserving resources, cutting back on waste and pollution, and promoting sustainable lifestyles and renewable energy. Assuring that everyone has access to necessities like food, water, healthcare, education, and a safe and secure environment is a key component of social sustainability. Protecting natural resources, lowering pollution and waste, and maintaining biodiversity are all aspects of environmental sustainability (Dincer, 2000).

The pursuit of sustainable development necessitates a comprehensive strategy that entails the cooperation and involvement of all parties, including communities, corporations, governments, and civil society organisations. It entails establishing targets and goals for sustainable development as well as putting those practises and policies into action. Additionally, fostering innovation and technology transfer as well as forming alliances and partnerships are all elements of sustainable development (Fawzy et al., 2020).

In conclusion, sustainable development is an all-encompassing method of growth that aims to strike a balance between social progress, economic expansion, and environmental preservation. It is crucial for building a sustainable future that guarantees a high standard of living for everyone, both now and in the future. All stakeholders must work together and participate in order to achieve sustainable development, and sustainable development-friendly policies and practises must be put into place (Nasir et al., 2023).

The key principles of sustainable development include:

1. Environmental Protection

Sustainable development recognizes the importance of preserving and protecting the natural environment. This involves reducing pollution, conserving resources, promoting biodiversity, and mitigating the impacts of climate change.

2. Social Equity

Sustainable development seeks to ensure social equity and justice by addressing poverty, inequality, and social exclusion. It aims to provide equal opportunities, access to basic services, and respect for human rights for all individuals (Hassan, Sameen, Salman et al, 2023).

3. Economic Prosperity

Sustainable development promotes economic growth and prosperity, but in a way that is sustainable and inclusive. It encourages the use of renewable resources, promotes sustainable business practices, and fosters innovation and technological advancements that benefit society and the environment (Hassan, Sameen, Salman et al, 2023).

4. Long-Term Planning

Sustainable development requires long-term planning and decision-making that takes into account the potential impacts on future generations. It emphasizes the need for integrated and holistic approaches to development that consider social, economic, and environmental factors (Hassan, Sameen, Salman et al, 2023).

5. Global Cooperation

Sustainable development recognizes that addressing global challenges requires international cooperation and collaboration. It emphasizes the importance of partnerships, knowledge sharing, and collective action to achieve sustainable development goals at the local, national, and global levels (Hassan, Sameen, Salman et al, 2023).

The participation of many stakeholders, including governments, corporations, civil society organisations, and individuals, is necessary to achieve sustainable development. It necessitates a change in attitudes, actions, and laws to support more sustainable practises and ways of living (Broman & Robèrt, 2017).

All parties, including governments, corporations, civil society organisations, and individuals, must take part in and work together to achieve sustainable development. It necessitates the creation and implementation of long-term-focused policies and plans that consider the requirements and goals of both the present and the future generations (Broman & Robèrt, 2017).

In order to solve some of the most pressing issues facing the world today, including environmental degradation, poverty, inequality, and climate change, sustainable development is essential. It is also essential for promoting economic growth and development that is inclusive and equitable, and that benefits all members of society. By pursuing sustainable development, we can create a more prosperous, just, and sustainable world for ourselves and for future generations (Hassan, Sameen, Salman et al, 2023).

Research Objectives

In order to combat climate change and environmental deterioration, there is a critical need for sustainable energy solutions. This is the background and reason for researching the role of artificial intelligence (AI) in renewable energy. Due to their abundance and cleanliness, renewable energy sources like solar, wind, and hydropower have drawn a lot of interest as alternatives to fossil fuels. However, there are a number of difficulties in integrating renewable energy into the current power grids, including issues with system stability, unpredictability, and intermittency (Ashfaq et al., 2022).

Artificial intelligence has emerged as a promising method to overcome these issues and improve renewable energy systems due to its capacity to analyse enormous volumes of data, identify patterns, and make defensible conclusions. By leveraging AI techniques, renewable energy generation can be enhanced, forecasting accuracy can be improved, and grid integration can be optimized. Additionally, AI-driven energy management systems can enable efficient utilization of renewable energy resources and promote demand response strategies (Chami et al., 2019).

The goal of this book chapter is to draw attention to how AI has the potential to alter the renewable energy industry. The chapter's goal is to show how AI may boost sustainability and hasten the transition to a low-carbon future by looking at its different uses and advantages. Moreover, it seeks to explore the existing limitations and ethical considerations associated with AI in renewable energy, along with potential future directions for research and development in this field (Hassan, Sameen, Salman et al, 2023).

Through this chapter, readers will gain insights into the role of AI in renewable energy and understand its significance in shaping the future of sustainable energy systems.

THE ROLE OF AI IN RENEWABLE ENERGY SYSTEMS

Artificial intelligence (AI) plays a crucial role in advancing renewable energy systems. By leveraging AI techniques, renewable energy generation, forecasting, grid optimization, energy management, and demand response can be significantly improved. AI facilitates the integration of energy from renewable sources into current power systems and addresses the issues brought on by their intermittent and unpredictable character (Panwar et al., 2011). It facilitates more accurate forecasting of renewable energy generation by analyzing weather data and load prediction models. AI also optimizes grid stability and power flow control, ensuring seamless integration of energy from renewable source into the power infrastructure. Additionally, AI-based energy management systems support demand response techniques for balancing energy supply and demand as well as efficient utilisation of renewable energy sources. The use of AI in renewable energy systems helps hasten the transition to a clean and sustainable energy future (Hassan, Sameen, Salman et al, 2023).

Enhancing Renewable Energy Generation

Enhancing renewable energy generation is the process of enhancing and optimising the generation of clean energy from renewable sources like solar, wind, hydro, or geothermal energy. It involves utilizing advanced technologies, techniques, and strategies to maximize the efficiency, output, and reliability of renewable energy systems (Appiah-Otoo et al., 2022). This can include optimizing the positioning and tracking of solar panels, maximizing wind turbine performance, optimizing the operation of hydropower plants, and implementing intelligent control systems. Enhancing renewable energy production aims to make renewable energy sources more efficient and viable overall, ultimately leading to a more sustainable and environmentally friendly energy future (Hassan, Abdulateef, Hafedh et al, 2023).

Solar Power and AI

Solar power is the use of photovoltaic (PV) panels or solar thermal systems to convert sunlight into electrical energy. It is a sustainable and renewable energy source that offers numerous environmental benefits and helps reduce greenhouse gas emissions. Artificial intelligence (AI) has emerged as a powerful tool to enhance the efficiency, effectiveness, and integration of solar power systems (Hassan, Sameen, Salman et al, 2023).

AI in solar power involves the application of advanced algorithms and data analytics to optimize various aspects of solar energy generation. In order to accurately anticipate solar irradiance, AI systems may examine weather patterns, historical data, and current conditions. This enables better resource allocation and planning. By optimizing the positioning and tracking of solar panels, AI can ensure maximum sunlight exposure and energy capture throughout the day. AI can improve the monitoring and maintenance of solar power systems. Through real-time data analysis, AI algorithms can detect performance issues, equipment failures, or shading problems that may impact solar panel efficiency. This enables proactive maintenance and prompt corrective actions, minimizing downtime and optimizing energy generation (Hassan, Abbas, Tabar et al, 2023).

AI-driven solar power systems can also enhance grid integration and stability. AI algorithms can monitor power output, demand, and grid conditions to enable better control and coordination of solar power generation. It can facilitate intelligent power flow control, allowing solar energy to be efficiently integrated into existing power grids while maintaining stability and reliability. The integration of AI in solar power systems offers tremendous potential to increase energy production, improve efficiency, optimize maintenance, and enable seamless integration with the power grid, thereby accelerating the adoption of clean and sustainable solar energy (Abbas et al., 2023).

Wind Power and AI

By deploying wind turbines, wind power is the process of turning wind energy into electricity. It is a clean, renewable energy source that has drawn a lot of interest as a long-term replacement for fossil fuels. The development of artificial intelligence (AI) has made it possible to maximise and optimise the effectiveness of wind power systems (Lu & Zhou, 2018).

Applying cutting-edge algorithms and data analytics to improve various aspects of wind energy generation is known as artificial intelligence (AI) in wind power. To predict wind conditions precisely, AI systems can examine both historical and current weather data, including wind speed, direction, and turbulence. This enables better planning and resource allocation for wind farms, ensuring optimal positioning and operation of wind turbines. By leveraging AI, wind power systems can optimize turbine performance. AI algorithms can adjust turbine settings in real-time based on wind conditions, such as changing the blade pitch or yaw angle, to maximize power output and efficiency. AI can also facilitate predictive maintenance by analyzing sensor data and identifying potential turbine issues before they escalate, reducing downtime and optimizing energy generation (Chami et al., 2019).

Furthermore, AI-driven control systems can improve grid integration and stability of wind power. AI algorithms can monitor power output, demand, and grid conditions to enable better coordination and control of wind turbine operation. This includes intelligent power flow control, where AI algorithms optimize the power dispatch and distribution to ensure grid stability and minimize the impact of fluctuations in wind power generation (Xiao et al., 2015).

Integrating AI into wind power systems offers significant advantages, including increased energy production, improved operational efficiency, enhanced maintenance, and better grid integration. Wind power can be used more successfully by utilising AI's data analysis, optimisation, and real-time control skills, which will help create a cleaner and more sustainable energy future (Hassan, Abdulateef, Hafedh et al, 2023).

Hydropower and AI

Hydropower refers to the conversion of the energy of flowing or falling water into electricity. It is a renewable and widely used source of energy that offers significant advantages in terms of reliability, scalability, and low carbon emissions. Hydropower systems could undergo a transformation thanks to artificial intelligence (AI), which can boost productivity, streamline processes, and guarantee the sustainable use of water resources (Hassan, Abdulateef, Hafedh et al, 2023).

AI in hydropower involves the application of advanced algorithms and data analytics to enhance various aspects of hydropower generation. AI algorithms can analyze historical and real-time data related to water flow, reservoir levels, weather conditions, and energy demand to optimize the operation of hydropower plants. This includes determining the optimal release of water, adjusting turbine operations, and optimizing power generation to maximize energy output while considering environmental factors. Furthermore, AI can contribute to the efficient management of water resources. By analyzing water inflow patterns, weather forecasts, and demand forecasts, AI algorithms can aid in making informed decisions regarding water storage, release, and distribution. This ensures sustainable utilization of water resources for hydropower generation while maintaining ecosystem balance and addressing water scarcity concerns (Johansson et al., 1992).

AI-driven control systems can also improve grid integration and stability of hydropower. AI algorithms can monitor power output, demand, and grid conditions to enable better coordination and control of hydropower plant operation. This includes optimizing power flow, regulating frequency, and ensuring grid stability during varying load conditions (Jha et al., 2017).

Integrating AI into hydropower systems holds great potential for increased energy generation, improved operational efficiency, sustainable water management, and enhanced grid integration. By leveraging AI's capabilities in data analysis, optimization, and real-time control, hydropower can play a vital role in meeting the growing energy demands while minimizing environmental impacts and ensuring long-term sustainability (Fazal et al., 2023).

Intelligent Forecasting for Renewable Energy

To effectively estimate and forecast the generation of renewable energy, intelligent forecasting for renewable energy uses cutting-edge technologies, data analytics, and artificial intelligence (AI) methodologies. It is essential for maximising the use and integration of renewable energy sources in the electrical grid (Jha et al., 2017).

Intelligent forecasting for renewable energy provides several benefits. It enables grid operators to anticipate fluctuations in renewable energy generation, helping them balance energy supply and demand more efficiently. It also assists energy market participants in making informed decisions regarding energy trading and pricing. Additionally, by offering insightful information for grid management and stability, intelligent forecasting supports the integration of renewable energy into the electrical system (Jha et al., 2017).

Weather Data and AI

Weather data is information that offers measurements and observations of atmospheric conditions and is gathered from a variety of sources, including weather stations, satellites, radars, and sensors. The parameters in this information include those relating to temperature, humidity, wind direction and speed, precipitation, and air pressure. Artificial intelligence (AI) methods can be used to analyse and decipher weather data, allowing for more precise climate modelling, forecasting, and decision-making across a range of industries and sectors (Anđelković & Bajatović, 2020).

Large amounts of weather data may be processed by AI algorithms, which can then spot relationships, trends, and patterns that human observers would not instantly see. AI can gain useful insights from weather data and increase the precision of weather forecasts by utilising machine learning and data analytics. This allows meteorologists and forecasters to provide more reliable and timely forecasts, which have significant implications for sectors such as agriculture, transportation, energy, and emergency management (Chami et al., 2019).

AI-powered weather models can assimilate real-time weather data and generate predictions based on historical trends, atmospheric physics, and statistical patterns. These models can handle complex interactions between different atmospheric variables, enabling more precise and localized forecasts. AI algorithms can also learn from past weather patterns and historical data to provide probabilistic forecasts, helping users assess risks and make informed decisions based on the likelihood of specific weather events occurring. Moreover, the integration of AI with weather data facilitates the development of early warning systems for severe weather events like hurricanes, storms, or floods. By analyzing historical data, current atmospheric conditions, and real-time observations, AI algorithms can identify potential risks and issue timely alerts to populations at risk, allowing for proactive measures such as evacuations and resource allocation (He et al., 2022).

The combination of weather data and AI offers immense potential for improving weather forecasting, climate modeling, risk management, and decision-making. By harnessing the power of AI to analyze and interpret weather data, we can enhance our understanding of weather patterns, mitigate the impacts of extreme weather events, and make more informed choices in various sectors that rely on accurate and timely weather information (Chami et al., 2019).

Load Prediction and AI

Load prediction refers to the process of estimating the electricity demand or load on a power grid at a specific time or within a certain time frame. It plays a crucial role in energy management, grid stability, and efficient allocation of resources. Artificial intelligence (AI) techniques can be applied to load prediction, enabling accurate and reliable forecasts that assist in optimizing energy generation, transmission, and distribution (Hou et al., 2022).

AI algorithms can analyze historical load data, weather conditions, time of day, day of the week, and other relevant factors to identify patterns and trends in electricity consumption. By leveraging machine learning and data analytics, AI models can learn from past load patterns and make predictions for future load demand. These AI-powered load prediction models can handle the complexities of load variation, taking into account factors like seasonality, holidays, and special events. As fresh data becomes available, they can also update projections in real-time to account for shifting patterns (Zhang et al., 2021).

The integration of AI with load prediction offers several benefits. It enables utilities and grid operators to optimize energy generation and distribution, ensuring that sufficient electricity is available to meet the demand without overloading the grid or resorting to excessive reserve capacity. This enhances the electricity system's effectiveness and dependability. Demand response programmes, where users can modify their electricity usage depending on anticipated high or low load periods, are also made possible by AI-driven load prediction. With the help of AI, grid operators and consumers can communicate more effectively, encouraging peak load management and energy conservation (Fazal et al., 2023).

The combination of load prediction and AI empowers utilities, grid operators, and energy consumers to make informed decisions, allocate resources efficiently, and maintain grid stability. By leveraging AI algorithms to analyze historical data and identify patterns, load prediction models enable effective energy management and support the transition to a more sustainable and resilient power grid (Jha et al., 2017).

Optimizing Renewable Energy Grid Integration

Optimizing renewable energy grid integration refers to the process of efficiently integrating and managing renewable energy sources into existing power grids. To enable the seamless and reliable integration of renewable energy generation into the grid infrastructure, it includes utilising a variety of techniques, technologies, and operational practises (Wu et al., 2023).

The goal of optimizing renewable energy grid integration is to maximize the utilization of renewable energy resources while maintaining grid stability, reliability, and security. It involves addressing challenges such as intermittency, variability, and uncertainty associated with renewable energy sources. By employing advanced control systems and real-time monitoring, grid operators can actively manage and balance the supply and demand of renewable energy. This includes adjusting power flows, coordinating energy storage systems, and implementing intelligent grid management strategies (Pang et al., 2022).

Optimizing renewable energy grid integration also involves the development and implementation of forecasting techniques to accurately predict renewable energy generation. Accurate forecasts for the availability and output of renewable energy sources can be created using weather data, historical patterns, and cutting-edge algorithms. Furthermore, grid optimization techniques can be applied to enhance the flexibility and adaptability of the power grid. This may include the deployment of advanced sensors, smart meters, and communication technologies to improve grid monitoring, fault detection, and power quality management (Okonkwo et al., 2021).

Optimizing renewable energy grid integration aims to create a symbiotic relationship between renewable energy sources and the existing power grid. It requires the deployment of advanced technologies, data-driven decision-making, and efficient grid management practices to maximize the benefits of renewable energy while ensuring a reliable and stable electricity supply (Jha et al., 2017).

Grid Stability and AI

Grid stability refers to an electrical power grid's capacity to maintain a steady and balanced flow of electricity while making sure that the frequency and voltage levels stay within allowable bounds. It is essential for both the grid's effective and secure operation and the provision of high-quality power to consumers. By analysing data, forecasting system behaviour, and enabling proactive control and management, artificial intelligence (AI) can significantly contribute to improving grid stability (Fazal et al., 2023).

AI algorithms can monitor real-time data from sensors and smart grid devices to detect anomalies, voltage fluctuations, and other factors that may impact grid stability. By analyzing this data, AI systems can identify potential issues and trigger appropriate responses to mitigate or prevent disruptions. AI-powered predictive analytics can forecast system behavior, load patterns, and renewable energy generation to anticipate potential grid stability challenges. This enables grid operators to proactively adjust generation, optimize load distribution, and implement measures to maintain stability. Moreover, AI can facilitate intelligent control and coordination of power generation, transmission, and distribution (Okonkwo et al., 2021). AI algorithms can optimize power flow, balance supply and demand, and dynamically adjust parameters such as voltage regulation and reactive power control to maintain stability. By providing real-time monitoring, control, and coordination of these resources, artificial intelligence (AI) can also help with the integration of distributed energy resources (DERs), such as solar panels and wind turbines. This makes it possible for them to be properly integrated into the grid without endangering its stability (Hassan, Sameen, Salman et al, 2023).

AI technologies can enhance grid stability by leveraging data analytics, predictive capabilities, and intelligent control systems. By enabling real-time monitoring, proactive response, and optimized coordination, AI empowers grid operators to maintain a stable and reliable power grid, even in the presence of varying loads, intermittent (Okonkwo et al., 2021).

Power Flow Control and AI

Power flow control refers to the management and regulation of electrical power flow within a power grid to ensure efficient and reliable energy transmission. It involves controlling the flow of electricity in real-time to maintain grid stability, minimize losses, and optimize the utilization of transmission infrastructure. In order to improve power flow regulation, artificial intelligence (AI) approaches can be used to analyse data, improve system performance, and facilitate wise decision-making (Dasgupta et al., 2020).

Huge volumes of data about grid conditions, electricity demand, generation capacity, and transmission limits can be processed by AI systems. By analyzing this data, AI systems can predict and optimize power flow patterns, considering factors such as load balancing, voltage stability, and line capacity. By leveraging machine learning and data analytics, AI can learn from historical power flow data and identify patterns and trends. This enables the development of accurate models and algorithms for power flow control, allowing grid operators to make informed decisions in real-time (Şerban & Lytras, 2020).

AI-driven power flow control systems can adjust parameters such as reactive power, voltage regulation, and line impedance to optimize power flow and minimize losses. These systems can also manage dynamic changes in grid conditions, such as fluctuations in renewable energy generation, to maintain grid stability and reliable power supply. Additionally, AI can make it easier to integrate cutting-edge innovations like HVDC (High-Voltage Direct Current) systems and FACTS (Flexible AC Transmission Systems), which improve power flow control capabilities. AI algorithms can optimize the operation of these systems, enabling efficient transmission of electricity over long distances and minimizing transmission losses (Dasgupta et al., 2020).

AI in power flow control provides the ability to analyze real-time data, optimize power flow patterns, and make intelligent decisions to ensure efficient and reliable electricity transmission. By leveraging AI algorithms and data-driven techniques, power flow control systems can enhance grid stability, minimize losses, and maximize the utilization of transmission infrastructure, ultimately contributing to a more efficient and resilient power grid (Hassan, Sameen, Salman et al, 2023).

AI-BASED ENERGY MANAGEMENT AND DEMAND RESPONSE

AI-based energy management refers to the use of artificial intelligence (AI) techniques to optimize energy consumption, production, and distribution in various sectors. It involves analyzing real-time data, predicting energy demand, and making intelligent decisions to achieve energy efficiency and cost savings. Demand response, on the other hand, is a strategy that involves adjusting electricity usage based on supply and demand conditions. By integrating AI with energy management and demand response, smart systems can optimize energy usage, balance load, and enable consumers to respond to price signals or grid conditions, ultimately promoting a more sustainable and resilient energy system (Alamir et al., 2023).

Energy Management Systems and AI

Energy Management Systems (EMS) refer to software-based systems designed to monitor, control, and optimize energy consumption within various settings such as buildings, industrial facilities, and power grids. These systems help organizations and grid operators effectively manage energy resources, reduce costs, and enhance efficiency. Artificial intelligence (AI) plays a significant role in enhancing EMS by leveraging advanced algorithms and data analytics (Alamir et al., 2023).

AI enables EMS to analyze vast amounts of energy data, including historical usage patterns, real-time sensor data, weather conditions, and occupancy information. By processing this data, AI algorithms can identify energy-saving opportunities, detect anomalies, and predict energy demand more accurately (Hassan, Sameen, Salman et al, 2023).

AI-based EMS can optimize energy usage by automatically adjusting parameters such as lighting, heating, cooling, and equipment operation based on real-time data and user preferences. They can also identify and prioritize energy-saving measures, provide energy consumption forecasts, and suggest optimal schedules for energy-intensive processes. Additionally, AI-powered EMS can interact with demand response programmes, energy storage systems, and renewable energy sources. They can dynamically manage energy flows, balance supply and demand, and optimize the use of distributed energy resources. By integrating AI into EMS, organizations can achieve significant energy savings, reduce carbon emissions, and improve operational efficiency. AI-enabled EMS helps create smarter and more sustainable energy ecosystems, enabling better energy management decisions and contributing to a greener and more resilient future (Şerban & Lytras, 2020).

Demand Response and AI

Demand response refers to a strategy in the energy sector that involves adjusting electricity consumption in response to supply and demand conditions. It aims to balance the electricity grid by shifting or reducing energy usage during peak demand periods or in times of grid stress. By analysing data, forecasting demand patterns, and allowing wise decision-making, artificial intelligence (AI) plays a critical role in enabling successful demand response (Alamir et al., 2023).

In order to predict electricity demand with accuracy, AI algorithms can evaluate real-time data from a variety of sources, including smart metres, IoT devices, and weather forecasts. By leveraging machine learning and data analytics, AI systems can identify consumption patterns, detect anomalies, and forecast future demand with high precision. With AI-powered demand response, grid operators and energy aggregators can communicate price signals or demand requests to consumers or connected devices in real-time. AI algorithms can optimize energy usage, schedule load shifts, and automate energy-saving measures based on predefined preferences and predefined strategies (Rohit et al., 2017).

Furthermore, AI can enable intelligent control and coordination of flexible loads, energy storage systems, and distributed energy resources. It can prioritize and optimize the participation of these resources in demand response programs to ensure grid stability and maximize the economic and environmental benefits (Praveen et al., 2023).

By integrating AI with demand response, the energy industry can achieve better grid reliability, reduce peak demand, and avoid the need for additional generation capacity. AI-enabled demand response promotes a more efficient, flexible, and sustainable electricity system by empowering consumers to actively participate in balancing the supply and demand dynamics of the grid (Fazal et al., 2023).

BENEFITS AND CHALLENGES OF AI IN RENEWABLE ENERGY

Advantages of AI in Renewable Energy

Renewable energy is one of many areas that artificial intelligence (AI) is revolutionising. AI has many benefits that can considerably improve the effectiveness, dependability, and integration of renewable energy systems in light of the growing focus on sustainable energy sources. We shall examine the main benefits of AI in renewable energy in this essay (Khan et al., 2023).

1. Improved Energy Forecasting

For the best management of renewable energy generation and grid integration, accurate energy forecasting is essential. In order to generate more precise and trustworthy energy generation projections, AI systems may examine enormous volumes of historical data, weather patterns, and other pertinent elements. AI systems can recognise complex patterns and connections by utilising machine learning and data analytics techniques, leading to more accurate projections of energy supply and demand. This aids grid managers and energy suppliers in planning maintenance tasks, resource allocation, and the effective grid integration of renewable energy sources. Effective energy trading and proactive decision-making to reduce potential grid imbalances are both made possible by improved forecasting (Appiah-Otoo et al., 2022).

2. Enhanced Energy Efficiency

Energy efficiency is a key focus in renewable energy systems. AI can play a pivotal role in optimizing energy consumption and distribution. By analyzing real-time data from sensors, smart meters, and other IoT devices, AI algorithms can identify energy-saving opportunities, detect anomalies, and automatically adjust parameters for maximum efficiency. For instance, based on occupancy, weather, and energy demand, AI can optimise the functioning of heating, ventilation, and air conditioning (HVAC) systems in buildings. By dynamically adjusting temperature settings and airflow, AI-enabled systems can reduce energy waste and provide optimal comfort. AI can also optimize energy usage in industrial processes, such as manufacturing and transportation, leading to significant energy savings and cost reduction (Broman & Robèrt, 2017).

3. Optimal Resource Management

Solar panels, wind turbines, and energy storage devices are examples of renewable energy sources that require efficient management to reach their full potential. AI can optimize the management of these resources by analyzing data on weather conditions, energy demand, grid stability, and other relevant factors. By understanding these variables, AI algorithms can determine the optimal use and allocation of renewable resources (Jha et al., 2017). For example, AI can dynamically adjust the tilt angle and orientation of solar panels to maximize energy generation based on the position of the sun. AI can also optimize the operation of wind turbines by adjusting blade angles and turbine speed to capture the maximum amount of wind energy. In order to maintain effective energy use and grid stability, AI algorithms can also optimise the charging and discharging cycles of energy storage devices. AI helps to maximise energy output, reduce waste, and enhance the overall effectiveness of renewable energy systems by optimising resource management (Johansson et al., 1992).

4. Grid Stability and Resilience

The addition of renewable energy sources to the current power grid poses particular difficulties in terms of grid stability, unpredictability, and intermittency. AI can address these challenges by actively monitoring and managing energy supply and demand, anticipating grid imbalances, and coordinating the integration of renewable sources. Intelligent load balancing, voltage control, and dynamic adjustments are made possible by AI to guarantee the power grid's dependable and steady operation. AI systems, for instance, may analyse real-time data from sensors and smart grid devices to find anomalies, voltage variations, and other elements that might affect the stability of the grid. By analyzing this data, AI systems can identify potential issues and trigger appropriate responses to mitigate or prevent disruptions. AI can also optimize power flow and transmission parameters to maintain grid stability and minimize losses. By integrating AI with renewable energy systems and grid management, the power grid becomes more resilient, adaptive, and capable of efficiently integrating renewable energy sources (Kalogirou, 2001).

5. Cost Reduction

In the field of renewable energy, AI-based optimisation strategies can result in significant cost savings. AI algorithms can reduce operational costs by optimising energy generation, transmission, and distribution. For example, AI can optimize the scheduling and dispatching of renewable energy generation sources based on real-time energy demand and market conditions, thereby reducing the reliance on costly conventional energy sources. AI can also optimize maintenance schedules for renewable energy systems, enabling preventive maintenance and minimizing downtime. Furthermore, AI can facilitate intelligent load management and demand response programs, allowing for efficient utilization of energy resources and reducing the need for additional generation capacity. By reducing operational costs, AI contributes to making renewable energy more economically viable and competitive with traditional energy sources (Chalk & Miller, 2006).

6. Scalability and Flexibility

Renewable energy systems need to be scalable and flexible to adapt to changing energy demands, resource availability, and grid conditions. AI-based systems can meet these requirements effectively. Real-time data analysis, technological adaptation, and energy system evolution are all possible with AI algorithms. For instance, AI can dynamically modify energy production and consumption to balance supply and demand as the use of renewable energy sources rises. Using AI, new technologies like energy storage, electric vehicles, and smart grids may be integrated more effectively. AI helps ensure the dependable and effective functioning of renewable energy systems in both centralised and decentralised settings by offering scalability and flexibility (Appiah-Otoo et al., 2022).

7. Integration of Distributed Energy Resources (DERs)

The landscape of renewable energy is increasingly dominated by distributed energy resources (DERs), such as rooftop solar panels, home energy storage systems, and electric car charging stations. In order to successfully integrate these DERs into the grid, AI is essential. AI algorithms can provide real-time monitoring, control, and coordination of these resources, optimizing their contribution to the overall energy system. For example, AI can prioritize the use of locally generated solar energy to power electric vehicles or manage the charging and discharging cycles of residential energy storage systems based on grid conditions and user preferences. By integrating DERs through AI, the renewable energy system becomes more decentralized, resilient, and capable of harnessing the full potential of distributed energy resources (Khan et al., 2023).

There are many benefits to artificial intelligence in the field of renewable energy. AI facilitates the transition to a sustainable and reliable energy future by enhancing energy forecasting, increasing energy efficiency, optimising resource management, ensuring grid stability, lowering costs, enabling scalability and flexibility, and facilitating the integration of distributed energy resources. AI's potential in the field of renewable energy is anticipated to increase as it develops and advances, helping to create a cleaner, greener, and more effective energy ecosystem (Friedlingstein et al., 2019).

Challenges and Limitations

While Artificial Intelligence (AI) offers numerous advantages in the renewable energy sector, it also faces certain challenges and limitations. Understanding and addressing these challenges is crucial for the effective implementation and utilization of AI technologies in renewable energy systems. In this chapter, we'll look at some of the main obstacles and restrictions to using AI in the field of renewable energy (Hassan, Abdulateef, Hafedh et al, 2023).

1. Data Availability and Quality

For precise analysis, prediction, and decision-making, AI systems primarily rely on high-quality data. However, in the renewable energy domain, data quality and availability can pose challenges. Weather data, for instance, is critical for predicting energy generation from solar and wind sources. However, weather data can be inconsistent, incomplete, or not easily accessible in some regions. Additionally, historical energy generation and consumption data may be limited, particularly in emerging renewable energy markets. Insufficient or low-quality data can hinder the performance and reliability of AI algorithms, leading to less accurate predictions and suboptimal decision-making (Henderson & Loreau, 2023).

2. Data Integration and Compatibility

Renewable energy systems generate large volumes of data from various sources, such as smart meters, sensors, and weather monitoring devices. Integrating and harmonizing this heterogeneous data can be a significant challenge. Data may be stored in different formats, structures, or systems, making it difficult to consolidate and analyze effectively. Data compatibility issues can arise when attempting to integrate data from different vendors or technologies. Ensuring seamless data integration and compatibility is essential for the successful implementation of AI in renewable energy systems (Anđelković & Bajatović, 2020).

3. Model Complexity and Interpretability

Deep learning and neural network-based AI systems in particular can be extremely sophisticated and challenging to understand. Because these models function as "black boxes," it is difficult to comprehend the underlying thought process. Lack of interpretability raises concerns regarding transparency, accountability, and trust in the decision-making of AI systems. In the renewable energy sector, where critical decisions impact energy generation, distribution, and grid stability, interpretability is essential. Developing AI models that are both accurate and interpretable remains an ongoing research challenge (Fazal et al., 2023).

4. Scalability and Computational Requirements

As renewable energy systems continue to grow in scale and complexity, the scalability of AI algorithms becomes crucial. AI models need to be able to handle large datasets and compute-intensive tasks. Training complex AI models may require significant computational resources, including processing power and memory. Deploying and running AI models in real-time applications can also pose challenges, as they need to process data rapidly and make timely decisions. Ensuring the scalability and computational efficiency of AI algorithms is vital to support the growing demand for renewable energy systems (Dincer, 2000).

5. Limited Domain Knowledge and Expertise

Domain knowledge and proficiency in both AI and renewable energy systems are necessary for the successful use of AI in renewable energy. However, there is a lack of experts who have a thorough understanding of both disciplines. AI experts may lack sufficient knowledge of the intricacies of renewable energy systems, including their technical, economic, and regulatory aspects. Likewise, renewable energy professionals may have limited familiarity with AI algorithms and their potential applications. To properly utilise AI's capabilities in renewable energy systems, the knowledge gap between AI and experts in the field must be closed (Henderson & Loreau, 2023).

6. Cost and Investment

Infrastructure, data collecting, software development, and system integration can all have considerable upfront expenses when implementing AI technology in renewable energy systems. Small-scale renewable energy projects or organizations with limited resources may find it challenging to invest in AI technologies. The cost-effectiveness of implementing AI in renewable energy systems must be carefully evaluated, considering factors such as the scale of the project, available resources, and potential long-term benefits. Identifying cost-effective AI solutions and fostering partnerships between AI developers, renewable energy companies, and research institutions can help overcome the financial challenges associated with implementing AI in the renewable energy sector (Rohit et al., 2017).

7. Regulatory and Policy Frameworks

For efficient integration of AI and renewable energy technology, regulatory and policy frameworks may offer difficulties. Regulatory barriers, such as complex permitting processes, data privacy regulations, and restrictions on data sharing, can impede the development and deployment of AI solutions in the renewable energy sector. Insufficiently addressing the unique needs and prospects of AI in renewable energy may stifle innovation and restrict the use of AI technologies. To overcome these obstacles, it is essential to create clear and encouraging legal frameworks that encourage innovation, data exchange, and collaboration (Hassan, Sameen, Salman et al, 2023).

8. Ethical Considerations

The application of AI in renewable energy poses moral questions about algorithmic biases, data security, and privacy. Huge volumes of data, particularly sensitive and personal data, are used in AI systems. It is crucial to ensure the security and privacy of this data. Additionally, biases can inadvertently be introduced into AI models, leading to unfair or discriminatory outcomes. Addressing ethical considerations requires the development of robust data governance frameworks, algorithmic transparency, and ethical guidelines specific to AI in renewable energy (Şerban & Lytras, 2020).

While AI offers significant advantages in the renewable energy sector, it is not without challenges and limitations. Overcoming these challenges requires collaboration between AI researchers, renewable energy experts, policymakers, and industry stakeholders. Addressing data quality and availability issues, improving data integration and compatibility, enhancing model interpretability, ensuring scalability and computational efficiency, fostering domain knowledge and expertise, managing costs and investments, developing supportive regulatory frameworks, and addressing ethical considerations are key steps towards harnessing the full potential of AI in renewable energy. By addressing these challenges, AI can become a transformative force in accelerating the transition towards a sustainable and clean energy future (Fazal et al., 2023).

REAL-WORLD APPLICATIONS

AI Applications in Solar Farms

Solar energy is a renewable energy source that is expanding quickly and is essential to the transition to a clean and sustainable energy future. A variety of applications for artificial intelligence (AI) technologies are available that can improve the operation, effectiveness, and administration of solar farms. We will examine the numerous AI uses in solar farms in this post, along with any potential advantages (Appiah-Otoo et al., 2022).

1. Solar Panel Performance Optimization

AI can optimize the performance of solar panels by analyzing real-time data, weather conditions, and historical patterns. Machine learning algorithms can predict the optimal tilt angle and orientation of solar panels to maximize energy generation based on the sun's position throughout the day. AI can also identify and diagnose issues such as panel degradation, dust accumulation, or shading, enabling timely maintenance and improving overall efficiency (Abbas et al., 2023).

2. Intelligent Solar Farm Design

AI can assist in the design and layout of solar farms by analyzing various factors such as land topography, shading analysis, and energy yield predictions. Machine learning algorithms can optimize the arrangement of solar panels to maximize energy generation while considering factors like available space, terrain, and geographical constraints. The arrangement and configuration of solar panels can be decided by solar farm developers using AI, which will boost energy output and improve project profitability (Abbas et al., 2023).

3. Predictive Maintenance and Fault Detection

AI algorithms can analyze sensor data from solar panels, inverters, and other equipment to detect anomalies and predict maintenance requirements. By identifying potential faults or malfunctions in real-time, AI enables proactive maintenance, reducing downtime and optimizing system performance. Predictive maintenance can help prevent costly repairs, extend the lifespan of equipment, and ensure reliable operation of the solar farm (Kalogirou, 2001).

4. Solar Irradiance Forecasting

Accurate solar irradiance forecasting is crucial for managing the variability of solar energy generation. AI algorithms can accurately anticipate sun irradiance levels by analysing historical meteorological information, satellite photos, and atmospheric variables. By supplying early warning of variations in solar energy generation, these forecasts aid operators of solar farms in optimising energy production and managing grid integration (Hassan, Tabar, Sameen et al, 2023).

5. Energy Demand Forecasting and Grid Integration

AI can forecast energy demand patterns by analyzing historical data, weather conditions, and other relevant factors. This information enables solar farm operators to align energy generation with demand, optimize energy dispatch, and support grid integration. By accurately predicting energy demand, AI can facilitate the efficient use of solar energy, reduce reliance on fossil fuel-based backup systems, and support the stability of the electricity grid (Hassan, Abbas, Tabar et al, 2023).

6. Optimization of Energy Storage Systems

Integrating energy storage systems with solar farms is crucial for ensuring reliable and consistent power supply. Based on solar energy production, energy demand, and power costs, AI algorithms can optimise the charging and discharging cycles of energy storage devices. By intelligently managing energy storage, AI can maximize self-consumption of solar energy, minimize grid reliance during peak demand periods, and enable cost-effective operation of the storage system (Hassan, Tabar, Sameen et al, 2023).

7. Intelligent Grid Management and Virtual Power Plants

AI can enable the integration of solar farms into virtual power plants (VPPs), where multiple distributed energy resources are aggregated and managed as a single entity. AI algorithms optimize the dispatch of energy from solar farms within the VPP, considering factors such as energy prices, grid constraints, and demand response signals. This intelligent grid management improves the stability and reliability of the electricity grid while maximizing the economic benefits of solar energy generation (Şerban & Lytras, 2020).

8. Data Analytics and Performance Monitoring

AI-based data analytics platforms can process and analyze large volumes of data from solar farms, providing real-time insights into system performance, energy generation, and operational efficiency. These analytics tools enable operators to monitor key performance indicators, identify inefficiencies, and make data-driven decisions for improving solar farm operations. AI can also generate actionable insights and recommendations for optimizing system performance and reducing operational costs (Hassan, Tabar, Sameen et al, 2023).

AI applications have the potential to revolutionize the solar energy industry by enhancing the performance, efficiency, and management of solar farms. From optimizing solar panel performance and intelligent farm design to predictive maintenance, energy forecasting, and grid integration, AI offers numerous benefits for solar farm operators. Solar farms may boost energy production, lower operating costs, boost system dependability, and aid in the worldwide transition to a clean and sustainable energy future by utilising AI (Jha et al., 2017).

AI-Driven Wind Turbine Control Systems

A major contributor to the worldwide shift to a clean and sustainable energy future is wind energy, which is a fast expanding renewable energy source. To maximise energy production and optimise the integration of wind power into the electrical grid, wind turbine performance and efficiency are essential. Artificial Intelligence (AI) technologies have emerged as powerful tools for enhancing the control systems of wind turbines. In this article, we will explore the concept of AI-driven wind turbine control systems, their applications, and the potential benefits they offer (Xiao et al., 2015).

1. Intelligent Control Strategies

AI-driven wind turbine control systems utilize machine learning algorithms to develop intelligent control strategies. Traditional control systems operate based on predefined rules and models, while AI-driven systems learn from data and adapt their control actions accordingly. By analyzing real-time data from sensors, weather conditions, and historical patterns, AI algorithms can optimize the control parameters of wind turbines to maximize energy capture and minimize structural loads. These intelligent control strategies improve the overall efficiency and performance of wind turbines (Lu & Zhou, 2018).

2. Condition Monitoring and Predictive Maintenance

AI can enable real-time condition monitoring of wind turbines by analyzing sensor data to detect anomalies and identify potential faults. Machine learning algorithms can learn the normal operating patterns of wind turbines and detect deviations from these patterns, indicating potential issues. By providing early warning signals, AI-driven systems facilitate proactive maintenance, reducing downtime and optimizing the availability and reliability of wind turbines. Predictive maintenance based on AI analytics can help prevent costly repairs, extend the lifespan of components, and enhance the overall performance of wind farms (Xiao et al., 2015).

3. Wind Farm Optimization

AI-driven control systems can optimize the operation of entire wind farms rather than individual turbines. By analyzing real-time data from multiple turbines within a wind farm, AI algorithms can optimize the allocation of power among turbines, considering factors such as wind conditions, wake effects, and grid constraints. This optimization improves the overall power output, reduces turbulence, and minimizes the impact of wake losses. AI-driven wind farm optimization leads to increased energy generation, improved grid integration, and enhanced economic viability (Lu & Zhou, 2018).

4. Power Curve Optimization

The link between wind speed and power output is depicted by a wind turbines power curve. AI-driven control systems can optimize the power curve by continuously adjusting the turbine's control parameters based on real-time wind conditions. Machine learning algorithms learn from historical data to predict the optimal power curve for different wind speeds, allowing the turbine to operate at its maximum efficiency. Power curve optimization enhances energy capture, reduces mechanical stress on the turbine, and increases the overall energy production of wind farms (Zhang et al., 2021).

5. Grid-Friendly Operation

AI-driven control systems enable wind turbines to operate in a grid-friendly manner, addressing the challenges associated with intermittent wind power generation. By analyzing real-time grid conditions, market prices, and demand forecasts, AI algorithms can optimize the power output of wind turbines to match grid requirements. This grid-friendly operation facilitates the stable integration of wind power into the electricity grid, improves grid stability, and reduces the need for fossil fuel-based backup systems. AI-driven control systems also support grid ancillary services such as frequency regulation and voltage control (Zhang et al., 2021).

6. Turbine Health Monitoring

AI-driven control systems can monitor the health of individual wind turbines by analyzing sensor data and detecting potential mechanical issues or component degradation. By continuously monitoring key performance indicators, such as rotor speed, temperature, and vibration levels, AI algorithms can identify early signs of turbine wear, misalignment, or bearing failures. This proactive monitoring allows operators to take timely maintenance actions, prevent catastrophic failures, and optimize the operational lifespan of wind turbines (Xiao et al., 2015).

7. Adaptive Control in Complex Environments

Wind turbines operate in complex environments with varying wind conditions and dynamic loads. AI-driven control systems offer adaptive control capabilities that adjust turbine operations in real-time based on environmental changes. By continuously analyzing sensor data and environmental inputs, AI algorithms can adapt the turbine's control parameters to optimize energy capture and respond to changing wind conditions. This adaptive control improves the turbine's efficiency, reduces mechanical stress, and ensures safe operation under different operating conditions (Dasgupta et al., 2020).

By optimising energy production, strengthening turbine performance, and enhancing grid integration, AI-driven wind turbine control systems have the potential to completely transform the wind energy sector. AI has many advantages for both grid operators and operators of wind turbines, from sophisticated control techniques and condition monitoring to wind farm optimisation and grid-friendly operation. Wind farms can boost energy output, lower maintenance costs, lengthen turbine lifespan, and support the worldwide transition to a clean and sustainable energy future by utilising AI (Dasgupta et al., 2020).

Grid Optimization With AI

The process of enhancing the effectiveness, dependability, and resilience of electrical power grids is referred to as grid optimisation. The need for grid optimisation is crucial as electricity consumption rises and renewable energy sources are increasingly integrated. Artificial intelligence (AI) has become a potent tool for managing numerous difficulties in electricity generation, transmission, distribution, and demand management and optimising grid operations. In this post, we'll examine the idea of grid optimisation with AI, its uses, and any advantages it might have (Chalk & Miller, 2006).

1. Load Forecasting and Demand Response

In order to anticipate electricity demand effectively, AI systems can examine historical data, weather trends, and other important elements. Grid operators can effectively plan and deploy resources, optimise generation schedules, and lessen the danger of supply-demand imbalances thanks to load forecasting. Additionally, by offering insights into customer behaviour and preferences, AI can help demand response programmes. Demand response enables grid operators to reward customers for adjusting their electricity usage at peak times, which reduces load, boosts grid stability, and saves money (Hou et al., 2022).

2. Renewable Energy Integration and Forecasting

The integration of renewable energy sources, such as solar and wind, poses challenges due to their intermittent and variable nature. AI can optimize the grid's incorporation of renewable energy by analyzing real-time data, weather conditions, and historical patterns. AI algorithms can forecast renewable energy generation accurately, enabling grid operators to balance supply and demand, optimize energy dispatch, and mitigate the impact of fluctuations in renewable energy production. With this optimisation, grid stability is increased, backup systems that rely on fossil fuels are used less frequently, and the move towards sustainable energy is aided (Ashfaq et al., 2022).

3. Grid Fault Detection and Self-Healing

AI-driven grid optimization systems can detect faults and abnormalities in real-time by analyzing data from sensors, smart meters, and other monitoring devices. Machine learning algorithms can learn from historical data to identify patterns associated with grid failures, such as equipment malfunctions or transmission line faults. AI-driven fault detection enables grid operators to respond quickly to grid disturbances, isolate affected areas, and restore power efficiently. Additionally, AI can support self-healing capabilities by automatically rerouting power, optimizing grid reconfiguration, and minimizing the impact of disruptions on consumers (Chalk & Miller, 2006).

4. Grid Stability and Voltage Control

Maintaining grid stability and voltage control is crucial for reliable and efficient power delivery. In order to manage reactive power flow, optimise voltage levels, and control system frequency, AI algorithms can analyse real-time data and grid conditions. By continuously monitoring grid parameters and dynamically adjusting control actions, AI-driven systems can enhance grid stability, reduce power losses, and improve the overall quality of electricity supply (Kalogirou, 2001).

5. Asset Management and Maintenance Optimization

AI can optimize asset management and maintenance activities by analyzing sensor data, historical performance, and maintenance records. Machine learning algorithms can predict the likelihood of equipment failures, prioritize maintenance tasks, and optimize maintenance schedules. By implementing condition-based maintenance strategies, grid operators can reduce downtime, extend equipment lifespan, and minimize maintenance costs. Additionally, AI-driven asset management enables proactive decision-making, enhancing grid reliability and operational efficiency (Alamir et al., 2023).

6. Energy Trading and Market Optimization

AI can optimize energy trading and market operations by analyzing market data, pricing signals, and demand-supply dynamics. AI algorithms can identify market opportunities, forecast energy prices, and optimize trading strategies. Grid operators can use AI-driven market optimization to maximize revenue, minimize energy procurement costs, and participate effectively in energy markets. This optimization improves market efficiency, facilitates the integration of renewable energy, and supports grid stability (Ashfaq et al., 2022).

7. Grid Planning and Expansion

AI-driven grid optimization systems can assist in grid planning and expansion by analyzing geographical data, population growth, and energy demand forecasts. Machine learning algorithms can optimize the placement of transmission lines, substations, and renewable energy installations to minimize grid losses, reduce infrastructure costs, and ensure reliable power supply. AI-driven grid planning enables proactive decision-making, efficient resource allocation, and scalability to accommodate future energy needs (Kalogirou, 2001).

Grid optimization with AI offers significant opportunities to enhance the efficiency, reliability, and sustainability of electrical power grids. From load forecasting and demand response to renewable energy integration, fault detection, and asset management, AI-driven systems can address various challenges faced by grid operators. Grids can be optimised for maximum performance, effective energy use, lower prices, and improved grid stability by utilising AI. Grid optimisation with AI integration opens the door to a smarter, more reliable, and sustainable energy future (Okonkwo et al., 2021).

FUTURE DIRECTIONS AND CONCLUSION

Potential Future Trends

The world is witnessing rapid advancements in technology and evolving societal needs, which shape the trajectory of various industries. In this article, we will explore potential future trends across different sectors and domains, considering the latest developments and emerging possibilities (Beynaghi et al., 2016).

1. Machine Learning (ML) and Artificial Intelligence (AI)

ML and AI are expected to continue their exponential growth, permeating various industries and domains. Intelligent robotics, personalised healthcare, smart cities, and other advanced applications will be made possible by improvements in AI algorithms, neural networks, and computer capacity. AI and ML will play a pivotal role in data analysis, pattern recognition, decision-making, and automation, revolutionizing industries and improving efficiency (Pandey, 2018).

2. Internet of Things (IoT)

The IoT will continue to expand, connecting an increasing number of devices and objects. This interconnected network will enable seamless communication, data sharing, and automation across various sectors, including transportation, healthcare, manufacturing, and agriculture. IoT devices will gather and transmit real-time data, enabling smart monitoring, predictive maintenance, and optimized resource allocation. The integration of IoT with AI will unlock new possibilities for intelligent automation and decision-making (Khan et al., 2023).

3. Renewable Energy and Sustainable Solutions

With the growing concern for environmental sustainability, there will be an increased focus on renewable energy sources such as wind, hydropower and solar. Advances in energy storage technologies, such as batteries and fuel cells, will increase the effectiveness and dependability of renewable energy systems. Moreover, the integration of AI and IoT will optimize the management and grid integration of renewable energy, facilitating a transition towards a greener and more sustainable energy landscape (Beynaghi et al., 2016).

4. Blockchain and Decentralized Systems

Blockchain technology, originally known for cryptocurrencies, will find applications beyond finance. Supply chain management, healthcare, voting systems, and identity verification are just a few of the industries that it can be used in because of its decentralised and secure nature. Blockchain can enable transparent and tamper-proof transactions, enhance data privacy, and streamline processes that require trust and accountability (Leng et al., 2020).

5. Virtual Reality (VR) and Augmented Reality (AR)

VR and AR technologies will continue to evolve, transforming industries such as gaming, entertainment, education, and healthcare. AR can overlay digital information onto the real world, enhancing productivity and providing immersive experiences. VR creates virtual environments that simulate real-world scenarios, enabling training, simulations, and virtual tourism. These technologies will have broader applications, ranging from remote collaboration to immersive training experiences and virtual tourism (Khan et al., 2023).

6. Genetic Engineering and Biotechnology

Advancements in genetic engineering and biotechnology will revolutionize healthcare, the agriculture sector, and environmental preservation. Precision medicine, gene editing, and regenerative medicine will lead to more personalized and effective treatments for diseases. Genetic engineering in agriculture will result in improved crop yields, disease resistance, and nutritional value. Moreover, biotechnology will contribute to sustainable solutions, such as biofuels, biodegradable materials, and environmental remediation (Şerban & Lytras, 2020).

7. Cybersecurity and Privacy

As technology evolves, the importance of cybersecurity and privacy will continue to grow. The increasing interconnectivity and reliance on digital systems will demand robust security measures to protect sensitive data and infrastructure. AI will play a vital role in threat detection, anomaly detection, and real-time response to cyber threats. Privacy regulations and frameworks will evolve to safeguard individuals' data and ensure responsible use of emerging technologies (Jha et al., 2017).

These potential future trends demonstrate the transformative power of technology across various domains. As AI, IoT, renewable energy, blockchain, AR/VR, biotechnology, and cybersecurity continue to advance, they will shape industries, societies, and economies. Embracing these trends and harnessing their potential will pave the way for a more connected, sustainable, and technologically empowered future (Pandey, 2018).

CONCLUSION

In conclusion, the rapid advancements in technology, particularly in the fields of artificial intelligence, renewable energy, IoT, blockchain, biotechnology, and augmented/virtual reality, are shaping the future landscape across various industries and sectors. These emerging trends offer immense opportunities for innovation, efficiency, and sustainability.

Artificial intelligence and machine learning are revolutionizing decision-making, automation, and data analysis, paving the way for intelligent systems and applications. The integration of AI with IoT is enabling seamless connectivity, real-time data monitoring, and predictive analytics, transforming industries such as healthcare, manufacturing, transportation, and agriculture.

Renewable energy is gaining prominence as a viable and sustainable solution to address the challenges of climate change. Advancements in energy storage technologies, along with the optimization of renewable energy integration using AI and IoT, are propelling the transition towards a greener and cleaner energy landscape.

Blockchain technology is disrupting traditional systems by providing secure, decentralized, and transparent solutions for several industries, including financing, management of supply chain, and identity verification. Its potential for enhancing trust, security, and accountability is driving innovation and transforming industries.

Augmented reality and virtual reality technologies are reshaping the way we interact with the digital world, offering immersive experiences, training simulations, and improved productivity in sectors such as gaming, education, and healthcare.

Genetic engineering and biotechnology will revolutionize healthcare, the agriculture sector, and environmental preservation. Precision medicine, gene editing, and biodegradable materials are revolutionizing healthcare treatments, agricultural practices, and sustainable solutions.

As technology continues to advance, cybersecurity and privacy become paramount. Robust security measures, powered by AI, are crucial for protecting sensitive data and infrastructure. Privacy regulations and frameworks are evolving to ensure responsible use of emerging technologies and safeguard individuals' data.

In embracing these potential future trends, industries, societies, and economies can harness the transformative power of technology to drive innovation, improve efficiency, and foster sustainability. By embracing these trends, we can shape a future that is technologically empowered, interconnected, and sustainable.

REFERENCES

Abbas, M. K., Hassan, Q., Tabar, V. S., Tohidi, S., Jaszczur, M., Abdulrahman, I. S., & Salman, H. M. (2023). Techno-economic analysis for clean hydrogen production using solar energy under varied climate conditions. *International Journal of Hydrogen Energy*, *48*(8), 2929–2948. doi:10.1016/j.ijhydene.2022.10.073

Alamir, N., Kamel, S., Megahed, T. F., Hori, M., & Abdelkader, S. M. (2023). Developing Hybrid Demand Response Technique for Energy Management in Microgrid Based on Pelican Optimization Algorithm. *Electric Power Systems Research*, *214*, 108905. doi:10.1016/j.epsr.2022.108905

Anđelković, A. S., & Bajatović, D. (2020). Integration of weather forecast and artificial intelligence for a short-term city-scale natural gas consumption prediction. *Journal of Cleaner Production*, *266*, 122096. doi:10.1016/j.jclepro.2020.122096

Appiah-Otoo, I., Song, N., Acheampong, A. O., & Yao, X. (2022). Crowdfunding and renewable energy development: What does the data say? *International Journal of Energy Research*, *46*(2), 1837–1852. doi:10.1002/er.7301

Ashfaq, A., Kamran, M., Rehman, F., Sarfaraz, N., Ilyas, H. U., & Riaz, H. H. (2022, March). Role of Artificial Intelligence in Renewable Energy and its Scope in Future. In *2022 5th International Conference on Energy Conservation and Efficiency (ICECE)* (pp. 1-6). IEEE. 10.1109/ICECE54634.2022.9758957

Beynaghi, A., Trencher, G., Moztarzadeh, F., Mozafari, M., Maknoon, R., & Leal Filho, W. (2016). Future sustainability scenarios for universities: Moving beyond the United Nations Decade of Education for Sustainable Development. *Journal of Cleaner Production*, *112*, 3464–3478. doi:10.1016/j.jclepro.2015.10.117

Broman, G. I., & Robèrt, K. H. (2017). A framework for strategic sustainable development. *Journal of Cleaner Production*, *140*, 17–31. doi:10.1016/j.jclepro.2015.10.121

Chalk, S. G., & Miller, J. F. (2006). Key challenges and recent progress in batteries, fuel cells, and hydrogen storage for clean energy systems. *Journal of Power Sources*, *159*(1), 73–80. doi:10.1016/j.jpowsour.2006.04.058

Chami, R., Cosimano, T., Fullenkamp, C., & Oztosun, S. (2019). Nature's solution to climate change. *Finance and Development Magazine*, *56*, 34–38.

Dasgupta, K., Roy, P. K., & Mukherjee, V. (2020). Power flow based hydro-thermal-wind scheduling of hybrid power system using sine cosine algorithm. *Electric Power Systems Research*, *178*, 106018. doi:10.1016/j.epsr.2019.106018

Dincer, I. (2000). Renewable energy and sustainable development: A crucial review. *Renewable & Sustainable Energy Reviews*, *4*(2), 157–175. doi:10.1016/S1364-0321(99)00011-8

Fawzy, S., Osman, A. I., Doran, J., & Rooney, D. W. (2020). Strategies for mitigation of climate change: A review. *Environmental Chemistry Letters*, *18*(6), 2069–2094. doi:10.100710311-020-01059-w

Fazal, S., Haque, M. E., Arif, M. T., Gargoom, A., & Oo, A. M. T. (2023). Grid integration impacts and control strategies for renewable based microgrid. *Sustainable Energy Technologies and Assessments*, *56*, 103069. doi:10.1016/j.seta.2023.103069

Friedlingstein, P., Jones, M. W., O'sullivan, M., Andrew, R. M., Hauck, J., Peters, G. P., Peters, W., Pongratz, J., Sitch, S., Le Quéré, C., Bakker, D. C. E., Canadell, J. G., Ciais, P., Jackson, R. B., Anthoni, P., Barbero, L., Bastos, A., Bastrikov, V., Becker, M., ... Zaehle, S. (2019). Global carbon budget 2019. *Earth System Science Data, 11*(4), 1783–1838. doi:10.5194/essd-11-1783-2019

Hák, T., Janoušková, S., & Moldan, B. (2016). Sustainable Development Goals: A need for relevant indicators. *Ecological Indicators, 60*, 565–573. doi:10.1016/j.ecolind.2015.08.003

Hassan, Q., Abbas, M. K., Tabar, V. S., Tohidi, S., Sameen, A. Z., & Salman, H. M. (2023). Techno-economic assessment of battery storage with photovoltaics for maximum self-consumption. *Energy Harvesting and Systems*, (0).

Hassan, Q., Abdulateef, A. M., Hafedh, S. A., Al-samari, A., Abdulateef, J., Sameen, A. Z., Salman, H. M., Al-Jiboory, A. K., Wieteska, S., & Jaszczur, M. (2023). Renewable energy-to-green hydrogen: A review of main resources routes, processes and evaluation. *International Journal of Hydrogen Energy, 48*(46), 17383–17408. doi:10.1016/j.ijhydene.2023.01.175

Hassan, Q., Sameen, A. Z., Salman, H. M., Al-Jiboory, A. K., & Jaszczur, M. (2023). *The role of renewable energy and artificial intelligence towards environmental sustainability and net zero*. Academic Press.

Hassan, Q., Tabar, V. S., Sameen, A. Z., Salman, H. M., & Jaszczur, M. (2023). A review of green hydrogen production based on solar energy: Techniques and methods. *Energy Harvesting and Systems*, (0).

He, B., Ye, L., Pei, M., Lu, P., Dai, B., Li, Z., & Wang, K. (2022). A combined model for short-term wind power forecasting based on the analysis of numerical weather prediction data. *Energy Reports, 8*, 929–939. doi:10.1016/j.egyr.2021.10.102

Henderson, K., & Loreau, M. (2023). A model of Sustainable Development Goals: Challenges and opportunities in promoting human well-being and environmental sustainability. *Ecological Modelling, 475*, 110164. doi:10.1016/j.ecolmodel.2022.110164

Hou, H., Liu, C., Wang, Q., Wu, X., Tang, J., Shi, Y., & Xie, C. (2022). Review of load forecasting based on artificial intelligence methodologies, models, and challenges. *Electric Power Systems Research, 210*, 108067. doi:10.1016/j.epsr.2022.108067

Jha, S. K., Bilalovic, J., Jha, A., Patel, N., & Zhang, H. (2017). Renewable energy: Present research and future scope of Artificial Intelligence. *Renewable & Sustainable Energy Reviews, 77*, 297–317. doi:10.1016/j.rser.2017.04.018

. Johansson, T. B., Kelly, H., Reddy, A. K., & Williams, R. H. (1992). Renewable fuels and electricity for a growing world economy: Defining and achieving the potential. *Energy Studies Review, 4*(3).

Kalogirou, S. A. (2001). Artificial neural networks in renewable energy systems applications: A review. *Renewable & Sustainable Energy Reviews*, *5*(4), 373–401. doi:10.1016/S1364-0321(01)00006-5

Khan, O., Parvez, M., Kumari, P., Parvez, S., & Ahmad, S. (2023). The future of pharmacy: How AI is revolutionizing the industry. *Intelligent Pharmacy*, *1*(1), 32–40. doi:10.1016/j.ipha.2023.04.008

Leng, J., Ruan, G., Jiang, P., Xu, K., Liu, Q., Zhou, X., & Liu, C. (2020). Blockchain-empowered sustainable manufacturing and product lifecycle management in industry 4.0: A survey. *Renewable & Sustainable Energy Reviews*, *132*, 110112. doi:10.1016/j.rser.2020.110112

Lu, Z., & Zhou, S. (2018). *Integration of large scale wind energy with electrical power systems in China*. John Wiley & Sons. doi:10.1002/9781118910054

Nasir, O., Javed, R. T., Gupta, S., Vinuesa, R., & Qadir, J. (2023). Artificial intelligence and sustainable development goals nexus via four vantage points. *Technology in Society*, *72*, 102171. doi:10.1016/j.techsoc.2022.102171

Okonkwo, E. C., Wole-Osho, I., Bamisile, O., Abid, M., & Al-Ansari, T. (2021). Grid integration of renewable energy in Qatar: Potentials and limitations. *Energy*, *235*, 121310. doi:10.1016/j.energy.2021.121310

Pandey, G. (2018). Challenges and future prospects of agri-nanotechnology for sustainable agriculture in India. *Environmental Technology & Innovation*, *11*, 299–307. doi:10.1016/j.eti.2018.06.012

Pang, Y., Pan, L., Zhang, J., Chen, J., Dong, Y., & Sun, H. (2022). Integrated sizing and scheduling of an off-grid integrated energy system for an isolated renewable energy hydrogen refueling station. *Applied Energy*, *323*, 119573. doi:10.1016/j.apenergy.2022.119573

Panwar, N. L., Kaushik, S. C., & Kothari, S. (2011). Role of renewable energy sources in environmental protection: A review. *Renewable & Sustainable Energy Reviews*, *15*(3), 1513–1524. doi:10.1016/j.rser.2010.11.037

Praveen, P., Khan, A., Verma, A. R., Kumar, M., & Peoples, C. (2023). Smart Village Infrastructure and Rural Communities. In Smart Village Infrastructure and Sustainable Rural Communities (pp. 1-15). IGI Global. doi:10.4018/978-1-6684-6418-2.ch001

Rohit, A. K., Devi, K. P., & Rangnekar, S. (2017). An overview of energy storage and its importance in Indian renewable energy sector: Part I–Technologies and Comparison. *Journal of Energy Storage*, *13*, 10–23. doi:10.1016/j.est.2017.06.005

Şerban, A. C., & Lytras, M. D. (2020). Artificial intelligence for smart renewable energy sector in europe—Smart energy infrastructures for next generation smart cities. *IEEE Access : Practical Innovations, Open Solutions*, *8*, 77364–77377. doi:10.1109/ACCESS.2020.2990123

Wu, X., Liao, B., Su, Y., & Li, S. (2023). Multi-objective and multi-algorithm operation optimization of integrated energy system considering ground source energy and solar energy. *International Journal of Electrical Power & Energy Systems*, *144*, 108529. doi:10.1016/j.ijepes.2022.108529

Xiao, L., Wang, J., Dong, Y., & Wu, J. (2015). Combined forecasting models for wind energy forecasting: A case study in China. *Renewable & Sustainable Energy Reviews*, *44*, 271–288. doi:10.1016/j.rser.2014.12.012

Zhang, L., Wen, J., Li, Y., Chen, J., Ye, Y., Fu, Y., & Livingood, W. (2021). A review of machine learning in building load prediction. *Applied Energy*, *285*, 116452. doi:10.1016/j.apenergy.2021.116452

KEY TERMS AND DEFINITIONS

Artificial Intelligence: Artificial Intelligence (AI) refers to the simulation of human intelligence in machines that can perform tasks, learn from experience, and make decisions. It encompasses various technologies like machine learning, natural language processing, and computer vision, enabling systems to analyze data, recognize patterns, and automate processes.

Demand Response: Demand Response is a strategy that aims to balance electricity supply and demand by incentivizing consumers to adjust their energy usage during peak periods. AI in Demand Response leverages data analysis and predictive modeling to optimize load management, improve grid stability, and maximize energy efficiency.

Energy Management: Energy Management involves monitoring, controlling, and optimizing energy consumption to ensure efficient use of resources. AI in Energy Management utilizes data analytics, machine learning, and automation to optimize energy systems, enhance renewable energy integration, and enable intelligent decision-making for improved energy efficiency and cost savings.

Energy Systems: Energy Systems refer to the interconnected components and processes involved in generating, transmitting, distributing, and consuming energy. AI in Energy Systems leverages advanced algorithms, data analysis, and automation to optimize renewable energy generation, grid integration, demand management, and overall system efficiency for a sustainable and reliable energy future.

Forecasting: Forecasting in the role of AI in Renewable Energy (RE) involves using advanced algorithms and data analysis techniques to predict future energy generation, demand, and market conditions. AI-powered forecasting improves the accuracy and reliability of predictions, enabling optimal decision-making, resource planning, and grid management in the RE sector.

Grid Optimization: Grid Optimization in the role of AI in Renewable Energy (RE) involves using AI algorithms and techniques to improve the efficiency, reliability, and stability of the electrical power grid. AI-driven grid optimization enables intelligent energy management, optimal grid operation, and effective integration of renewable energy sources for a sustainable and resilient grid infrastructure.

Renewable Energy: Renewable Energy refers to energy derived from naturally replenishing sources such as sunlight, wind, and water. In the role of AI in Renewable Energy (RE), AI technologies are utilized to optimize renewable energy generation, improve resource forecasting, enhance grid integration, and enable intelligent energy management for a sustainable and efficient energy system.

Sustainability: Sustainability, in the role of AI in Renewable Energy (RE), refers to the application of AI technologies to promote and achieve environmentally sustainable practices in the generation, distribution, and consumption of renewable energy. AI in RE enables efficient resource utilization, optimized energy management, reduced carbon footprint, and enhanced decision-making for a sustainable energy future that balances environmental, social, and economic considerations.

Chapter 9
Ethics and Scientific Integrity in Research:
The Undeniable Process in the Development of the Higher Education System

Jamileh Alamolhoda
Shahid Beheshti University, Iran

Esmaeil Jafari
https://orcid.org/0000-0003-3950-8795
Shahid Beheshti University, Iran

ABSTRACT

The improving the state of research ethics among the scientific community requires attention to the necessary infrastructure and quality in research. This chapter considers the training of ethical principles in the research process in curricula as a neglected aspect of research activities and it considers ethics and scientific integrity as the most important aspect of promoting academic excellence. The first step to prevent ethical abuse in research and to internalize scientific excellence in universities considers the formation of values or the formulation of an ethics and scientific integrity charter as the main solution. Therefore, this chapter first considers the reason for ethics and scientific integrity in research and tries to establish a link between these. Then it considers the position of ethics and scientific integrity in research in higher education centers. In the following, the higher education system is faced with two ethical discourses. Finally, it deals with standardization and drawing a vision of the ethics in the higher education system.

DOI: 10.4018/978-1-6684-9601-5.ch009

INTRODUCTION

Higher education is the manifestation of the growth and development of information and knowledge in societies and provides the power of societies in the field of international competition. Due to its inherent importance, this growing phenomenon is increasingly being noticed by societies. At the same time, due to the diversity of the needs of societies, it includes various functions such as education, research and human resource development (Kitagawa & Oba, 2010). Among the various functions of higher education, the function of research has become very important; Because the material and spiritual growth and prosperity of human civilizations and societies depends on research. However, if the research is properly guided and is not placed on the path of spiritual and ethical development, it is considered a factor of destruction. Therefore, there should be a guide or instruction that protects this function from deviation, and generally this responsibility rests with scientific ethics in general and ethics in research in particular (Vallance, 2004). In this regard, although many common elements of ethics and integrity are shared between different fields, there are also specific and unique challenges for each field in terms of how to deal with ethics and scientific integrity in research (Iphofen, 2020). In the context of very rapid scientific-technological development, this gradually covered wider areas and included issues related to consent, privacy, data management, fraud and plagiarism (Sleem et al, 2010).

On the other hand, the introduction of information and communication technology (ICT) in the educational system and research in this field raised a set of ethical risks related to the access and exchange of information in the form of ethical issues in the learning environment (Jafari & Alamolhoda, 2021). In this regard, these general issues should be understood and revealed in specific cases. However, while one should be open to the possibility of using good ideas and practices across disciplines, It should equally be able to resist over-weak generalizations or intentional or reckless "ethical imperialism" (Schrag, 2010).

Although ethical issues have been raised in the higher education system since the 1970s, issues that we now call "scientific ethics and integrity" have become common concerns in the scientific and decision-making communities in the higher education system. Therefore, the concept of ethical research in the higher education system has been expanded and includes questions. Faced with a number of proven cases of misconduct, the international community is faced with the need to regulate in order to maintain public trust and create appropriate frameworks. Therefore, at the international level, scientific integrity, which initially focused more on forgery and plagiarism, gradually included questions whose scope is epistemological (especially in the quality of knowledge produced, social responsibility and values and goals of science). Today, we can confirm that ethical issues and scientific integrity in research are closely related and should be considered together. Seeing these two elements together is mostly due to the fact that the increasing entanglement between universities and industry in the context of international competition creates more incentives for research misconduct (Jafari, 2022).

As mentioned, the majority of people involved in the higher education system conduct advanced research that covers a range of topics in natural, social, formal and applied sciences. However, the nature of the higher education system, which focuses on human dignity and values, sometimes conflicts with the ethics of what can be considered purely academic. This is primarily due to ambiguity or apparent conflict with current ethical standards and codes (Holder-Webb & Cohen, 2012). In this regard, this chapter focuses on the importance of ethics in research in the field of higher education system. It will then address the dichotomy between ethics and scientific integrity through a comprehensive analysis of the relevant literature. It will examine the importance and reasons for concern regarding the ethical issues faced in research in the higher education system. Because how to deal with these issues can help clarify the key issues in this chapter. In addition, it can help develop a framework that higher education stakeholders can use to maintain academic ethics and integrity while examining issues and proposing appropriate solutions for problems that have not yet been addressed in the field of higher education.

The issues raised in this chapter are deliberately structured in order to raise a correct understanding of ethics and scientific integrity. The main purpose of this chapter is to provide advice and guidance in a wide range of fields and professions in the higher education system. It is intended to help all readers apply their approach to difficult questions of ethical principles, values, and standards. In this regard, researchers can usually be fruitful in helping to resolve and identify questions such as "How can we try to be as ethical and honest as possible".

The Duality of Ethics and Scientific Integrity

No activity can be completed without understanding the main issues of ethics and scientific integrity. Unfortunately, they can rarely be acted upon in a separate way; Because, issues related to ethics and scientific integrity significantly overlap. These terms are a combination of principles (respecting privacy, protecting vulnerable people) and ways of realizing those principles (securing informed consent, anonymization, minimizing deception, maintaining confidentiality, etc.). Therefore, although these principles may be considered conceptually distinct, in practice they are related to each other in various ways and sometimes cause conflicts and dilemmas. For example, addressing the concept of "privacy" requires an assessment of what constitutes public space (in the real and virtual sence).

There is no universal consensus on the meaning of these two terms. Some scholars use "scientific integrity" as an overarching term (Neves & do Céu, 2018). It seems that "ethics" is used as a comprehensive term, which means scientific ethics is a field of study, which has sub-branches, similar to biological ethics or environmental ethics. Academic integrity, if used as a general term, then usually refers to ethics as an ideal that the educational environment should serve. An analysis of the scientific literature on scientific integrity shows that integrity is very similar to a person's etichs (Kaiser, 2014).

Scientific integrity is partly related to the element of public trust and everything in the environment (stakeholders and university) and the bureaucracy of the university is the guardian of this trust. However, one of the criticisms leveled at the university is the mechanization of relationships and the lack of attention to the charismatic personality and outstanding characteristics of academic stakeholders. In the absence of human relations and especially with the presence of young people who are called generation Z, the university has suffered a ethical collapse. Bureaucracy and strict rules of the university have helped to create a machine atmosphere of the university and scientific relations are formed in the form of administrative or business ethics relations. Therefore, authority lacks qualities such as knowledge, kindness and respect and is only the result of values related to trade or values related to governance or administrative discipline. The presence of generation Z in higher education and their authority-avoidance and disobedience character, which has challenged both the university bureaucracy and the principles of academic ethics, it is a new to return to an old issue, namely the relationship between ethics and scientific integrity with research and science (Iphofen, 2020).

In addition, the separation of ethical values from scientific and research activity can be seen as a result of the business view of related science. Emphasizing the application of science is the result of moving away from the inherent virtue of science and looking at it as a tool to achieve personal and organizational goals. In this type of view, science is not a virtue or a part of happiness, but it is only considered a tool to achieve prosperity and peace, but it is only considered a tool to achieve prosperity and peace, which will lead to the meaninglessness of life and the dominance of technology over science and scientific research. This makes learning to be evaluated only according to the values of employment and in proportion to the employment crisis. Therefore, the university focuses on the employment of graduates and follows the employment policies and becomes subject to the market. In this case, the university is unable to be active in the organization of people's knowledge and value system. The market of cultural goods and entertainment devices, especially cinema and media, has taken responsibility in the absence of an effective university and the passivity of science, and of course, it has not done well.

Therefore, considering the importance of this duality, it is obvious that it is necessary to recognize and express this duality. This issue may happen especially in

Iran due to the dependence of scientific institutions on the government and the lack of relative independence of the scientific institution from the political institution and other institutions. Also, it may be more important when wrong and expedient and opportunistic beliefs are carried out by the government towards their goals in research activities (Mirzaie & Qarakhani, 2013). In this regard, Neves (2018) states that the scientific integrity is not to move towards the development of a specific action, but towards defense or protection against potentially harmful external actions.

Reasons for Concern about Ethics and Scientific Integrity

It may be considered why ethics and integrity are important in the higher education system and even in the medical education system. It is not simply for any of these reasons that ethics and integrity are important. Rather, health, well-being, freedom, and human rights—even the survival of humans, other species, and the planet as we know it—depend on the integrity and ethics of research (WHO, 2015). This is because there should never be challenges to scientific findings that might cause someone to doubt the validity of those findings and those who produced those results. It is entirely possible to draw erroneous conclusions from careful research efforts, but this should never be the result of bad faith, a conflict of interest, or simple incompetence. Also, how these findings are disseminated, applied or used is important. Modern digital technology makes it possible to collect vast amounts of data about people, their thoughts, beliefs and actions. To preserve the integrity of science, it is important to be careful about how these data are collected, how they are used, and who is allowed to use them.

In the higher education system, the integration of technology and education has created unique dilemmas and ethical issues in learning environments. In fact, the emergence of the field of virtual education resulting from the use of ICT and the expansion of the use of the internet in the educational and research environment has led to the emergence of numerous emerging ethical issues that are sometimes more complex and unknown than the past ethical issues (Jafari & Alamolhoda, 2021). One of the common behaviors in ethical issues is scientific dishonesty, which has become a common issue in universities and in recent years, it has been recognized as one of the most researched aspects of education, which has taken on a wider form with the closure of universities in the era of covid-19 around the world. Academic dishonesty is easier in e-courses; Because the variety of facilities and the lack of supervision of students facilitate unethical behaviors (Sarwar et al., 2018). This is a worrying phenomenon; Because it has consequences both on the learning process of students during their academic studies and on the job market where they work after graduation. Investigating this phenomenon is necessary in the sense that learning will increasingly be done electronically and blended, especially in the post-Covid-19 era (Muhammad et al., 2021).

All of these require rigorous evidence that the ethics of doing so are fundamental. Valid and balanced research should not be undermined, and research based solely on blind bias and hollow challenges to existing evidence and literature should be the basis of activity. In general, the requirement of responsible behavior should apply to all those who comment on such issues. In medical research, the distortion of scientific results has been the result of the widespread influence of corporate financial interests, mainly pharmaceutical companies (Edwards, 2013). In research on social inequality, the quantifiable evidence base against which growing inequality is assessed is constantly challenged. For example, if there are social, ethnic, and cultural differences in what we consider "intelligence," we must be sure that the concept itself is valid and that such differences can be accurately measured. Otherwise, we should not address such inequalities unless they are shown to matter and make a difference to policy and social action.

These factors require that the researchers of the higher education system should ensure their scientific integrity as much as they can. The only challenges to scientific findings must be scientific and there is no valid reason to question the findings. Most importantly, policymakers should have no basis for questioning the reliability of findings, at least on the basis of ethical principles and scientific integrity. While science can be "wrong" based on accurate and reliable evidence, researchers must strive to never engage in ethically wrong activity. This is not to say that maintaining or not maintaining ethical and judgment are relatively simple matters. Also, it does not mean that scientific findings should be accepted uncritically. The reason we must think ethically is that the implications of scientific findings must be evaluated and considered in as balanced and discrete a manner as possible. Therefore, one should think deeply about ethics and scientific integrity. From the outset, he identified good reasons for doing research in a particular research area. Who can do that research, what their motivations are, how they do it. The more we engage with these issues, the less likely it is that actions based solely on ideology, prejudice, and opinion will easily influence higher education policymaking.

The Position of Research Ethics in Higher Education

Higher education is a professional system that covers human behavior and ethics; Therefore, the position of ethics in the higher education system has been the focus of many universities. Different researches in the field of ethics have confronted researchers and officials with many variables and dimensions in this regard, and since the mission of education is to promote ethics, the importance of ethics is obvious to researchers (Mesbahi & Abbaszadeh, 2013). Teaching ethics is one of the most important and necessary missions of universities (Rissanen & Löfström, 2014). Universities and higher education centers should be active in promoting ethical behavior, ethical principles in research, among researchers and students. Because researchers need to understand research ethics in their knowledge production and research (Resnik, 2020). The university is one of the most important organizations for the progress of any country; Therefore, it is necessary to have expert and ethical researchers. Universities and higher education centers should include research ethics in the educational program in addition to technical training and work skills, and the best way for this is to have ethical researchers (Sleem et al., 2010).

One of the most important challenges of the modern and advanced societies is to observe and adhere to ethical principles and rules. To improve the quality of research in universities and higher education centers, sensitivity to ethics adds to the credibility of research and guarantees the quality of research (Mousavi et al., 2015). It is necessary for faculty members, students, and administrators to be fully familiar with ethical codes and ethical standards related to their profession (Ishikawa, 2017). The goals of universities are to increase knowledge, commercialize knowledge and research, connect it with industry in the discussion of economic development and finally, the creation of knowledge-based companies, which requires high quality research and research ethics to achieve these goals.

The goals of universities and higher education centers are to increase knowledge, commercialize knowledge and research, connect it with industry in the development of the economy, and finally create knowledge-based companies, which requires high quality research and research ethics to achieve these goals. Adhering to scientific ethics and honesty in research leads to the correct progress of researchers. Today, in order to commercialize knowledge in universities and higher education centers, despite creating a favorable environment for economic and social development, it is necessary to establish knowledge-based companies. One of the important missions of higher education is to train students to comply with ethical standards and laws related to their field. In fact, teaching research ethics is one of the great academic missions (Heris et al., 2013). Ethics in research is a series of rules and guidelines that familiarize researchers with their professional standards (Baldini et al., 2006) to prevent possible deviations (Janvier & Farlow, 2015).

Ethics and Scientific Integrity: One Based on Rules or Based on Principles and Values

The higher education system and technology are more closely linked today than ever before. The latter, while being a type of incentive for higher education, leads to new technological developments. In fact, this shows the fact that technology has a pervasive and stronger impact on society and human life. This has led to a growing awareness that the higher education system cannot be considered beyond the realm of value judgments and thus ethics. Currently, there is no area that can be considered far removed from the real world and that can escape ethical considerations. Awareness of the relevance of ethics in operational research has been increasing in recent years (Brans, 2002). In this regard, in this section, two ethical principles are discussed that can help researchers and practitioners of the higher education system in research activities.

When we talk about ethics, we are actually talking about choosing between options. Because it is the possibility of choosing between different options that exposes us to ethical judgment. The fact that we can choose does not mean that choosing and judging is easy. However, proper reasoning is a good reason to believe that there is always a proper choice, even though some choices may not conform to what we consider to be the behavioral rules of the society in which we live. Hence, ethical discourse can be developed on two levels. The first is the level of ethical rules or norms. The second is the level of principles or values. The first level is a kind of approach to social responsibility towards research ethics, which is based on explaining requirements and creating norms among stakeholders. In this case, universities can only accept research activities from researchers and scientists who have certain ethical standards (Bevir, 2012). This approach expresses the fact that there is a set of behavioral rules that are provided from outside and depends on the specific cultural, social and political reality in which a person lives. The type of view of this principle on ethics is that all societies need a set of rules that are accepted in life and there is a general consensus. These are the rules and principles of morality codes that are sometimes defined generally and sometimes very precisely defined. However, as Preston (1997) states, if our morality discourse remains at the level of laws, we face the risk of fragmentation in our lives; Because the ethical standards applied by people in the workplace may differ from those in situations outside of the business environment.

At the second level, another step is taken with regard to rule-based ethics. At the second level, ethically is not based on the cultural, social and economic environment (from outside), but by the principles and values that are the result of our understanding as human beings and as members of a society. These principles and values are formed from the connection between the new generation and the past and future generations, and they give meaning to life and place us in the realm of freedom. In fact, this is the sense that Koch (2000) states that, first of all, a person should be considered as a citizen involved in a social context and then as an expert who insists on addressing this problem. Therefore, it can be said that the principle and basis of ethical discourse should be that responsibility towards others (values and principles), both human (past, present and future generations) and nature (social, political and economic environment), should always be considered.

The role of the higher education system is more evident here. To require all its members, based on rules and regulations or based on values and principles, to exercise professional judgment at all times to identify, assess, and address threats to basic ethical principles. These codes of ethics or values and principles determine the standards of behavior that the general public can expect from members of the higher education system, and adherence to this code restores trust in this system. Determining a code of conduct or the formation of ethics based on values and principles is a key requirement for the higher education system as a scientific and professional system, because the standards of professional behavior used provide the basis for self-regulation of this system (Herron & Gilbertson, 2004).

With all these interpretations, there are various reasons for choosing a code or standard of ethics based on principles and values instead of a code of ethics based on rules. The most obvious of these is the diverse nature of the scenarios in which the higher education system may find itself. Because it is impossible to create a code of conduct for each unique situation. Moreover, rule-based approaches often have negative consequences. First, the increase in rules and regulations often leads to an increase in administrative costs. The second thing is that outdated regulations can cause disregard for the law.

Too much focus on compliance to avoid penalties can divert management's focus from strategic expansion and limit the ability to identify growth opportunities. On the other hand, encouraging rules and regulations can lead to the inability to internalize ethical values. When rules challenge people too much, people look for ways to get around them. In this case, the focus shifts away from doing the right thing to whether the law can be interpreted "in my favor." Finally, often seen in organizations, excessive regulation can create a culture of blind conformity where authority is neither challenged nor questioned.

Therefore, the higher education system should try to internalize ethical values. Because rule-based regulation tends to create a culture of compliance rather than "doing the right thing". However, it should not be said that ethical standards based on rules are not useful. Because on the one hand, these ethical codes are prescriptive requirements and leave little room for misinterpretation. On the other hand, rule-based approaches are easier to implement and can provide comparability and standardization. As a result, higher education system administrators can make correct decisions.

It is in this situation that a person or an institution is tempted to ask the question: Which approach should scientists use to improve ethics and scientific integrity in research? However, this question presents a false dichotomy. Because both approaches can and should be followed. Both approaches are complementary. Because they focus on different aspects of ethical behavior. Principles and rules-based approaches emphasize the importance of following ethical rules, while values-based approaches emphasize the development of ethical character. Following the rules does not interfere with the development of ethical character, and character development does not prevent following the rules. Indeed, one might argue that one way to cultivate ethically is to understand and follow ethical rules. Although values-based and principles-based approaches are complementary, they have distinct strengths and weaknesses that affect their ability to serve as conceptual frameworks to promote research integrity. Therefore, they should be used in different research ethics. To understand this, it is necessary to consider the various ways that higher education institutions can promote research integrity. Therefore, both the rules-oriented approach and the value-oriented approach can play an important role in higher education.

Standardization of Ethics and Scientific Integrity

Today, it is difficult to find a term that adequately encompasses the diversity of global society. This is especially difficult considering the significant number of different religions and religions around the world. Without a common religious tradition, we cannot rely on religion to provide common standards of behavior. Of course, the main religions of the world agree on some general criteria. For example, everyone condemns murder, martyrdom, lying and stealing. However, this agreement is too general to decide most disputes. This is not a common standard. If the resolution of a particular case is disputed, different religions offer different and perhaps conflicting procedures for resolving the dispute. It is in this direction that these cases should be discussed in terms more or less independent of religion. So if we cannot rely on a common religious tradition for standards of common behavior, what can we rely on? In a short period of centralized law and decision-making, especially decision-making by the government, was the basis of movement and setting standards. However, whatever the explanation, we seem to have entered an era in which the voluntary decision of individuals has become more important. In this regard, we have to ask ourselves how a large number of people without strict regulations by law, without personal knowledge of each other, and without a common religious tradition can coordinate the complex behaviors that are necessary for a good life in this society? Ethics, as an institution that creates common standards as a decentralized process, can be the answer. In this regard, ethics has become an essential part of all activities and compliance with the rules is considered as a necessary requirement to achieve excellence. Compliance with commonly agreed ethical rules and procedures includes adherence to basic ethical principles embodied in legal norms and codes of conduct. Integrating ethical norms as a prerequisite is a difficult but efficient process for expanding universities' actions in the field of personal and public interest.

In recent years, the discussion of ethical standards has been one of the common topics in many work and professional fields, including higher education. Ethical standards are presented to express attitude and behavior in the face of education and are considered as behavioral expectations (Campbell, 2001). Ethical standards are intended to increase protection and minimize complaints and discomfort from interaction or information of individuals. Ethical analyses, ethical decision-making protocols and ethical standards have flourished in recent years (Reamer, 2017). These standards enable a person to be accountable to the ego, people and society. This part of this chapter is the latest frontier for original efforts to keep up with the times and develop ethics-based innovations to respond to individuals and societies. Research results (Reamer, 2017) show that ethical standards can help manage ethical challenges in universities. In this connection, the higher education system can prepare and design ethical and scientific standards at three levels.

At the individual level, ethics and scientific integrity can be defined as satisfaction and practical commitment to assigned tasks, with the condition that without a monitoring system, a person can perform tasks in the best possible way (Moradi et al., 2013). In this regard, teaching in any environment requires something more than course content. Teaching requires guidance and guidance for analysis, synthesis and application of value judgment, and achieving these skills requires high amounts of ethical responsibility. Professional obligation at the individual level means to pay attention to the conscience in doing the work, without having an external obligation or incurring legal punishment in case of violation. Adhering to duties such as respecting students and other colleagues, creating a safe and calm environment, encouraging fairness and justice, establishing appropriate relationships and refusing anything that may influence their decisions and actions is the main point in ethical standards at the individual level.

At the occupational level, ethical principles and professional behavior standards indicate fundamental values and professional requirements. The purpose of its formulation is to determine the criteria that faculty members, by observing and committing to them, will lead to the health and excellence of the profession. These principles and criteria in universities are a model for describing healthy behavior and evaluating people's professional behavior. If people deviate from these principles and criteria, there will be restrictions for them (Ahanchian and Soleimani, 2017). From the point of view of many experts, when talking about becoming a professional, attention should be paid to the discussion of adherence and ethical obligation of academic principles. Creation of insight and attitude in students and colleagues in the framework of the goals of higher education should be manifested. According to Keller (Keller, 2001), such people are able to overshadow the conditions and position of other factors in such an environment by emphasizing their commitment and professional ability.

At a higher level, when the topic of ethical standards regarding the system is raised, individuals and groups in the higher education system, in addition to having knowledge and skills, must have commitment and responsibility towards such a system and promote responsible behavior (Omoiee Milan et al., 2012). In this regard, commitment to the higher education system means that individuals and groups are determined based on appropriate procedures and statutes and regulations. On the other hand, move based on educational goals, policies and decisions. Also, individuals and groups can turn them into incentives for teaching and learning by validating the higher education system by valuing things. In fact, people's behavior is influenced by other stimuli such as systems or objects. In this regard, it is important to know how the object becomes a stimulus or motivation. In addition, validating the diverse perspectives of students and colleagues while working together to shape educational goals, policies, and decision-making is also very important in this field.

The Perspective and Future of Ethical Issues and Scientific Integrity

The existence of many ethical damages in the research field of our society is caused by two major facts. The first fact is that the personal and professional ethics of people are caused by the culture and atmosphere that governs the general and academic society. If we observe that many ethical harms have affected society's research, it indicates the fact that academic society in particular and society in general suffer from ethical problems. Issues such as emphasis on quantity and neglecting quality, lack of monitoring of the country's research and academic space have fueled many ethical issues. Because the points resulting from research activities for professors and students cause unhealthy competition in this field and do not reduce the quality of published articles. Another fact is related to the lack of value of scientific research in society. When research is not considered a value in a society and the conducted research remains stagnant without any use, the unethical motivation in researchers is unintentionally strengthened. Because there is no accountability and these two factors together cause immorality.

In research ethics, a better focus is needed on promoting ethical research rather than facilitating ethical approval. Research integrity and ethics play a prominent role in these initiatives. On the other hand, it leads to the training of necessary skills to reflect ethical issues in the field of research. This targets ethics and integrity for research in higher education. Also, skills to reflect on ethical issues and new teaching materials will be essential for this as well as moving towards a culture of ethical reflection throughout research activity. Ethical issues must be considered at all stages of research, from the design stage, during implementation, and to publication (O'Mathúna and Siriwardhana, 2017).

Other innovations in integrative research should be adapted for human environments. At the same time, researchers need support to address the ethical issues they encounter while conducting research. This requires committees for approval and compliance. At the same time, as researchers are helped to decide what should be done. Changes related to the ethical charter or protocol may also be necessary. However, researchers can benefit from discussing ethical issues with someone experienced in the field. Such individuals can serve as a ethical guardian or "counselor" available for such discussions. Such resources can help researchers conduct their research more ethically. Another aspect that expresses the perspective of ethics in the future and needs more attention is related to spiritual and religious issues. In some cultures, spiritual issues can be a source of discomfort or resilience (Wessells, 2013). Reluctance to include spiritual concerns can overlook important issues and needs. For example, research on a psychological program to help war-torn Angolan youth revealed two different worldviews. The program was based on a scientific and secular model, yet the children wanted to discuss spiritual issues. The program was originally designed to avoid spiritual issues. However, this "raised an important ethical issue. The imposition of this issue was something outside the cultural environment of the community that did not fit the local context" (Wessells, 2013). It will be important to conduct research in this area to ensure that all needs and resources are identified and that cultural blindness does not lead to ignoring important issues.

Finally, we can say that in order to be able to implement the ideals raised in this field, we must consider several components. First, any practical action in this direction needs to lay the foundation and prepare the necessary conditions for that action to occur. Therefore, the development of various infrastructures in the university community and the increase in quality and the provision of financial facilities for students and researchers are considered as default variables in this connection. In the macro section, if we consider the university society as a sub-scale of the macro society, the society's movement towards ethical values will lead to the development of compliance with these principles in the university as well. At the micro level, meeting the needs and providing the requirements of the researchers is an inevitable necessity. Second, through the development of legal requirements that are understood in different societies in the form of ethical codes, it is possible to take steps towards the development of compliance with ethical principles. Therefore, it is possible to move towards the development of compliance with ethical principles through the development of ethical models and codes that determine the ethical models related to research and researchers, as well as increasing the supervision of research. Finally, an all-out effort should be made to institutionalize ethical compliance in all matters. In this connection, the educational authorities of the country must have a comprehensive plan for this matter. On the other hand, in the process of university education, they should institutionalize ethics in doing all assigned tasks. In this case, by emphasizing and paying attention to these three basic elements, we can hope for the development of ethics in higher education in various dimensions, including research.

CONCLUSION

Considering the multiplicity of ethical dilemmas, trials and problems that educational researchers can face, it is very important to acknowledge the importance of ethical research codes. Also, their limitations to inform the daily discussions of education researchers have attracted attention. Checking the boxes of the research ethics approval forms and ending the matter can never be enough. Of course, this is only the beginning of the ethical debates of an educational researcher. Therefore, it is essential for researchers to have informed personal views regarding the ethical dilemmas they may face and the decisions they must make (Brooks et al., 2014). Most of the ethical issues raised in the higher education system tend to focus on human issues. While existing ethical frameworks and "codes of ethics" provided by leading organizations tend to focus on increasing ethical sensitivity or competence. However, the human-centered nature of some disciplines in the higher education system sometimes conflicts with the ethics of what might be considered purely scientific research. This is primarily due to ambiguity or apparent conflict with current research standards (eg, existing guideline frameworks or ethical codes). This requires seeing the ethical process in a more organic way. It is used to argue in support of a broader sense of ethical responsibility in the relationship between educational researchers and those charged with overseeing their research. What we have sought to highlight in this chapter is the importance of a general approach to this issue in the higher education system.

REFERENCES

Ahanchian, M. R., & Soleimani, E. (2017). The perception of faculty members on being professional in Iran's higher education system. *IRPHE*, *23*(2), 1–23.

Baldini, N., Grimaldi, R., & Sobrero, M. (2006). Institutional changes and the commercialization of academic knowledge: A study of Italian universities' patenting activities between 1965 and 2002. *Research Policy*, *35*(4), 518–532. doi:10.1016/j.respol.2006.01.004

Bevir, M. (2012). *Governance: A very short introduction*. Tehran: Kargadan Pub.

Brans, J. P. (2002). Or, Ethics and Decisions: The Oath of Prometheus. *European Journal of Operational Research*, *140*(2), 191–196. doi:10.1016/S0377-2217(02)00062-0

Brooks, R., Te Riele, K., & Maguire, M. (2014). Ethics and education research. *Sage (Atlanta, Ga.)*.

Campbell, E. (2001). Let right be done: Trying to put ethical standards into practice. *Journal of Education Policy*, *16*(5), 395–411. doi:10.1080/02680930110071011

Edwards, N. (2013). Bad Pharma. *Journal of the Royal Society of Medicine*, *106*(2), 72–73. doi:10.1177/0141076813477618

Heris, M. S., Piralaiy, E., & Fard, A. M. (2013). A study of status of the research ethics compliance among physical education academic members of Islamic Azad University in reigns 2 & 13. *European Journal of Experimental Biology*, *3*(1), 568–571.

Herron, T. L., & Gilbertson, D. L. (2004). Ethical principles vs. ethical rules: The moderating effect of moral development on audit independence judgments. *Business Ethics Quarterly*, *14*(3), 499–523. doi:10.5840/beq200414320

Holder-Webb, L., & Cohen, J. (2012). The cut and paste society: Isomorphism in codes of ethics. *Journal of Business Ethics*, *107*(4), 485–509. doi:10.100710551-011-1060-1

Iphofen, R. (2020). An introduction to research ethics and scientific integrity. Handbook of research ethics and scientific integrity, 3-13.

Ishikawa, S. (2017). Status and Problems of Humanities Education with Respect to Pharmacy Education. Yakugaku zasshi. *Journal of the Pharmaceutical Society of Japan*, *137*(4), 399–405. PMID:28381714

Jafari, E. (2022). Academic dishonesty in e-learning, a threat to the integrity of the medical education system under the COVID-19 epidemic. *Indian Journal of Medical Education*, *22*(23), 158–169.

Jafari, E., & Alamolhoda, J. (2021). Lived Experience of Faculty Members of Ethics in Virtual Education. *Technology. Knowledge and Learning*, *2021*, 1–9.

Janvier, A., & Farlow, B. (2015). The ethics of neonatal research: An ethicist's and a parents' perspective. *Seminars in Fetal & Neonatal Medicine*, *20*(6), 436–441. doi:10.1016/j.siny.2015.10.003 PMID:26497942

Keller, G. (2001). The new demographics of higher education. *Review of Higher Education*, *24*(3), 219–235. doi:10.1353/rhe.2001.0004

Kitagawa, F., & Oba, J. (2010). Managing differentiation of higher education system in Japan: Connecting excellence and diversity. *Higher Education*, *59*(4), 507–524. doi:10.100710734-009-9262-5

Koch, T. (2000). We live in the city not in a study. *ORMS Today*, *27*(2), 16–17.

Mâță, L., & Boghian, I. (2019). Perception of Teachers in Higher Education towards Ethical Issues of Information Technology Use. *Romanian Journal for Multidimensional Education*, *11*(1), 156–169. doi:10.18662/rrem/183

Mesbahi, M., & Abbaszadeh, A. (2013). System model of professional behavior in supreme education (Qualitative research). *Bioethics Journal*, *3*(10), 11–34.

Mirzaie, S. A., & Qarakhani, M. (2013). The Research Ethics In Iranian social Science. *Iranian Journal of Sociology*, *15*(2), 3–23.

Moradi, M., Khatooni, M., Jahani Hashemi, H., Zeighami, R., & Sheikh, M. (2013). Relationship between professional commitment and job satisfaction in Qazvin's pubic educational hospital nurses. *Medical Ethics (Burlington, Mass.)*, *7*(24), 55–78.

Mousavi, S., Mohsenpour, M., Borhani, F., & Ebadi, M. (2015). Ethical Sensitivity of nurses and nursing students working in Aja University of Medical Sciences. *Quarterly Journal of Medical Ethics*, *31*(9), 127–143.

Neves, P., & do Céu, M. (2018). On (scientific) integrity: Conceptual clarification. *Medicine, Health Care, and Philosophy*, *21*(2), 181–187. doi:10.100711019-017-9796-8 PMID:28905288

O'Mathúna, D., & Siriwardhana, C. (2017). Research ethics and evidence for humanitarian health. *Lancet*, *390*(10109), 2228–2229. doi:10.1016/S0140-6736(17)31276-X PMID:28602560

Omoiee Milan, M., Mehdinezhad, V., & Yaghoubi, N. (2011). Assessing Factors Affecting the Tendency to Use Electronic Learning Systems among Faculty Members. *Interdisciplinary Journal of Virtual Learning in Medical Sciences*, *2*(3), 28–38.

Preston, D. (1997). Can business ethics really exist? *Computers & Society*, *27*(1), 6–9. doi:10.1145/382069.382070

Reamer, F. G. (2017). Evolving ethical standards in the digital age. *Australian Social Work*, *70*(2), 148–159. doi:10.1080/0312407X.2016.1146314

Resnik, D. B. (2020). What is ethics in research and why it is important. National Institute of Environmental Health Sciences, 8(8).

Rissanen, M., & Löfström, E. (2014). Students' research ethics competences and the university as a learning environment. *International Journal for Educational Integrity*, *10*(2). Advance online publication. doi:10.21913/IJEI.v10i2.1004

Sarwar, S., Idris, Z. M., & Ali, S. M. (2018). Paid academic writing services: A perceptional study of business students. *International Journal of Experiential Learning & Case Studies*, *3*(1), 73–83. doi:10.22555/ijelcs.v3i1.2078

Schrag, Z. M. (2010). *Ethical imperialism: Institutional review boards and the social sciences, 1965–2009.* JHU Press. doi:10.1353/book.471

Sleem, H., Abdelhai, R. A. A., Al-Abdallat, I., Al-Naif, M., Gabr, H. M., Kehil, E. T., Sadiq, B. B., Yousri, R., Elsayed, D., Sulaiman, S., & Silverman, H. (2010). Development of an accessible self-assessment tool for research ethics committees in developing countries. *Journal of Empirical Research on Human Research Ethics; JERHRE*, *5*(3), 85–98. doi:10.1525/jer.2010.5.3.85 PMID:20831423

Sleem, H., Abdelhai, R. A. A., Al-Abdallat, I., Al-Naif, M., Gabr, H. M., Kehil, E. T., Sadiq, B. B., Yousri, R., Elsayed, D., Sulaiman, S., & Silverman, H. (2010). Development of an accessible self-assessment tool for research ethics committees in developing countries. *Journal of Empirical Research on Human Research Ethics; JERHRE*, *5*(3), 85–98. doi:10.1525/jer.2010.5.3.85 PMID:20831423

Sutrop, M., Parder, M. L., & Juurik, M. (2020). Research Ethics Codes and Guidelines. Handbook of Research Ethics and Scientific Integrity, 67-89.

Vallance, R. J. (2004, November). Formation in research ethics: Developing a teaching approach for the social sciences. *AARE Conference 28th Nove2nd Dec The University of Melbourne.*

Wessells, M. (2013). Reflections on Ethical and Practical Challenges of Conducting Research with Children in War Zones: Toward a Grounded Approach. *Research methods in conflict settings: A view from below*, 81.

World Health Organization. (2015). *Global health ethics key issues: Global network of WHO collaborating centres for bioethics.* World Health Organization.

Chapter 10
Unlocking the Power of Spatial Big Data for Sustainable Development:
From Capacity Building to Food Security and Food Traceability

Munir Ahmad
https://orcid.org/0000-0003-4836-6151
Survey of Pakistan, Pakistan

ABSTRACT

Food security is a fundamental human right. Enhancing agricultural practices, lowering food waste, and ensuring equitable resource allocation are among the actions taken to ensure food security. VGI has developed into a helpful tool for data collection, assisting in tracking resource availability, monitoring price fluctuations, mapping community food assets, and assisting with land management. VGI encourages fair pricing, capacity building, access to nutritious food, decision-making, locating food storage, and capacity building. VGI helps to address food waste, map crop production, compile production techniques, crowdsource food quality data, and track supply chains. Advanced technologies can be integrated with VGI to improve analysis and decision-making. Enhancing VGI programs and inclusivity can help people better understand and address issues related to food security.

DOI: 10.4018/978-1-6684-9601-5.ch010

INTRODUCTION

Food security, a vital concept, pertains to ensuring access to adequate, affordable, and nutritious food for all individuals. Achieving food security is not only a matter of human well-being but also a fundamental human right. Efforts to promote food security involve improving agricultural practices, minimizing food waste, and ensuring equitable distribution of food resources. By enhancing food safety measures, preventing fraudulent activities, and providing consumers with comprehensive information about the history of food products, food traceability strengthens consumer confidence.

VGI has gained widespread recognition as an effective method for data collection (Ahmad et al., 2022; Ahmad & Khiyal, 2023). For example, it has become widely accepted as a data collection technique for tracking the environment, and resources (Rollason et al., 2018). Similarly, Stream Tracker employs sensors, satellite data, and community volunteers to monitor the locations and flow patterns of streams (Stream Tracker, 2023). Furthermore, CrowdWater, a citizen science effort, aims to supplement existing observations by collecting independent and reliable data for flood and drought modeling (CrowdWater, 2023). Additionally, mobile applications have been utilized to gather comprehensive crowdsourced data for real-time monitoring of pricing changes in the food system (Adewopo et al., 2021; Arbia et al., 2023a). Another study showcased the development of an open framework called facilitated VGI, which merged local knowledge with expert guidance to collect data related to urban farming in Philadelphia (Quinn & Yapa, 2016). Members of the local food community were involved in the study (Fast & Rinner, 2018) to map the community food assets using VGI.

Addressing inequalities in land tenure is vital as they can undermine tenure security, limit land utilization, and jeopardize food security for those dependent on the land (Nara et al., 2020). Ensuring land rights and equitable access to resources is crucial for safeguarding food security. Participatory land administration, incorporating mobile applications and satellite imagery, has been utilized in data collection efforts focused on farms (Asiama et al., 2017).

Roughly half of the indicators associated with the Sustainable Development Goals (SDGs) are linked to population dynamics and geographical distribution. A fundamental principle of the SDGs is to ensure that no one is left behind, necessitating attention not only to population counts within administrative units but also to the specific locations where individuals reside. This underscores the importance of considering both the quantity and spatial aspects of human settlements when striving to achieve the SDGs (Qiu et al., 2022).

VGI is essential in advancing food security. It supports efforts to manage land and assists in tracking environmental resources, tracking price changes, mapping community food assets, and monitoring price fluctuations. It can improve our understanding, decision-making, and actions toward reaching food security goals by utilizing the potential of VGI. In addition, It can contribute to the larger agenda of sustainable development and the SDGs.

To this effect, the main objective of this chapter is to investigate how VGI can facilitate food security and food traceability issues. To achieve this objective, the structure of this chapter is as follows: the first section describes the connection between VGI and spatial big data. The next section is dedicated to the role of VGI in sustainable development. The role of VGI in food security and food traceability is presented in the next section, The last section served as the conclusion of the study.

VGI AND SPATIAL BIG DATA

Approximately 80% of the vast amount of big data available can be classified as spatial big data, which encompasses spatial information. The prevalence of spatial big data opens up exciting new possibilities for addressing challenges in urban and environmental sustainability through the application of advanced analytics tailored to spatial big data (Huang & Wang, 2020).

Finding relevant geographical information is becoming more and more crucial because of the growing amount of geospatial data that has been available with the advent of the "Big Data" era (Mobasheri, 2017). Spatial big data, which is essential for supporting both research endeavors and practical applications, can be obtained from VGI (Lawrence et al., 2020; G. Zhang, 2022). The utilization of VGI holds immense potential in uncovering the spatial and temporal dynamics of geographic phenomena (G. Zhang, 2020).

VGI encompasses user-generated digital geographic data, including text and multimedia (See et al., 2016). Enabled by the internet and communication technologies, VGI facilitates the production, collection, and dissemination of geographic information. It serves as a valuable tool for gaining deeper insights into the social, economic, and environmental conditions of specific regions. VGI is derived from diverse sources, such as unstructured social network data platforms like Twitter, Facebook, and Instagram, which contain explicit or implicit location information. Additionally, structured geographical data platforms like OpenStreetMap (OSM) contribute to the vast and varied pool of VGI. This rich resource enables a wide range of geographical analyses and applications. By complementing official data sources, VGI enhances our understanding of impacted populations, infrastructure conditions, and road networks (Goodchild, 2007; Soden & Palen, 2014).

VGI has been used in many applications as big data. For example, Jurado Rota et al., (2019) monitored visitor behavior and public utilization within protected areas using VGI as a source of big data. Similarly, Ben Rhaiem et al., (2022) used VGI in big spatiotemporal data analytics to support resilient smart cities. (Gao et al., 2014) argued for constructing gazetteers by mining the volunteered geospatial big data using a data-driven methodology.

VGI FOR SUSTAINABLE DEVELOPMENT

The notion of sustainable growth and development was introduced by Corbett & Rostow, (1960) and it brought forth the understanding of sustainability as an essential element of real development, encompassing various dimensions of society, including the environment, well-being, and economy. In 2015, the United Nations (UN) introduced the Sustainable Development Goals (SDGs) as a holistic framework to address global challenges and promote sustainable development worldwide. Consisting of 17 goals and 169 targets, the SDGs encompass a broad range of interconnected issues, encompassing areas such as poverty eradication, education access, gender equality, climate change mitigation, and beyond (UNDP, 2023).

Food is vital to life. Ensuring food security is vital to sustainable development. The United Nations Sustainable Development Goal (SDG) 2 focuses on food intending to achieve food security. The importance of food and agriculture systems in global food security cannot be overstated. These systems play crucial roles in providing safe, healthy, and nutritious food, supporting livelihoods, and contributing to economic development ("Food and Nutrition in Numbers: 2014," 2015). Food security encompasses dimensions of availability, access, stability, and utilization, as established by the World Food Summit of 1996.

To meet the global challenge of food insecurity, it is essential to know where food-deprived people are living. To answer this and many other questions, spatial data is essentially needed. Spatial data plays a vital role in supporting the study, monitoring, and achievement of SDGs (Scott & Rajabifard, 2017). Spatial data facilitates the analysis, mapping, and modeling of sustainable development issues, offering a framework for global collaboration and consensus decision-making (UN-GGIM, 2012).

In the current era of technological advancements, geospatial information is being produced not only by public sector organizations but also voluntarily by some communities and citizens. VGI is one such example. VGI serves as a valuable resource in supporting the SDGs by generating fresh geospatial data through online open platforms, thanks to the contributions of volunteers and humanitarian organizations. This data contributes to measuring and reporting progress on the SDGs, for example, Cramwinckel, (2019) investigated the role of open global geospatial data products in measuring SDG indicator 11.7.1: public open spaces. Increased awareness of hunger and food security issues has become prominent in the fields of Public Health and Medicine. Economic challenges have exacerbated food insecurity in populations and communities that were previously unaffected. These pockets of hunger often go unnoticed when relying on conventional measurement methods. To address this, a study conducted by Vaughan, (2011) utilized VGI analysis to enhance the effectiveness, outreach, and outcomes of public health programs.

The utilization of crowdsourcing can yield highly accurate maps compared to other available global land cover sources. This argument supports the notion that VGI possesses the potential to address food security concerns. In countries where the availability and accessibility of precise and locally consistent data are challenging, VGI emerges as a viable solution. By harnessing VGI, countries can leverage their power to support initiatives focused on addressing food security-related issues effectively.

VGI FOR FOOD SECURITY AND FOOD TRACEABILITY

OpenStreetMap is one of the popular representations of VGI. It presents a valuable platform to highlight and map essential resources related to food security. It provides a valuable opportunity to effectively map and locate fresh food resources (Wiki, 2020). Various types of tags (Figure 1) are available for a variety of categories such as grocery stores, food banks, farmers markets, community gardens, urban farms, community-supported agriculture initiatives, farming and gardening stores, compost material producers, restaurants, coffee shops, breweries, and recycling centers that

Figure 1. OSM tags related to food security
Source: Wiki (2020)

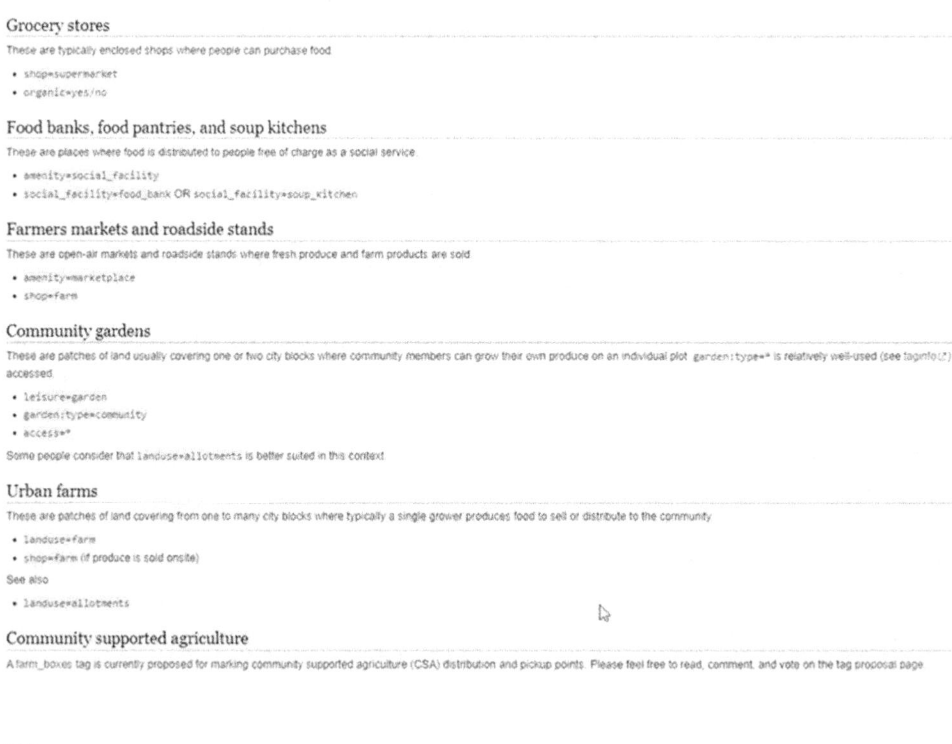

Grocery stores

These are typically enclosed shops where people can purchase food

- shop=supermarket
- organic=yes/no

Food banks, food pantries, and soup kitchens

These are places where food is distributed to people free of charge as a social service

- amenity=social_facility
- social_facility=food_bank OR social_facility=soup_kitchen

Farmers markets and roadside stands

These are open-air markets and roadside stands where fresh produce and farm products are sold

- amenity=marketplace
- shop=farm

Community gardens

These are patches of land usually covering one or two city blocks where community members can grow their own produce on an individual plot garden:type=* is relatively well-used (see taginfo) a accessed

- leisure=garden
- garden:type=community
- access=*

Some people consider that landuse=allotments is better suited in this context

Urban farms

These are patches of land covering from one to many city blocks where typically a single grower produces food to sell or distribute to the community

- landuse=farm
- shop=farm (if produce is sold onsite)

See also

- landuse=allotments

Community supported agriculture

A farm_boxes tag is currently proposed for marking community supported agriculture (CSA) distribution and pickup points. Please feel free to read, comment, and vote on the tag proposal page directly or indirectly support food security (Wiki, 2020). Based on the data available at OSM, specialized application has been developed with the titled as "Philly Fresh Food Map" (CWU, 2016; Quinn & Yapa, 2016).

VGI has found a diverse range of applications for food security and food traceability as can be seen in Figure 2.

Figure 2. VGI to support food security

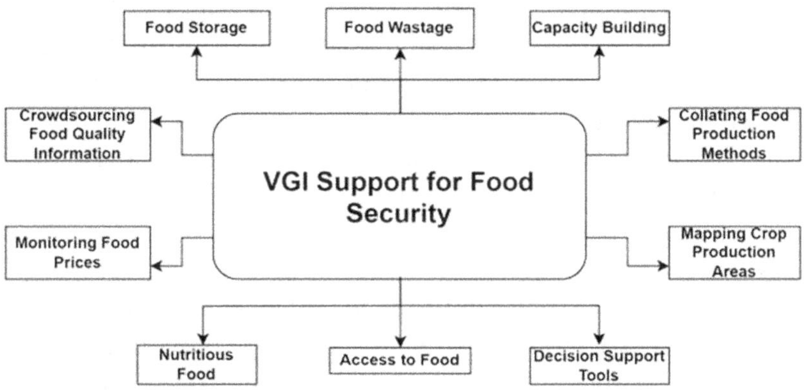

Monitoring Food Prices

The accessibility of food is significantly influenced by its prices. While numerous factors contribute to food prices, in some countries implementing government-set prices is also a big challenge. Accessibility and timeliness of price information are also crucial to comprehend changes in the market and the state of food security. Utilizing VGI can help in monitoring food prices by identifying geographic areas where prices exceed the government-regulated rates, potentially leading to reduced accessibility for consumers. Additionally, VGI can help identify locations where the government can provide low-priced food items, thereby enhancing accessibility for a wider range of communities. For example, mobile-based crowdsourced and voluntarily contributed price observations can help to support real-time data collection (Arbia et al., 2023a; Solano-Hermosilla et al., 2022; Zeug et al., 2017). Therefore, VGI holds the potential to facilitate the implementation of fair food pricing practices and enhance food accessibility to everyone throughout the country.

Nutritious Food

Unequal access to nutritious food presents a significant public health challenge, as limited availability of affordable, high-quality options can lead to inadequate diets and increased risks of health issues such as obesity, diabetes, and cardiovascular diseases (Dai & Wang, 2011). Addressing these disparities is crucial for endorsing overall well-being and reducing the burden of preventable health conditions.

Ensuring access to nutritious food is an essential component of achieving food security, as defined by the Food and Agriculture Organization (FAO). In this regard, VGI can play a valuable role. By integrating information about the availability of nutritious and non-nutritious food from various sources, VGI can help identify areas where improvements are needed. Furthermore, VGI can facilitate the integration of food choices among different community members, enabling the government to identify and address malnutrition situations, thus strengthening food security.

In cases like the study conducted by Widener & Li, (2014), advanced data-mining techniques were employed to extract geotagged information from Twitter, providing insights into the occurrence of healthy and unhealthy food across the United States. Additionally, Twitter data has been used to analyze the relationship between food access and dietary choices, highlighting the positive impact of shops offering fresh stock on users' nutritious choices (Chen & Yang, 2014). By leveraging VGI and data analysis, experts can identify areas with less nutritious food options and provide advisory guidelines to improve the situation.

Decision Support Tools

In decision-making processes involving socio-economic and technological issues, decision support tools play a crucial role (Nutley et al., 2013). This is equally applicable to the domain of food security, where effective decision-making is paramount. Geographical information serves as an essential component in most decision support tools, providing valuable insights (Barzehkar et al., 2021; Graymore et al., 2009; D. Zhang et al., 2015). VGI emerges as a valuable and freely available source of geographic data, particularly in local contexts (Sadeghi-Niaraki et al., 2020). Leveraging VGI, decision-making tools can be designed, developed, and implemented to address food security challenges, especially in situations where obtaining geographic information from the public sector remains challenging. By utilizing VGI, decision-makers can enhance their understanding of food security issues and make informed choices to address them effectively.

Food Storage

Food storage plays a critical role in ensuring a consistent and uninterrupted supply of food for all members of society (Habiba et al., 2015). In the event of disasters such as floods and earthquakes, the identification of food storage locations becomes vital for saving lives. VGI can serve as a valuable tool to rapidly and accurately identify and communicate these food storage sites in real time. Local volunteers residing in affected areas possess an intimate knowledge of the area's geography, making them well-positioned to identify, mark, and effectively communicate the locations of local food storage facilities. By leveraging VGI and the insights of local volunteers, timely and accurate information about food storage places can be shared, supporting disaster response efforts, and contributing to the well-being and survival of affected communities.

Capacity Building

While ignorance may provide temporary relief, it is insufficient for problem-solving. To effectively address issues, it is crucial for all relevant stakeholders to possess a comprehensive understanding of the realities and severity of the problems at hand. Food insecurity, a widespread global challenge, requires urgent capacity-building efforts. In this context, VGI can play a significant role. By engaging volunteers to gather and share real-time, up-to-date information, VGI can enhance the capacity of concerned stakeholders and promote situational awareness (Arbia et al., 2023b; Zeug et al., 2017). This enables more effective responses to food insecurity, empowering stakeholders to make informed decisions and take appropriate actions to combat this pressing issue.

Food Wastage

The non-availability or limited accessibility of food items in certain parts of the world can be attributed, in part, to food wastage. Several countries have recognized this issue and taken measures to address it. For instance, Italy and Germany have deployed social welfare inspectors to penalize customers who order excessive food quantities they cannot consume. The problem of food waste in restaurants is also conspicuous in some countries, contributing to the overall food insecurity situation. In response to this challenge, volunteers in certain regions are redistributing untouched, surplus food to individuals in need. Mobile applications can help in mitigating household food wastage as noted by (Farr-Wharton et al., 2014). Thus, VGI based mobile applications can play a significant role in addressing food waste by mapping and highlighting areas where food waste occurs, such as food processing facilities and supermarkets. This data can help in identifying locations where interventions can be implemented to reduce food waste and promote more sustainable practices. By leveraging VGI, efforts can be focused on identifying and integrating places where surplus food is available for distribution to those in need, thereby mitigating food waste and alleviating food insecurity.

Mapping Crop Production Areas

Crops play a fundamental role in ensuring food security but monitoring and obtaining precise information about them can be challenging. While satellite imagery has been extensively utilized worldwide for crop monitoring, it primarily offers a limited aerial perspective and necessitates on-ground surveys for accurate data. VGI emerges as a valuable resource for supporting field surveys, particularly in locations with difficult accessibility, because VGI is an optimal source of locally rich geographical information. VGI can be effectively employed to map and identify areas where food crops are cultivated, providing valuable insights into agricultural production and the specific types of crops being grown.

Spatial knowledge regarding the distribution of croplands is crucial for monitoring food security. In response to this need, a crowdsourcing approach involving volunteers was adopted by See et al., (2013) to map Google Earth imagery using Geo-Wiki tools in Ethiopia, catering to the specific requirements of the community. By leveraging VGI, decision-makers can enhance their understanding of crop cultivation areas and improve resource allocation strategies to ensure food security for vulnerable populations.

Collating Food Production Methods

VGI offers a valuable means of gathering data on diverse farming practices employed across various regions, encompassing factors such as irrigation methods, crop rotation, and fertilizer usage. This information serves as a foundation for identifying and promoting sustainable farming techniques, ultimately contributing to long-term food security initiatives. By harnessing VGI, stakeholders can access a wealth of knowledge, enabling them to make informed decisions that foster sustainable agriculture practices and enhance the resilience of food systems.

Crowdsourcing Food Quality Information

VGI can serve as a valuable tool for collecting information on various aspects of food products, including taste, freshness, and nutritional content (Chen & Yang, 2014). This data empowers consumers to make informed decisions regarding their food choices while promoting transparency within the food supply chain. The availability and quality of food are critical concerns in the context of food security, with considerations of time and spatial dimensions (Chen & Kwan, 2015). Social media platforms have emerged as essential channels for raising awareness about food security issues and reporting concerns related to food availability and quality. For example, Chen & Yang, (2014) explored the potential of geotagged tweets to examine users' diet choices. While there are limited examples of applications utilizing VGI for food security, the integration of VGI into these efforts holds great potential.

Tracing Food Products

VGI offers a valuable means to monitor and trace food products as they traverse the complex supply chain, thereby providing essential details regarding their origin, production methods, and distribution channels. By harnessing VGI, stakeholders can verify the authenticity and safety of food items, mitigating the risks associated with food fraud and ensuring consumer protection.

Access to Food

In a recent study conducted by Vonthron et al., (2023), a collaborative map leveraging VGI was employed to investigate and analyze disparities in food access. The primary objective of the investigation was to gain insights into the availability and distribution of food resources. Another study focused on the tagging of food resources in Philadelphia, USA (Quinn & Yapa, 2016). To address this objective, an open framework was developed to integrate VGI with expert opinions. This collaborative effort facilitated a comprehensive understanding of food access inequalities and provided a valuable resource for enhancing food security initiatives in the region.

CONCLUSION

Food security is a fundamental human right. Efforts to promote food security include enhancing agricultural practices, reducing food waste, and ensuring fair distribution of food resources. VGI has become a useful tool for data collecting with applications in monitoring price fluctuations, tracking environmental resource availability, mapping community food assets, and assisting with land management initiatives.

By assisting the fair food pricing practices, ensuring access to nutritious food, offering decision support tools, locating food storage facilities, and supporting capacity-building initiatives, VGI significantly supports food security. Additionally, it can help in the resolution of problems like food waste, mapping crop production regions, compiling food production techniques, crowdsourcing data on food quality, and tracking food goods along the supply chain. Stakeholders can improve resource allocation methods, encourage sustainable agricultural practices, and make well-informed decisions by utilizing VGI.

The analysis and use of VGI data can be improved by combining it with cutting-edge technologies like machine learning and artificial intelligence. In order to extract valuable insights from huge amounts of VGI data and facilitate more efficient monitoring and decision-making in the context of food security, this can involve developing algorithms and models. A more thorough understanding of food security challenges can be achieved by broadening the scope and impact of VGI programmes. This can involve engaging more people, using different data collection techniques, and looking into creative ways to include underrepresented groups and guarantee inclusivity in VGI initiatives.

REFERENCES

Adewopo, J. B., Solano-Hermosilla, G., Colen, L., & Micale, F. (2021). Using crowd-sourced data for real-time monitoring of food prices during the COVID-19 pandemic: Insights from a pilot project in northern Nigeria. *Global Food Security*, *29*, 100523. Advance online publication. doi:10.1016/j.gfs.2021.100523 PMID:34178595

Ahmad, M., Khayal, M. S. H., & Tahir, A. (2022). Analysis of Factors Affecting Adoption of Volunteered Geographic Information in the Context of National Spatial Data Infrastructure. *ISPRS International Journal of Geo-Information*, *11*(2). Advance online publication. doi:10.3390/ijgi11020120

Ahmad, M., & Khiyal, M. S. H. (2023). Assessment of land administration in Pakistan and the potential role of volunteered geographic information. In Handbook of Research on Driving Socioeconomic Development With Big Data. https://doi.org/ doi:10.4018/978-1-6684-5959-1.ch014

Arbia, G., Solano-Hermosilla, G., Nardelli, V., Micale, F., Genovese, G., Amerise, I. L., & Adewopo, J. (2023). From mobile crowdsourcing to crowd-trusted food price in Nigeria: Statistical pre-processing and post-sampling. *Scientific Data*, *10*(1), 446. doi:10.103841597-023-02211-1 PMID:37438443

Asiama, K., Bennett, R., & Zevenbergen, J. (2017). Participatory Land Administration on Customary Lands: A Practical VGI Experiment in Nanton, Ghana. *ISPRS International Journal of Geo-Information*, *6*(7), 186. Advance online publication. doi:10.3390/ijgi6070186

Barzehkar, M., Parnell, K. E., Soomere, T., Dragovich, D., & Engström, J. (2021). Decision support tools, systems and indices for sustainable coastal planning and management: A review. *Ocean and Coastal Management*, *212*, 105813. Advance online publication. doi:10.1016/j.ocecoaman.2021.105813

Ben Rhaiem, M. A., Selmi, M., Farah, I. R., & Bouzeghoub, A. (2022). A big spatiotemporal streaming data architecture for smart city crisis monitoring using VGI. *Proceedings - 2022 2nd International Conference of Smart Systems and Emerging Technologies, SMARTTECH 2022*. 10.1109/SMARTTECH54121.2022.00035

Chen, X., & Kwan, M. P. (2015). Contextual uncertainties, human mobility, and perceived food environment: The uncertain geographic context problem in food access research. American Journal of Public Health. doi:10.2105/AJPH.2015.302792

Chen, X., & Yang, X. (2014). Does food environment influence food choices? A geographical analysis through "tweets." *Applied Geography (Sevenoaks, England)*, *51*, 82–89. doi:10.1016/j.apgeog.2014.04.003

Corbett, D. C., & Rostow, W. W. (1960). The Stages of Economic Growth: A Non-Communist Manifesto. *International Journal (Toronto, Ont.)*, *16*(1), 83. Advance online publication. doi:10.2307/40198523

Cramwinckel, J. (2019). *The role of global open geospatial data in measuring SDG indicator 11.7. 1: Public open spaces*. Wageningen University.

CrowdWater. (2023). *CrowdWater*. https://crowdwater.ch/en/start/

CWU. (2016). *Philly Fresh Food Map (Powered by OpenStreetMap)*. http://www.gis.cwu.edu/phillyfood/

Dai, D., & Wang, F. (2011). Geographic disparities in accessibility to food stores in southwest Mississippi. *Environment and Planning. B, Planning & Design*, *38*(4), 659–677. Advance online publication. doi:10.1068/b36149

Farr-Wharton, G., Choi, J. H. J., & Foth, M. (2014). Food talks back: Exploring the role of mobile applications in reducing domestic food wastage. *Proceedings of the 26th Australian Computer-Human Interaction Conference, OzCHI 2014*. 10.1145/2686612.2686665

Fast, V., & Rinner, C. (2018). Toward a participatory VGI methodology: Crowdsourcing information on regional food assets. *International Journal of Geographical Information Science*, *32*(11), 2209–2224. doi:10.1080/13658816.2018.1480784

Food and nutrition in numbers: 2014. (2015). *Choice Reviews Online, 52*(12). doi:10.5860/CHOICE.191448

Gao, S., Li, L., Li, W., Janowicz, K., & Zhang, Y. (2014). Constructing gazetteers from volunteered Big Geo-Data based on Hadoop. *Computers, Environment and Urban Systems*. Advance online publication. doi:10.1016/j.compenvurbsys.2014.02.004

Goodchild, M. F. (2007). Citizens as sensors: The world of volunteered geography. GeoJournal. doi:10.100710708-007-9111-y

Graymore, M. L. M., Wallis, A. M., & Richards, A. J. (2009). An Index of Regional Sustainability: A GIS-based multiple criteria analysis decision support system for progressing sustainability. *Ecological Complexity*, *6*(4), 453–462. Advance online publication. doi:10.1016/j.ecocom.2009.08.006

Habiba, U., & Abedin, Md. (2015). Introduction and Overview of Food Security and Risk Reduction Issues. doi:10.1007/978-4-431-55411-0_1

Huang, B., & Wang, J. (2020). Big spatial data for urban and environmental sustainability. *Geo-Spatial Information Science*, *23*(2), 125–140. Advance online publication. doi:10.1080/10095020.2020.1754138

Jurado Rota, J., Pérez Albert, M. Y., & Serrano Giné, D. (2019). Visitor monitoring in protected areas: An approach to Natura 2000 sites using Volunteered Geographic Information (VGI). *Geografisk Tidsskrift - Danish Geografisk Tidskrift*, *119*(1), 69–83. Advance online publication. doi:10.1080/00167223.2019.1573409

Lawrence, H., Robertson, C., Feick, R., & Nelson, T. (2020). The spatial-comprehensiveness (S-COM) index: Identifying optimal spatial extents in volunteered geographic information point datasets. *ISPRS International Journal of Geo-Information*, *9*(9), 497. Advance online publication. doi:10.3390/ijgi9090497

Mobasheri, A. (2017). A rule-based spatial reasoning approach for openstreetmap data quality enrichment; case study of routing and navigation. *Sensors (Basel)*, *17*(11), 2498. Advance online publication. doi:10.339017112498 PMID:29088125

Nara, B. B., Lengoiboni, M., & Zevenbergen, J. (2020). Implications of customary land rights inequalities for food security: A study of smallholder farmers in Northwest Ghana. *Land (Basel)*, *9*(6), 178. Advance online publication. doi:10.3390/land9060178

Nutley, T., McNabb, S., & Salentine, S. (2013). Impact of a decision-support tool on decision making at the district level in Kenya. *Health Research Policy and Systems*, *11*(1), 34. Advance online publication. doi:10.1186/1478-4505-11-34 PMID:24011028

Qiu, Y., Zhao, X., Fan, D., Li, S., & Zhao, Y. (2022). Disaggregating population data for assessing progress of SDGs: methods and applications. International Journal of Digital Earth, 15(1). doi:10.1080/17538947.2021.2013553

Quinn, S., & Yapa, L. (2016). OpenStreetMap and Food Security: A Case Study in the City of Philadelphia. *The Professional Geographer*, *68*(2), 271–280. Advance online publication. doi:10.1080/00330124.2015.1065547

Rollason, E., Bracken, L. J., Hardy, R. J., & Large, A. R. G. (2018). The importance of volunteered geographic information for the validation of flood inundation models. *Journal of Hydrology (Amsterdam)*, *562*, 267–280. Advance online publication. doi:10.1016/j.jhydrol.2018.05.002

Sadeghi-Niaraki, A., Jelokhani-Niaraki, M., & Choi, S. M. (2020). *A volunteered geographic information-based environmental decision support system for waste management and decision making*. Sustainability. doi:10.3390u12156012

Scott, G., & Rajabifard, A. (2017). Sustainable development and geospatial information: A strategic framework for integrating a global policy agenda into national geospatial capabilities. *Geo-Spatial Information Science*, *20*(2), 59–76. doi:10.1080/10095020.2017.1325594

See, L., McCallum, I., Fritz, S., Perger, C., Kraxner, F., Obersteiner, M., Baruah, U. D., Mili, N., & Kalita, N. R. (2013). Mapping Cropland in Ethiopia Using Crowdsourcing. *International Journal of Geosciences*, *4*(6), 6–13. Advance online publication. doi:10.4236/ijg.2013.46A1002

See, L., Mooney, P., Foody, G., Bastin, L., Comber, A., Estima, J., Fritz, S., Kerle, N., Jiang, B., Laakso, M., Liu, H. Y., Milèinski, G., Nikšieč, M., Painho, M., Podör, A., Olteanu-Raimond, A. M. R., & Rutzinger, M. (2016). Crowdsourcing, citizen science or volunteered geographic information? The current state of crowdsourced geographic information. *ISPRS International Journal of Geo-Information*, *5*(5), 55. Advance online publication. doi:10.3390/ijgi5050055

Soden, R., & Palen, L. (2014). From Crowdsourced Mapping to Community Mapping: The Post-earthquake Work of OpenStreetMap Haiti. *COOP 2014 - Proceedings of the 11th International Conference on the Design of Cooperative Systems, 27-30 May 2014, Nice (France)*. 10.1007/978-3-319-06498-7_19

Solano-Hermosilla, G., Barreiro-Hurle, J., Adewopo, J. B., & Gorrín-González, C. (2022). Increasing engagement in price crowdsourcing initiatives: Using nudges in Nigeria. *World Development*, *152*, 105818. Advance online publication. doi:10.1016/j.worlddev.2022.105818 PMID:35370345

Stream Tracker. (2023). *Why Stream Tracker?* https://www.streamtracker.org/

UN-GGIM. (2012). *Monitoring sustainable development: Contribution of geospatial information to the rio+ 20 processes*. United Nations New York.

UNDP. (2023). *Sustainable Development Goals*. https://www.undp.org/sustainable-development-goals

Vaughan, L. (2011). Hunger In America: Utilizing GIS and Volunteered Geographic Information (VGI) to Improve Food Security. *139st APHA Annual Meeting and Exposition 2011*.

Vonthron, S., Perrin, C., & Soulard, C. T. (2023). Mapping the food environment: The reliability of volunteered geographical information and institutional data sources in France. *Transactions in GIS*, *27*(1), 134–151. Advance online publication. doi:10.1111/tgis.13013

Widener, M. J., & Li, W. (2014). Using geolocated Twitter data to monitor the prevalence of healthy and unhealthy food references across the US. *Applied Geography (Sevenoaks, England)*, *54*, 189–197. Advance online publication. doi:10.1016/j.apgeog.2014.07.017

Wiki. (2020). *Food Security*. https://wiki.openstreetmap.org/wiki/Food_security

Zeug, H., Zeug, G., Bielski, C., Solano-Hermosilla, G., & M'barek, R. (2017). *Innovative food price collection in developing countries. Focus on crowdsourcing in Africa*. Academic Press.

Zhang, D., Chen, X., & Yao, H. (2015). Development of a prototype web-based decision support system for watershed management. *Water (Basel)*, *7*(2), 780–793. Advance online publication. doi:10.3390/w7020780

Zhang, G. (2020). Spatial and temporal patterns in volunteer data contribution activities: A case study of eBird. *ISPRS International Journal of Geo-Information*, *9*(10), 597. Advance online publication. doi:10.3390/ijgi9100597

Zhang, G. (2022). Detecting and Visualizing Observation Hot-Spots in Massive Volunteer-Contributed Geographic Data across Spatial Scales Using GPU-Accelerated Kernel Density Estimation. *ISPRS International Journal of Geo-Information*, *11*(1), 55. Advance online publication. doi:10.3390/ijgi11010055

Chapter 11
Using AI Techniques to Improve the Power Quality of Standalone Hybrid Renewable Energy Systems

Hitesh Kumawat
Maharana Pratap University of Agriculture and Technology, India

Raunak Jangid
Geetanjali Institute of Technical Studies, India

ABSTRACT

As a developed and environmentally benign alternative for energy generation, renewable energy sources (RES) have attracted a lot of attention. The issue of energy quality, however, is one of the most difficult integration problems for RES. Voltage and frequency violations caused by waveform instability in the energy production from RES can harm equipment and lower the reliability of the power supply. In recent years, renewable energy sources like wind and solar power have drawn a lot of attention as environmentally friendly substitutes for traditional energy sources. Systems with artificial intelligence (AI) are becoming more and more common for use in automation, automated automation, and data analysis. In this research, the authors present a unique theory for an AI system that combines solar and wind energy to power AI applications. In order to improve the effectiveness and efficiency of AI systems, the suggested system makes use of the complementary properties of solar and wind energy, optimizing their generation and utilization.

DOI: 10.4018/978-1-6684-9601-5.ch011

INTRODUCTION

Renewable energy sources like PV, wind, hydro, etc., are now becoming increasingly popular due to the acute depletion of fossil fuels and pollution of the atmosphere. Solar and wind energy resources are abundantly available worldwide. Due to the fluctuating nature of renewable energy resources, power generation from renewable energy systems is intermittent. The combination of two or more energy sources with the storage system was inspired by these conditions (Arul et al., 2015). A remote hybrid system gives greater efficiency with low energy production costs, compared to a single source. It is essential to care for changes in the generated power which vary from time to time. In the literature, various types of HRES are introduced which work in grid-connected or stand- alone modes. HRES system energy management is completed using a PI controller (Ahmed et al., 2018). It is done by controlling a buck-boost bi-directional converter for battery charging and discharging. A current control strategy for power balance is introduced in the PI controller. The conventional controller strategy depends on mathematical system modeling. Hybrid systems are one effective solution for electrical energy generation, specifically for remote sites or for a micro-generation unit linked to a weak AC grid (Alexander & Thathan, 2015). Hybrid systems combine several conventional or renewable energy sources interconnected through a DC bus. For an isolated area, the association of electro- chemical storage with the hybrid system eliminates the diesel generator (Arezki & Boudour, 2014). In this condition, we describe an integrated generation system acquired via combining PV and wind turbines with storage batteries. This system is to overcome periods of inadequate generation and for system control. In hybrid systems, different topologies are available, depending on the interface converters between sources and the method of interconnection. According to this topology, the interconnection of the sources ensures maximum energy transfer, and the investigation of energy losses involves power conditioning converters as well as optimal control and energy management (Chauhan & Saini, 2014). The present work aims to develop the idea of hybrid system configuration, dynamic modeling, energy management and control strategies. For different operating conditions, the author proposes and investigates an efficient control strategy for a standalone hybrid solar-wind system. Due to the uninterrupted demand for energy, batteries are used as a backup in the hybrid system. So, the suggested system is efficient for working under variable weather and load conditions (Chishti et al., 2019).

REVIEW OF LITERATURE

The literature review provides an overview of power quality issues in HRES and the application of AI techniques for mitigating these issues. Various AI techniques such as artificial neural networks (ANN), fuzzy logic, and genetic algorithms have been used in previous studies to improve the power quality of HRES. These techniques have shown promising results in terms of reducing voltage fluctuations, improving power factor, and reducing current harmonics (Habib et al., 2019).

OBJECTIVES OF THE RESEARCH

There are four objectives of this chapter-

1. To develop dynamic model for hybrid, stand-alone micro grid consists solar-wind system.
2. To develop AI based intelligent control strategy for power quality improvement of hybrid standalone micro grid consists solar-wind system.
3. To simulate the developed control strategy using simulation software.
4. To investigate the performance for different load conditions.

DESCRIPTION OF PROPOSED HYBRID SYSTEM

We propose a standalone hybrid micro grid system combined with an intelligent control strategy to improve power quality. The micro grid consists of a 9.8 kW Photovoltaic (PV) system with Perturb and observe (P & O) MPPT algorithm to assure maximum power from PV panels and a 6.3Kw permanent magnet synchronous generator (PMSG) powered wind energy conversion system (WECS) with optimum torque (OT) control and maximum power point tracking (MPPT) to extract maximum power from wind. In emergency situations, a hybrid energy storage system (HESS) can provide constant power or deliver constant load power under variable load conditions (Tata et al., 2022). The HESS configuration is based on lithium-ion (Li-Ion) batteries and a super capacitor. A fully controlled AC/DC bridge rectifier is used to extract maximum power from the wind turbine. The converter gate pulse is generated by the optimum torque based MMPT technique. A DC/DC converter is used for the PV system and the gate pulse for the converter is developed by the P & O MPPT algorithm. Figure 1 illustrates the block diagram of the developed ANN Based Control Strategy for Power Quality Improvement of Hybrid RES (Roy et al., 2022).

In order to control the power from both the battery and super capacitor, two bidirectional DC/DC converters are made; these converters are directly connected to the 640 V DC bus. Whenever wind and solar power are sufficient and the generated power exceeds the load's needs, the battery is charged first (Chatterjee et al., 2023).

Figure 1. Block diagram of developed AI-based control strategy for power quality improvement of hybrid RES

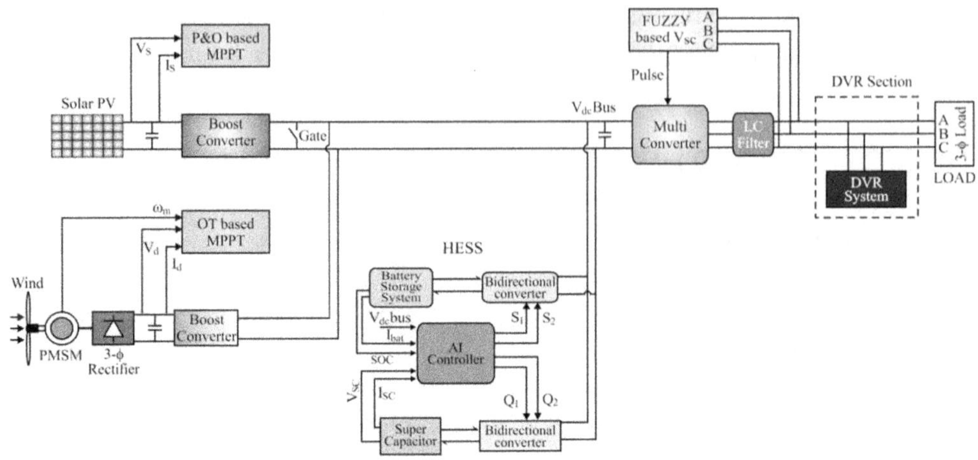

CONTROL OF SYSTEM COMPONENTS

ANN-based control is quick, robust and stable. It has the potential to improve the micro grid's power quality and efficiency. The proposed control strategy is intended for two conditions: change in wind velocity, change in solar irradiation. The block diagram of the proposed ANN-based control strategy for management of hybrid storage is shown in Figure 2 (Kumar et al., 2016).

Figure 2. Proposed AI-based control strategy for management of hybrid storage

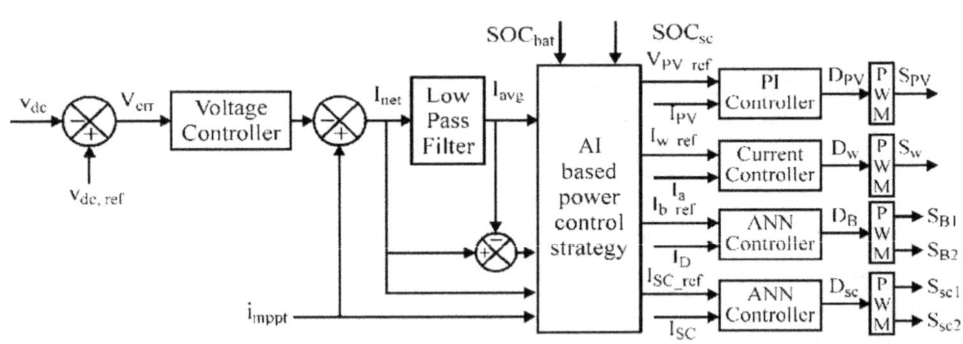

PERFORMANCE EVALUATION OF ANN TRAINING DATA FOR HYBRID RES

In this paper we compare the error histogram with 20 bins, the performance, regression and training set and develop Levenberg-based Artificial Intelligence Technique based MPPT algorithms for hybrid renewable energy. It is evident from comparative results that developed systems produce better results (Roy et al., 2021).

Figure 3 shows that R = 1, which corresponds to the most accurate output prediction based on input information. Furthermore, it correlates the Hybrid RES output voltage with its target voltage. To calculate an error, subtract the output from the target. As shown in Figure 3, the LM algorithm is used to train the data with negligible error, resulting in an output value that follows the target value perfectly (Rai et al., 2013).

Figure 3. Regression plot of the ANN model

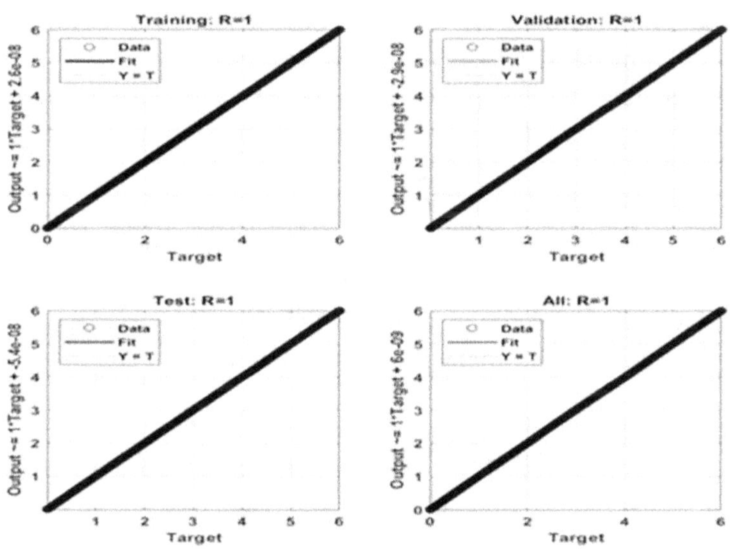

Additionally, Figure 4 shows that the ANN algorithm provides zero error in the training, validation, and test phases. Based on the error histogram in Figure 4, the leftmost bin shows a total error of –4.1e-05, while the rightmost bin exhibits a total error of 3.21e-05 (Al Hadi et al., 2023).

The bins represent the number of vertical bars in the error histogram. Each vertical bar represents the number of samples within a particular bin from the selected dataset. Each bin is divided into 20 smaller bins with a bin width of 0.0000147. Based on the validation dataset, the middle bin has an error of 1.21e-06 for 150

Figure 4. Error histogram plot of ANN training data for hybrid RES

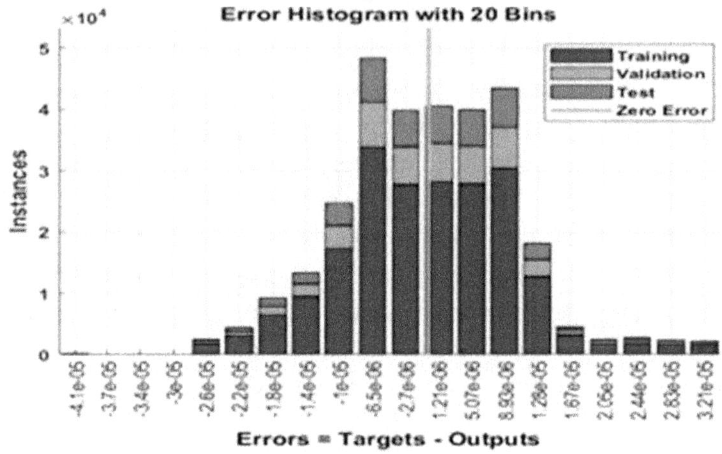

samples. The error histogram converges to zero after 20 bins, which indicates that ANNs are applicable to MPPTs (Zhang et al., 2019).

Figure 5 and Figure 6 illustrate the training state and performance phase of ANN for handling the selected dataset. The gradient, momentum parameter (Mu) and validation check of the trained dataset are shown in Figure 5. The simulation shows that the gradient is 1.2437e-07 at 1000 epochs, which indicates negligible deviation from the conditioned data. Simulated results indicate that the summed error is the mean per input vector and a zero-output decision. It is clear from the very small Mu value, gradient and validation checks of the trained dataset that the Levenberg Marquardt algorithm can be applied to MPPT (Zanib et al., 2022).

Figure 5. Training state plot of ANN MPPT

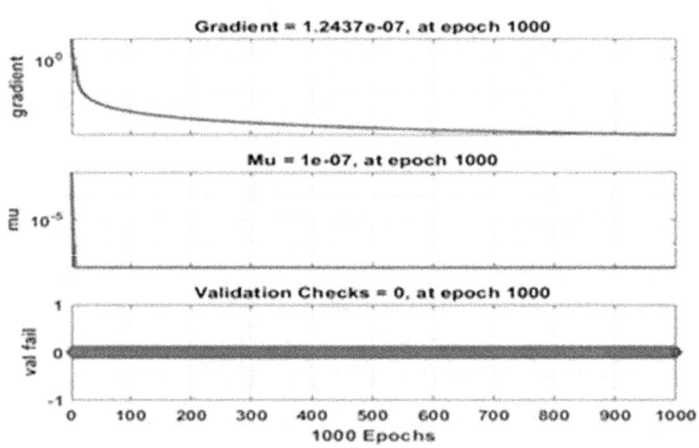

Figure 6. Performance test of ANN algorithm

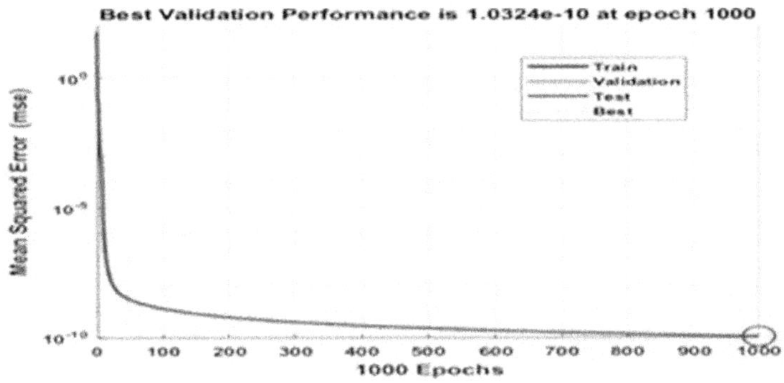

A convergent dataset of 1000 epochs is shown in Figure 6 where the mean squared error is shown for each epoch. It is therefore at 1000 epochs that the trained dataset achieves its most effective validation performance. Simulation results indicate that 1.0324e-10 is the optimal validation performance at 1000 epochs. According to Levenberg Marquardt's MPPT prediction, "nearly zero" validation performance indicates negligible error (Chawda et al., 2020).

SIMULATION RESULTS AND DISCUSSION

This paper presents the detailed dynamic model in MATLAB/Simulink software for simulation of the hybrid PV - wind system developed with artificial neural network-based DVR for identification, classification and mitigation of power quality issues during faulty and sag/swell conditions in an isolated area and presents several case studies designed to investigate the application of the proposed hybrid system for variable wind, irradiance level, and load conditions at faulty and sag/swell conditions (Swapna & Gayatri, 2021).

Case 1: Simulation Results During Asymmetrical Fault at Constant Wind Speed, Constant Load and Constant Irradiance

The simulation results for Line-to-Line fault are shown in the following figures.

Figure 7 shows the plot of irradiance and temperature kept constant at 1000W/m^2 and 25°C respectively. Fig. 8 shows the plot of PV voltage and PV current. PV array voltage is constant at 1011V and PV current becomes stable at 70A after a spike at 110A (Bellini et al., 2009).

In Figure 9 PV array power is constant at 53780W for the whole simulation time. In Figure 10 PV output power is shown compared to irradiance where both are constant throughout the time period (Xue et al., 2022).

Figure 7. Simulation results at constant irradiance and temperature of 1000w/m² and 25°C respectively, waveform of irradiance, and temperature

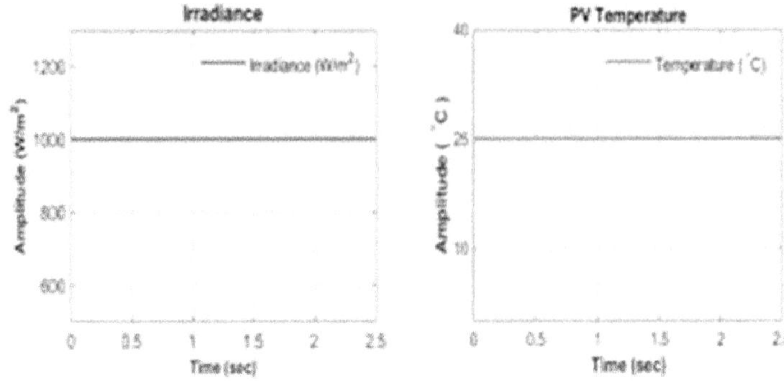

Figure 8. Simulation results at constant irradiance and temperature of 1000w/m2 and 25°C respectively, waveform of PV voltage, and PV current

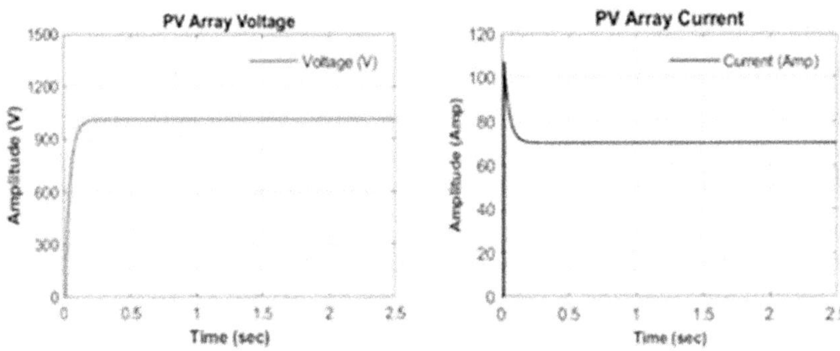

Figure 9. Simulation results at constant irradiance and temperature of 1000w/m2 and 25°C,respectively, waveform of PV power

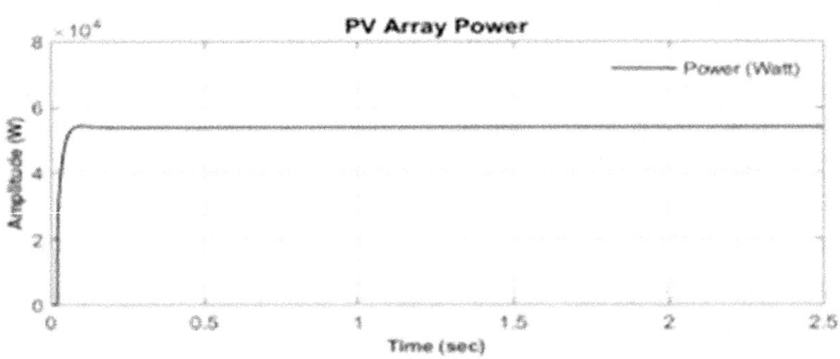

Figure 10. Simulation results at constant irradiance and temperature of 1000w/m2 and 25°C respectively, waveform of PV output power, and irradiance

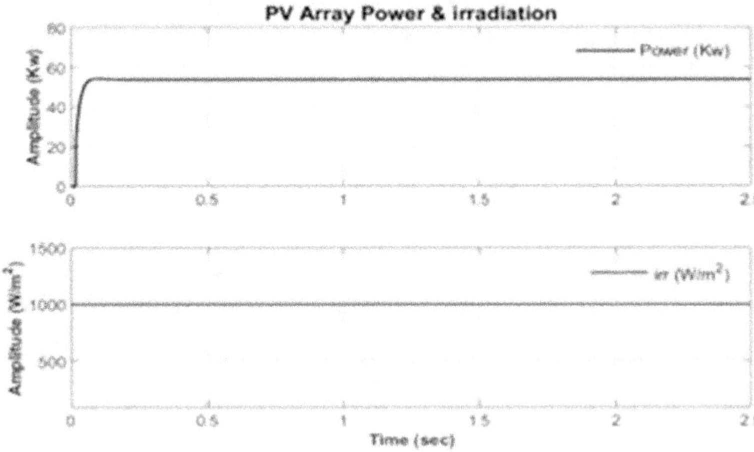

In Figure 11 Input wind speed is constant at 12m/s for whole simulation time and PMSG rotor speed is constant for time t=0.2sec to t=1.2sec at 1930rpm, until there is a line-to- line fault in the system at t=1.2sec which continues till t=1.6sec hence the rotor speed slightly increases during the asymmetrical fault. After that, the system gains stability at time t=1.6secs and the rotor speed decreases back to its initial value (Liu et al., 2021).

Figure 11. Simulation results at constant wind speed, 12m/s and load respectively, waveform of input wind speed, and rotor speed

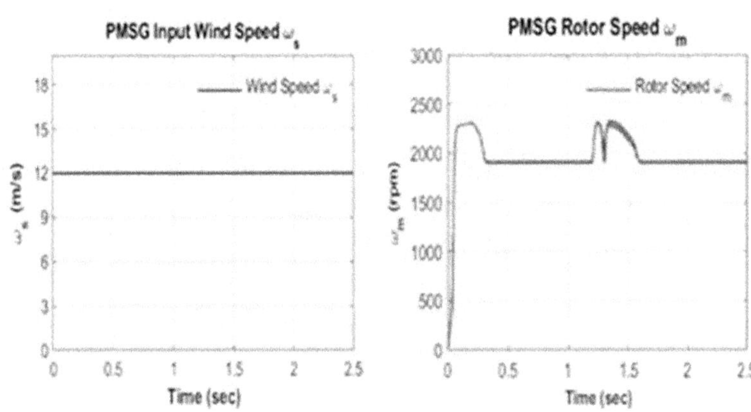

In Figure 12 PMSG output power and PMSG Electromagnetic and Mechanical torque are shown, where these are constant from time t=0.5sec to t=1.2sec at 40.8KW & 345.9N-m respectively. At time t=1.2sec, both torque parameters and output power decrease in the same pattern at 18KW & 130N-m. When t=1.5sec becomes 1.8sec, output power and torque parameters reach their initial values and become constants (Zhang et al., 2021).

Figure 12. Simulation results at constant wind speed, 12m/s and load respectively, waveform of PMSG output power, PMSG electromagnetic torque T_e, and mechanical torque T_m

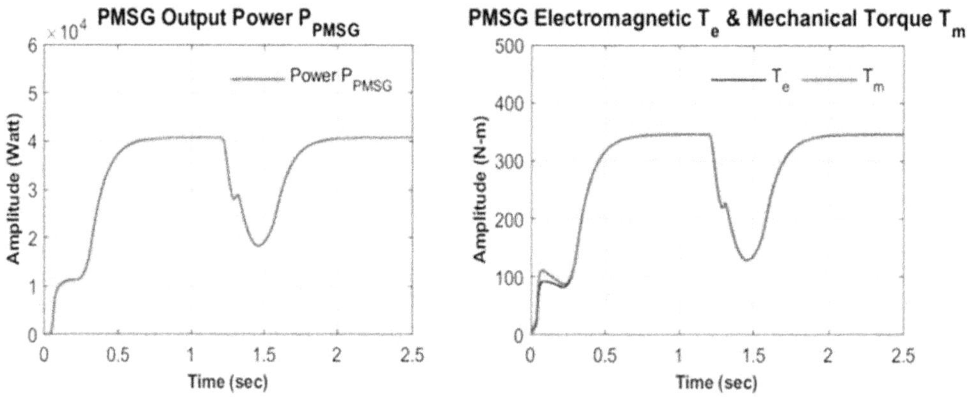

In Figures 13 and 14, Both PMSG phase voltage and phase current are constant from time t=0.3sec to t=1.2sec, but due to line-to-line fault phase voltage increases slightly and phase current decreases and again both become constant at t=1.6sec at their initial constant value (Sahoo et al., 2023).

Figure 13. Simulation results at constant wind speed, 12m/s and load respectively, waveform of PMSG phase-to-phase voltage Vph

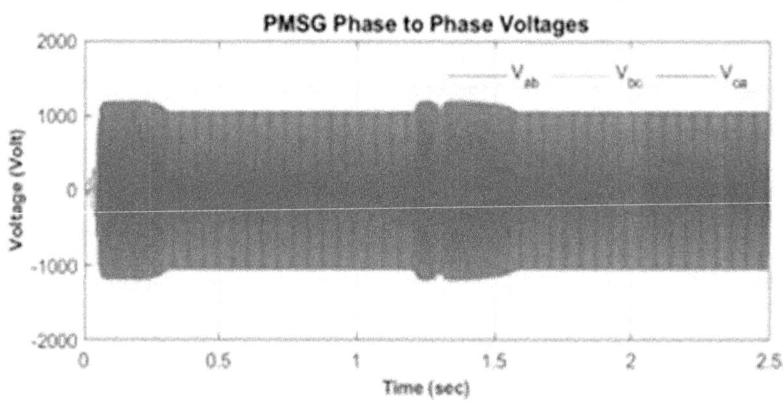

Figure 14. Simulation results at constant wind speed, 12m/s and load respectively, waveform of PMSG phase current Iph

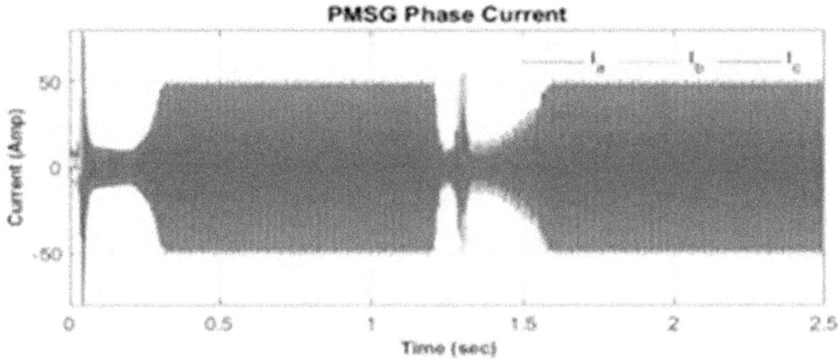

In the Figure 15 and Figure 16, Both PMSG phase RMS voltage and RMS current are constant from time t=0.3sec to t=1.2sec at 732V and 32.89A respectively. However, due to line-to-line fault phase RMS voltage increases slightly and phase RMS current decreases before both become constant at t=1.6sec at their initial nominal value (Calik et al., 2021).

Figure 17 shows the PMSG output power plot against input wind speed. This is constant at 12m/s throughout the time period. However, output power is constant from t=0.5sec to t=1.2sec at 40.8KW, then decreases during line-to-line faults and again increases to reach the initial constant value (Liu et al., 2021).

According to Figure 18 phase-to-phase voltage of the generation side is constant till time t=1.2sec, after that it decreases a little due to line-to-line fault, then at t=1.3sec system gains stability and voltage again becomes constant to its initial value, which can be clearly observed in the zoom in window shown in Figure 19 (Fekry et al., 2020).

According to Figure 20 phase-to-phase current of generation side is constant till t=1.2sec, then increases due to line-to-line fault, then at t=1.3sec system gains stability and current again becomes constant to its initial value, which can be clearly observed in the zoom in window shown in Figure 21 (Alibegovic & Freiland Paulsson, 2021).

Even during the line-to-line fault in the system, the phase-to-phase voltage of load is constant throughout the simulation time, because of the ANN tuned DVR used in the network which compensates for distortions in generation side voltage to load side voltage, as shown in Figure 22. This can also be observed in the zoom-in window provided in Figure 23 for load phase voltage (Aworo, 2019).

In Figure 24 the phase current of load is constant at 92A, but it experiences a slight distortion from time t=1.2 sec to t=1.3 sec which can be clearly observed in the zoom in window provided in Figure 25 (Nam-In & Kumsuwan, 2020).

Figure 15. Simulation results at constant wind speed, 12m/s and load respectively, waveform of PMSG phase-to-phase RMS voltage

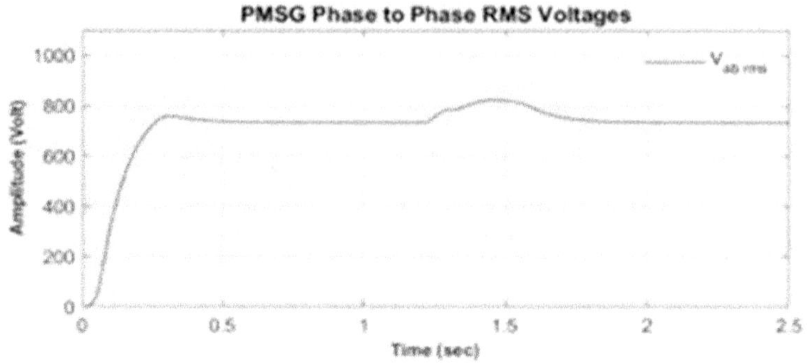

Figure 16. Simulation results at constant wind speed, 12m/s and load respectively, waveform of PMSG phase RMS current

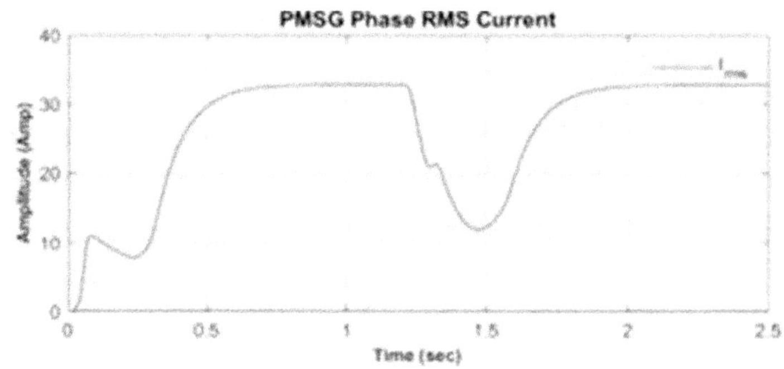

Figure 17. Simulation results at constant wind speed, 12m/s and load respectively, waveform of PMSG output power, and wind speed

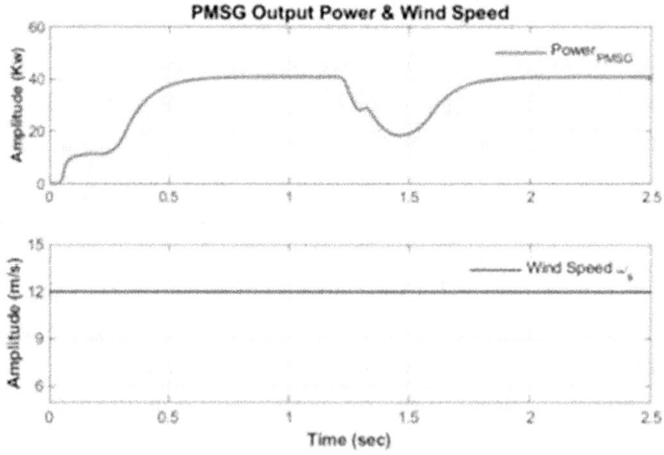

Figure 18. Simulation results at constant irradiance of 1000 w/m2, constant temperature of 25°C and wind speed 12m/s respectively, waveform of phase-to-phase voltages of generation side

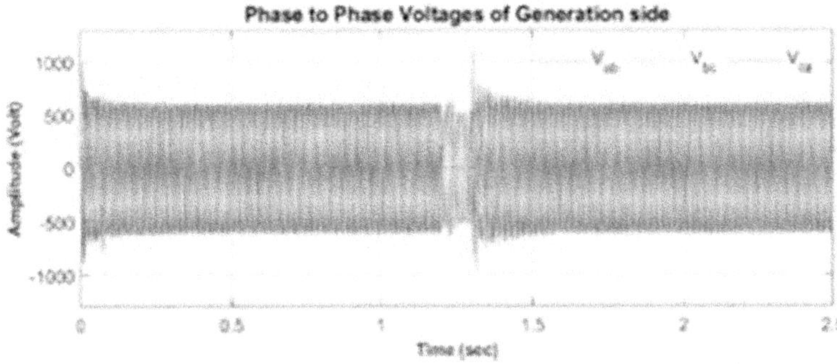

Figure 19. Simulation results at constant irradiance of 1000 w/m2, constant temperature of 25°C and wind speed 12m/s respectively, waveform of phase-to-phase voltage of generation side during line-to-line fault

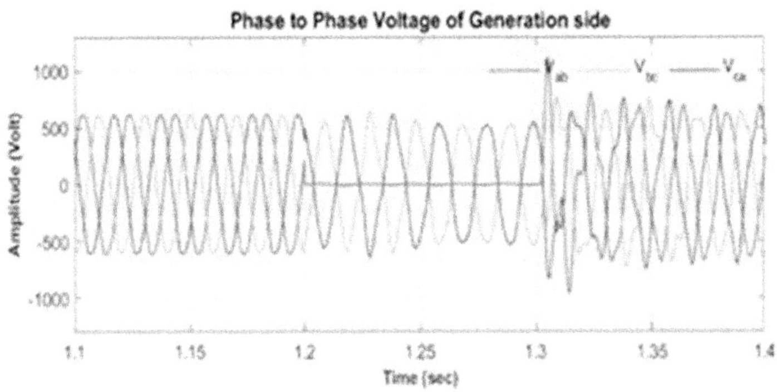

Figure 20. Simulation results at constant irradiance of 1000 w/m2, constant temperature of 25°C and wind speed 12m/s respectively, waveform of phase current of generation side

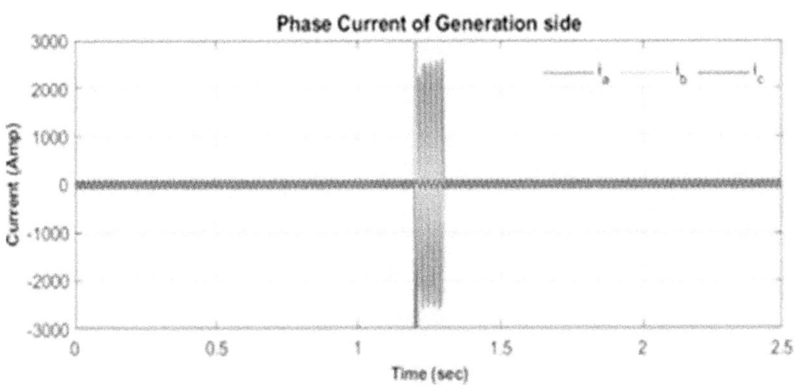

Figure 21. Simulation results at constant irradiance of 1000 w/m2, constant temperature of 25°C and wind speed 12m/s respectively, waveform of phase current of generation side during line-to-line fault

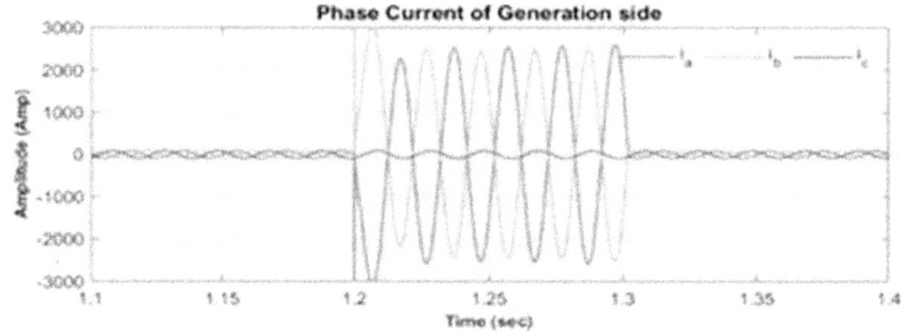

Figure 22. Simulation results at constant irradiance of 1000 w/m2, constant temperature of 25°C and wind speed 12m/s respectively, waveform of phase-to-phase voltage of load

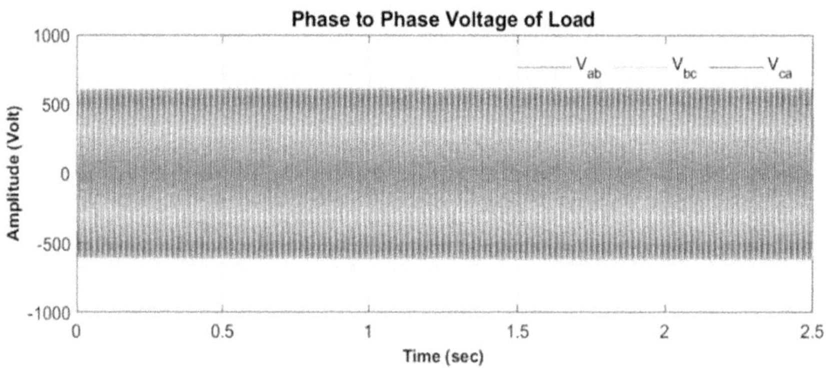

Figure 23. Simulation results at constant irradiance of 1000 w/m2, constant temperature of 25°C and wind speed 12m/s respectively, waveform of phase-to-phase voltage of load during line-to-line fault

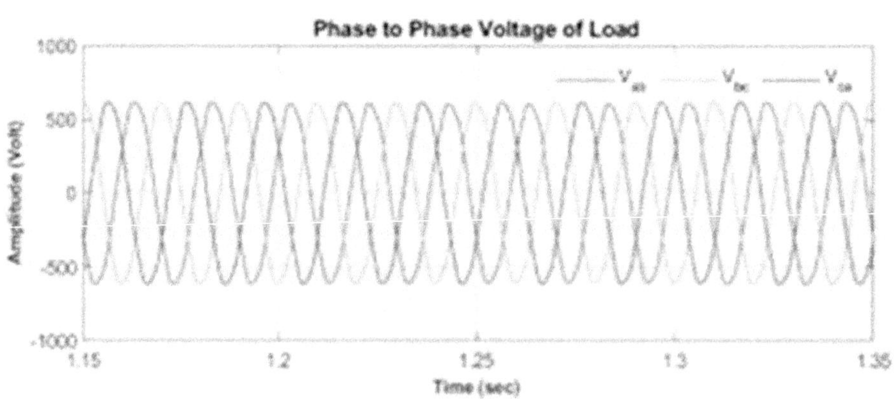

Figure 24. Simulation results at constant irradiance of 1000 w/m2, constant temperature of 25°C and wind speed 12m/s respectively, waveform of phase current of load

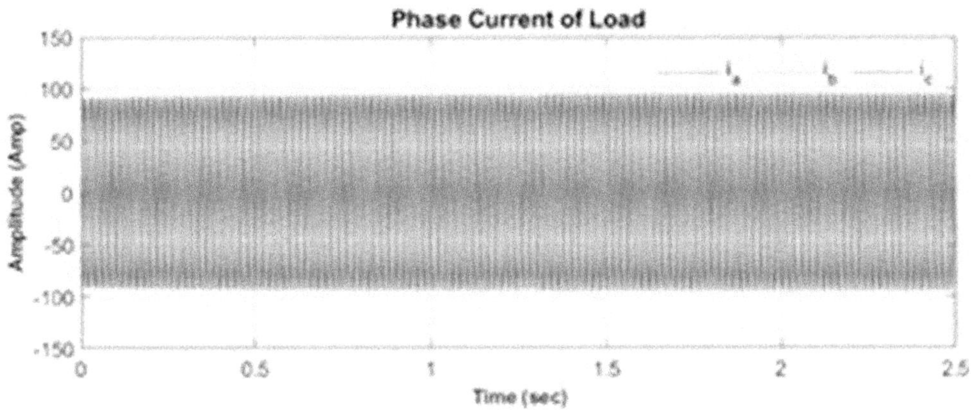

Figure 25. Simulation results at constant irradiance of 1000 w/m2, constant temperature of 25°C and wind speed 12m/s respectively, waveform of phase current of load during line-to-line fault

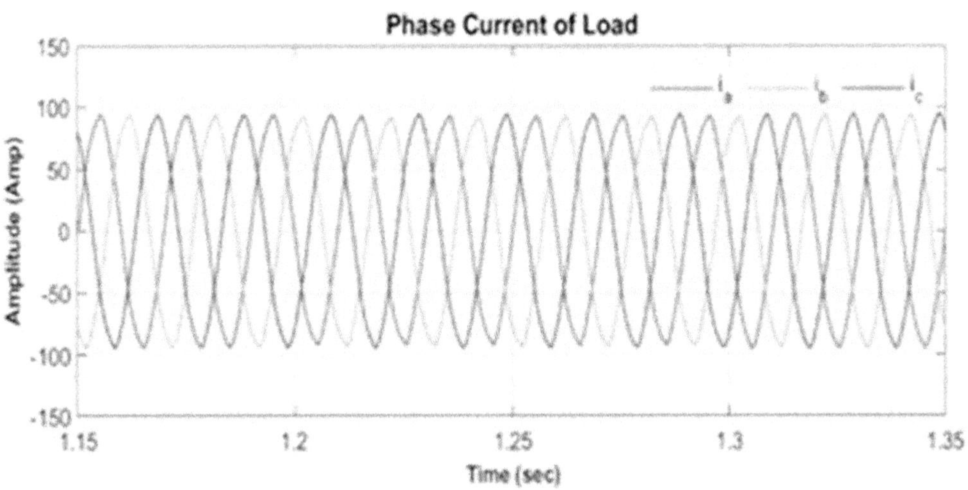

In Figure 26 both RMS phase voltage and RMS phase current are constant throughout the simulation time at 440V and 65.65A respectively. However, they experience slight changes when a line-to-line ground fault occurs i.e., time t=1.2sec to t=1.3sec (Bento et al., 2019).

According to Figure 28, we can clearly observe the distortions caused by line-to-line faults on the generation side which decreased phase-to-phase voltage, while

Figure 26. Simulation results at constant irradiance of 1000 w/m2, constant temperature of 25°C and wind speed 12m/s respectively, waveform of phase-to-phase RMS voltage of load and RMS phase current of load

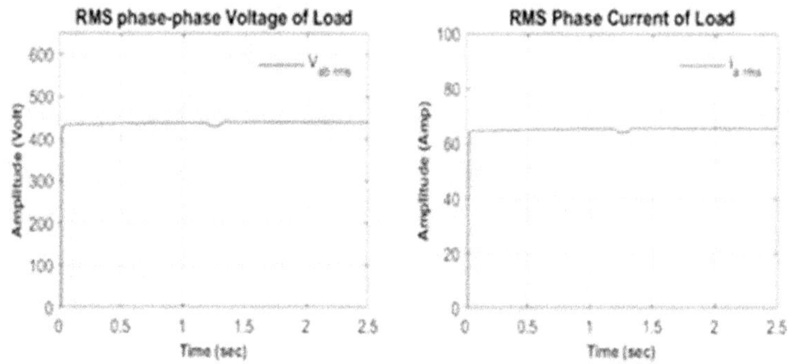

Figure 27. Simulation results at constant irradiance of 1000 w/m2, constant temperature of 25°C and wind speed 12m/s respectively, waveform of phase-to-phase voltage of generation side in PU

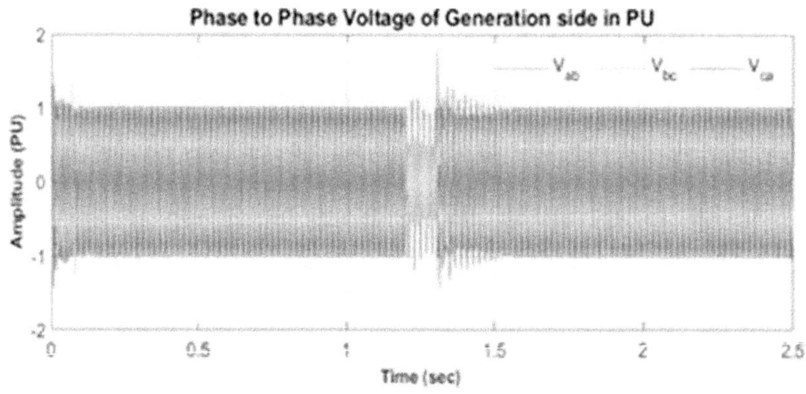

Figure 28. Simulation results at constant irradiance of 1000 w/m2, constant temperature of 25°C and wind speed 12m/s respectively, waveform of phase-to-phase voltage of load side in PU

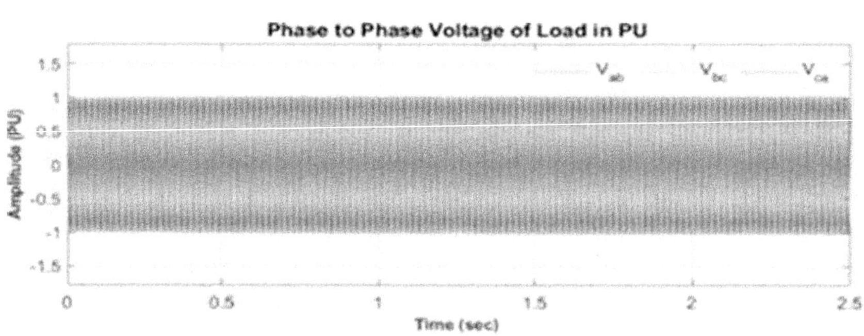

asymmetrical faults do not affect the phase-to-phase voltage on the load side. This is due to the fact that an ANN tuned DVR is used in the system, which isolates the load from line-to-line faults (Zichang et al., 2020).

In Figure 29, it can be observed that the ANN tuned DVR injects phase voltages into the load while isolating it from the generation side, ensuring that the supply of phase voltage to the load remains uninterrupted at the time of the fault (Tanaka et al., 2018).

Figure 29. Simulation results at constant irradiance of 1000 w/m2, constant temperature of 25°C and wind speed 12m/s respectively, waveform of injected phase-to-phase voltage in PU

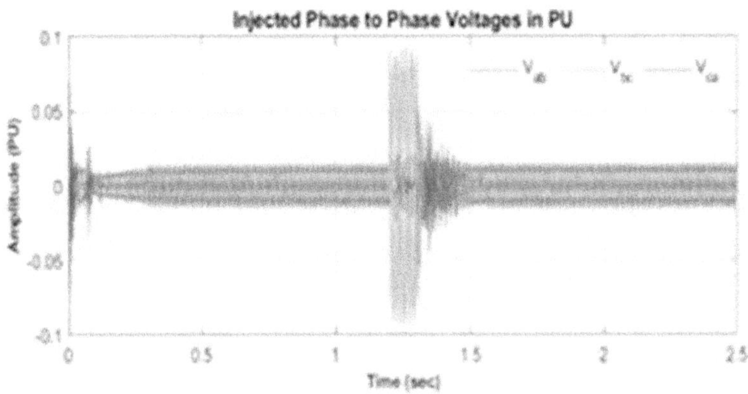

In Figure 30 and Figure 31 phase current of generation side and phase current of load side is shown per unit system (Li et al., 2018).

Figure 30. Simulation results at constant irradiance of 1000 w/m2, constant temperature of 25°C and wind speed 12m/s respectively, waveform of phase current of generation side in PU

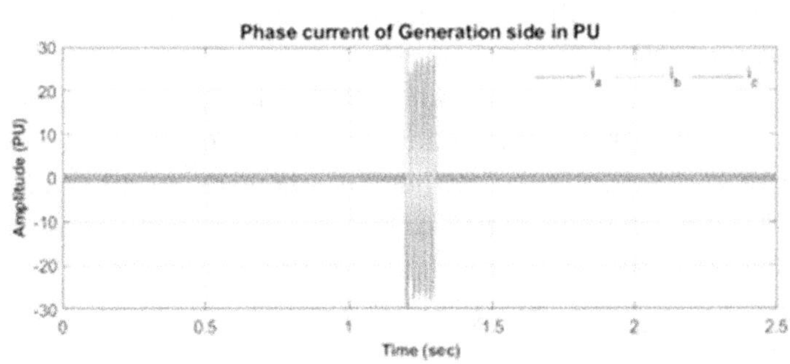

Figure 31. Simulation results at constant irradiance of 1000 w/m2, constant temperature of 25°C and wind speed 12m/s respectively, waveform of phase current of load side in PU

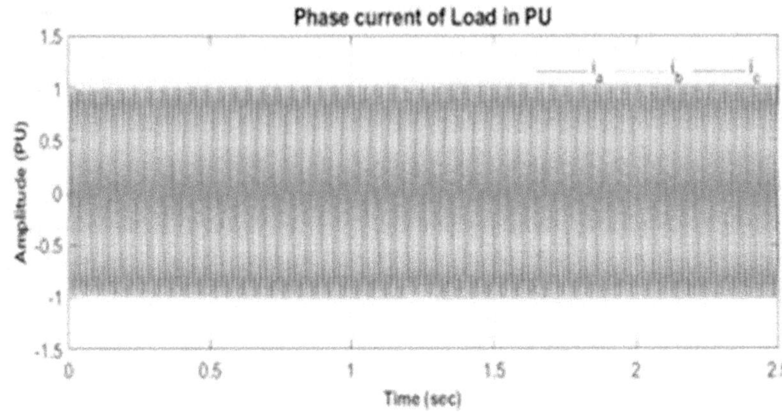

Case 2: Simulation results during symmetrical fault at constant wind speed, constant load and constant irradiance

The simulation results for three phase Line to ground fault are shown in the following figures.

Figure 32 shows the plot of irradiance and temperature which is kept constant at 1000W/m^2 and 25°C respectively. Figure 33 shows the plot of PV voltage and PV current where PV array voltage is constant at 1011V and PV current becomes constant at 70A after a spike at 110A (Khasim et al., 2021).

In Figure 34 PV array power is constant at 53780W for whole simulation time. In Figure 35 PV output power is shown compared to irradiance where both are constant throughout the time period (Bhukya et al., 2019).

In Figure 36 Input wind speed is constant at 12m/s for whole simulation time and PMSG rotor speed is constant from time t=0.2sec to t=1sec at 1930rpm, until there is a three- phase line to ground fault in the system at time t=1.2sec which continues till t=1.6sec hence the rotor speed slightly increases during the symmetrical fault. After that system gains stability at t=1.6secs, & the rotor speed decreases to initial constant value again (Dida, 2017).

In Figure 37 PMSG output power and PMSG Electromagnetic and Mechanical torque are shown, where these are constant from time t=0.5sec to t=1.2sec at 40.8KW & 345.9N- m respectively. At t=1.2sec both the torque parameters and output power decreases in a same pattern at 11KW & 80N-m respectively. From time t=1.5sec to t=1.8sec Output power and both torque parameters increases & reaches initial values and become constant again (Dida, 2017).

Figure 32. Simulation results at constant irradiance and temperature of 1000w/m2 and 25°C respectively, waveform of irradiance and temperature

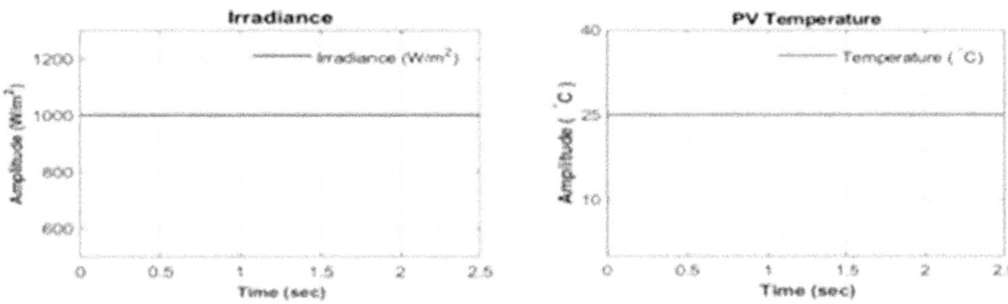

Figure 33. Simulation results at constant irradiance and temperature of 1000w/m² and 25°C,respectively, waveform of PV voltage and PV current

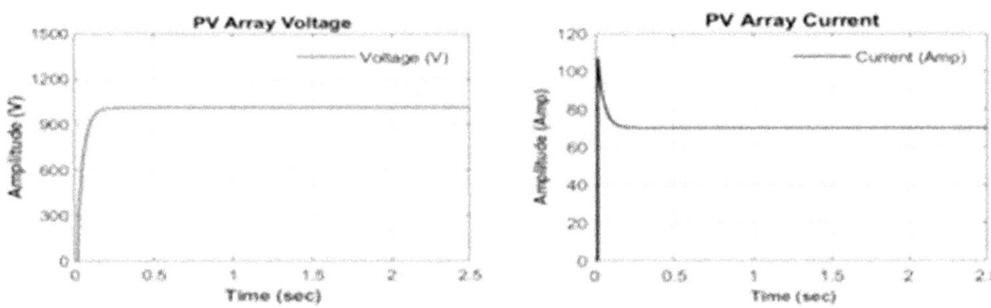

Figure 34. Simulation results at constant irradiance and temperature of 1000w/m² and 25°C respectively, waveform of PV power

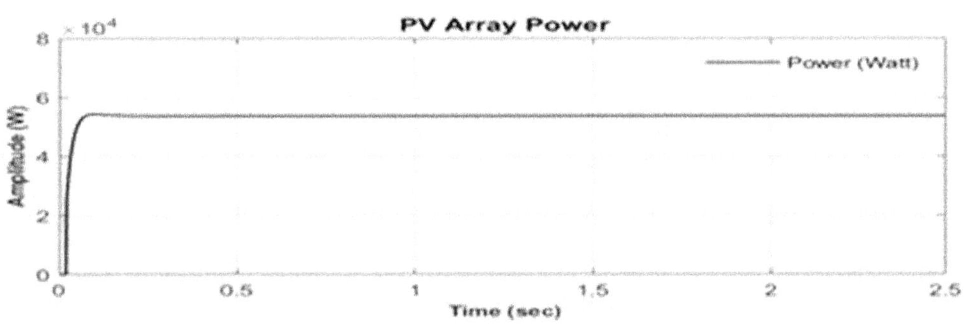

In Figures 38 and 39, both PMSG phase voltage and phase current are constant from t=0.3sec to t=1.2sec, but due to three phase line to ground fault voltage increases slightly and current decreases and again both becomes constant at t=1.6sec at their initial value (Barrett, 2020).

Figure 35. Simulation results at constant irradiance and temperature of 1000w/m² and 25°C respectively, waveform of PV output power and irradiance

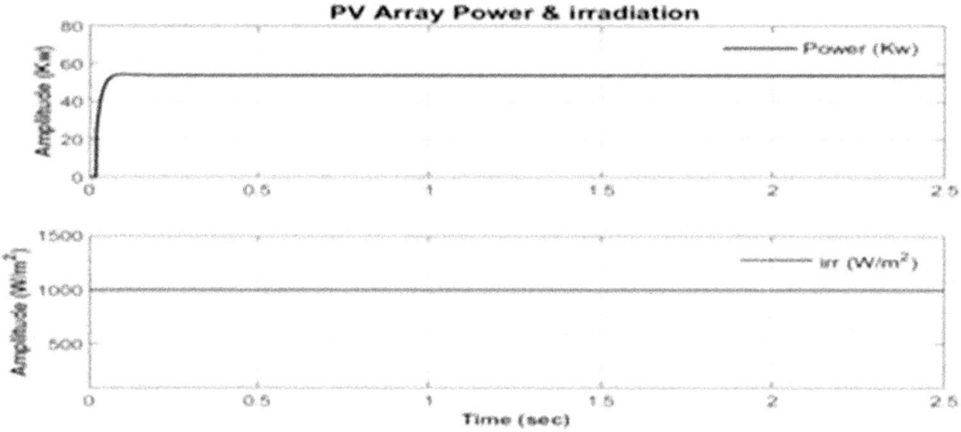

Figure 36. Simulation results at constant wind speed, 12m/s and load respectively, waveform of input wind speed and rotor seed

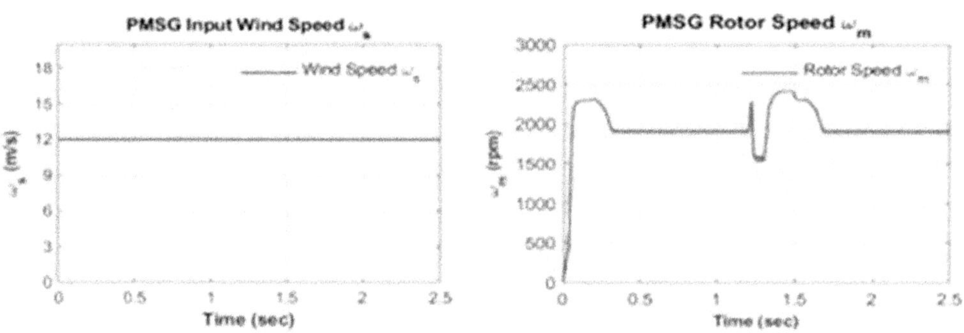

Figure 37. Simulation results at constant wind speed 12m/s and load respectively, waveform of PMSG output power, PMSG electromagnetic torque T_e and mechanical torque T_m

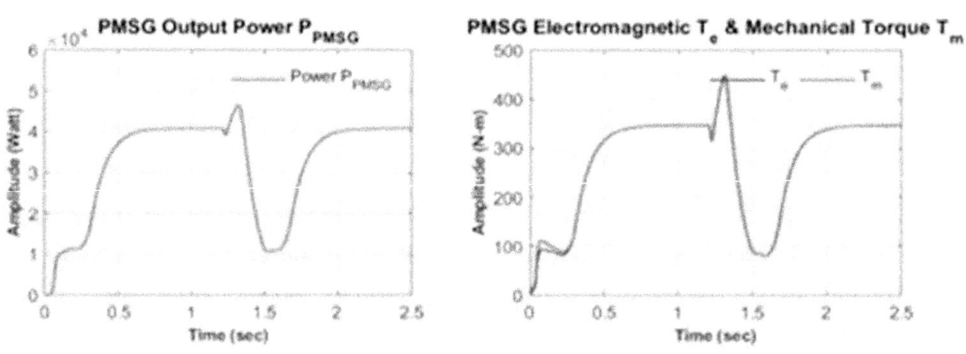

Figure 38. Simulation results at constant wind speed, 12m/s and load respectively, waveform of PMSG phase-to-phase voltage V_{ph}

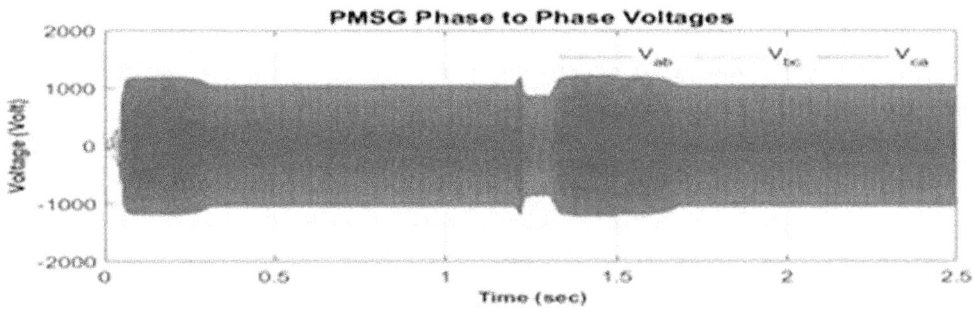

Figure 39. Simulation results at constant wind speed, 12m/s and load respectively, Waveform of PMSG phase current I_{ph}

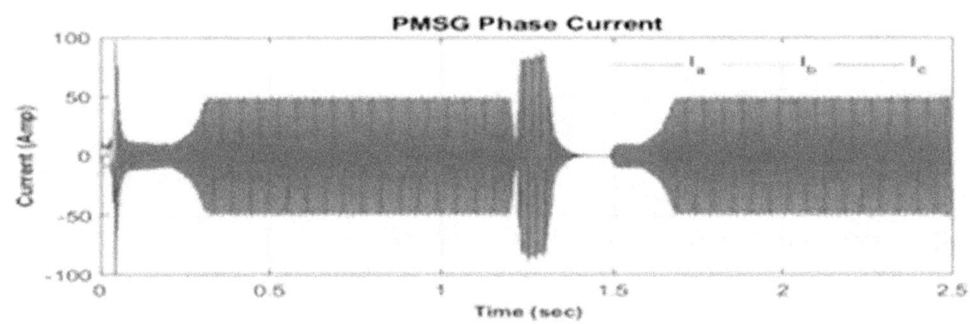

In the Figure 40 and Figure 41, Both PMSG phase RMS voltage and RMS current are constant from time t=0.3sec to t=1.2sec at 732V and 32.89A respectively, but due

Figure 40. Simulation results at constant wind speed, 12m/s and load respectively, waveform of PMSG phase-to-phase RMS voltage

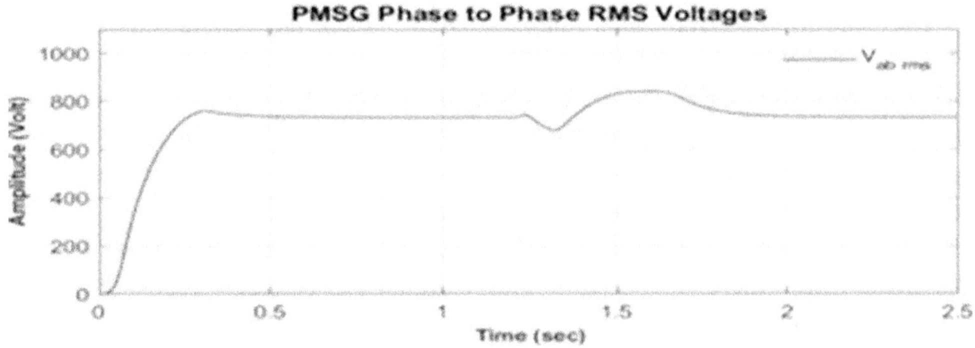

to line-to-line fault voltage increases slightly and current decreases and again both becomes constant at t=1.6sec at their initial constant value (Bahloul et al., 2023).

Figure 41. Simulation results at constant wind speed, 12m/s and load respectively, waveform of PMSG phase RMS current

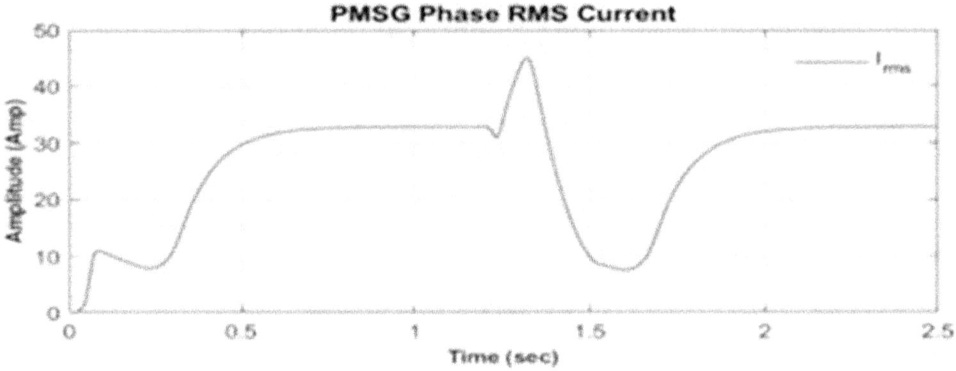

Figure 42 shows the PMSG output power plot with respect to input wind speed which is constant at 12m/s throughout the time period, but output power is constant from time t=0.5sec to t=1.2sec at 40.8KW, after that it decreases during three phase line to ground fault and again increases to reach the initial constant value (Fouad et al., 2018).

According to Figure 43 phase-to-phase voltage of generation side is constant till time t=1.2sec, after that it decreases to zero due to three phase line to ground fault, then at t=1.3sec system gains stability and voltage again becomes constant to its initial value, which can be clearly observed in the zoom in window shown in Figure 44 (Moawwad et al., 2016).

According to Figure 45 phase-to-phase current of generation side is constant till t=1.2sec, after that it increases due to line-to-line fault, then at t=1.3sec system gains stability and current again becomes constant to its initial value, which can be clearly observed in the zoom in window shown in Figure 46 (Chettibi et al., 2016).

Even during the three-phase line to ground fault in the system, the phase-to-phase voltage of load is constant throughout the simulation time, because of the ANN tuned DVR used in the network which compensate the effect of distortions in generation side voltage to load side voltage, as shown in Figure 47 & this can also be observed in the zoom in window provided in Figure 48 for phase voltage of load (Sadeghkhani et al., 2016).

In Figure 49 phase current of load is constant at 92A, it experiences a slight distortion from time t=1.2 sec to t=1.3sec which can be clearly observed in the zoom in window provided in Figure 50 (Gawande & Dambhare, 2016).

Figure 42. Simulation results at constant wind speed, 12m/s and load respectively, waveform of PMSG output power and wind speed

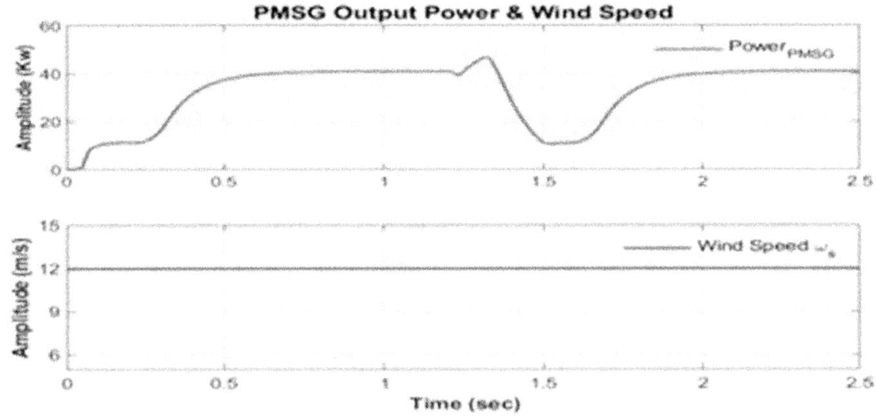

Figure 43. Simulation results at constant irradiance of 1000 w/m², constant temperature of 25°C and wind speed 12m/s respectively, waveform of phase-to-phase voltages of generation side

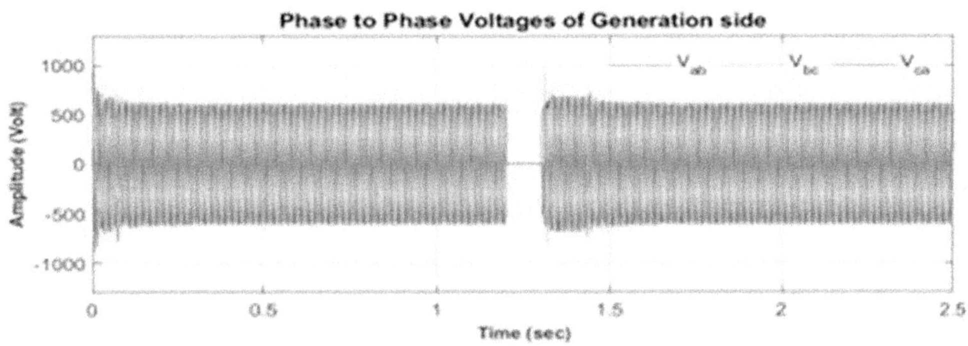

Figure 44. Simulation results at constant irradiance of 1000 w/m², constant temperature of 25°C and wind speed 12m/s respectively, waveform of phase-to-phase voltage of generation side during three phase line to ground fault

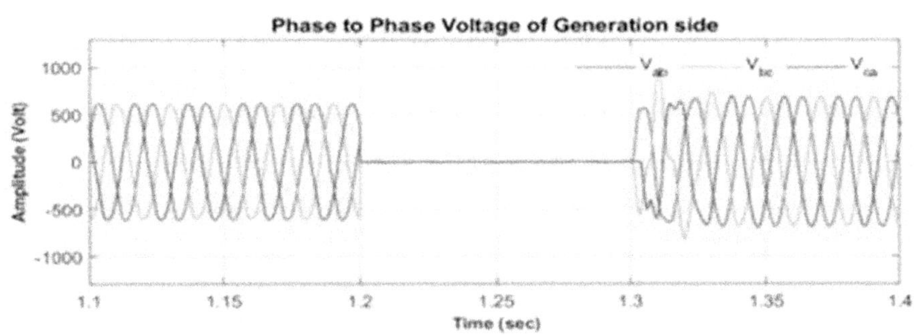

Figure 45. Simulation results at constant irradiance of 1000 w/m², constant temperature of 25°C and wind speed 12m/s respectively, waveform of phase current of generation side

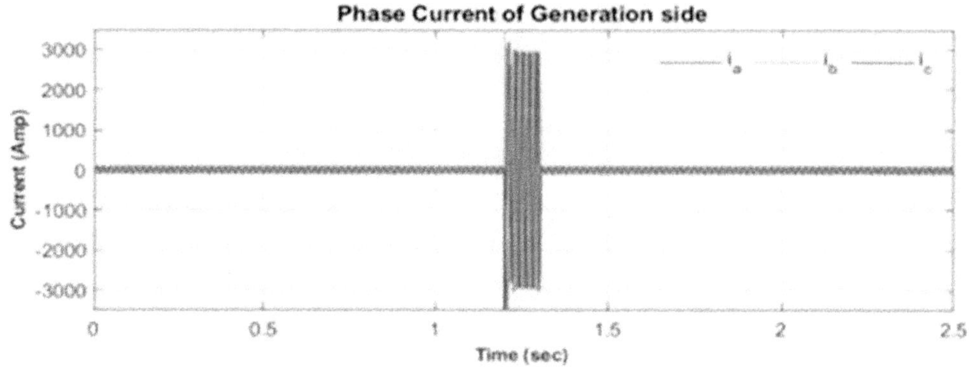

Figure 46. Simulation results at constant irradiance of 1000 w/m², constant temperature of 25°C and wind speed 12m/s respectively, waveform of phase current of generation side during three phase line to ground fault

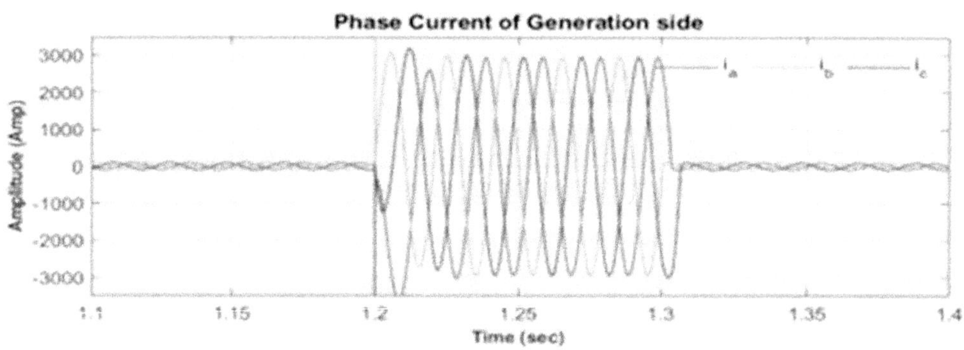

Figure 47. Simulation results at constant irradiance of 1000 w/m², constant temperature of 25°C and wind speed 12m/s respectively, waveform of phase-to-phase voltage of load

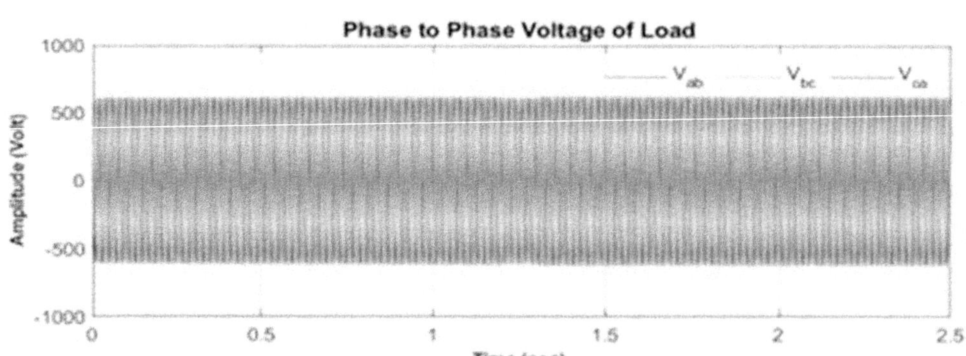

Figure 48. Simulation results at constant irradiance of 1000 w/m², constant temperature of 25°C and wind speed 12m/s respectively, waveform of phase-to-phase voltage of load during three phase line to ground fault

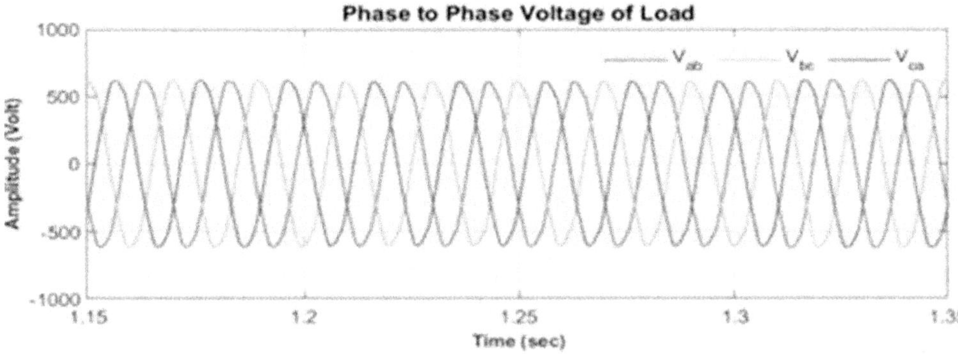

Figure 49. Simulation results at constant irradiance of 1000 w/m², constant temperature of 25°C and wind speed 12m/s respectively, waveform of phase current of load

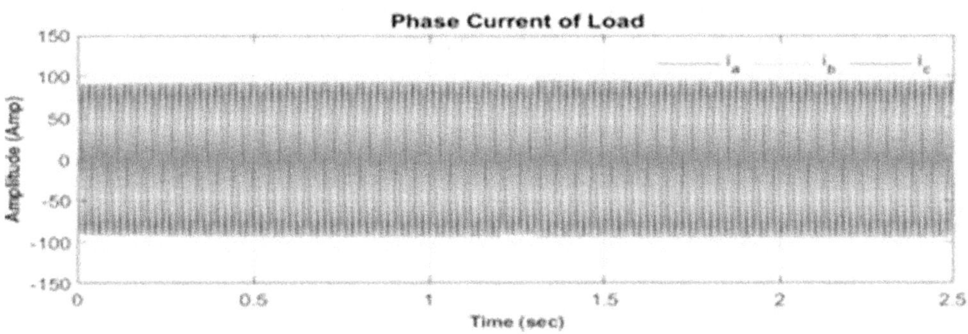

Figure 50. Simulation results at constant irradiance of 1000 w/m², constant temperature of 25°C and wind speed 12m/s respectively, waveform of phase current of load three phase line to ground fault

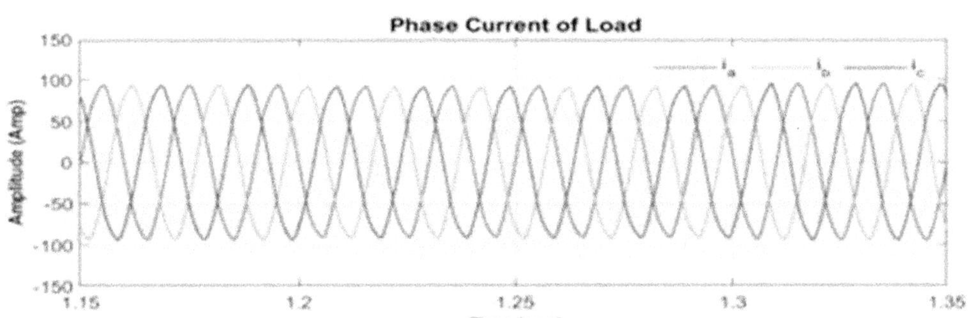

In Figure 51 both RMS phase voltage and phase current are constant throughout the time period at 440V and 65.65A respectively, experience slight changes when three phase line to ground fault occurs i.e., time t=1.2sec to t=1.3sec. This can also be observed in the zoom-in window provided in Figure 23 for load phase voltage (Li et al., 2019).

In Figure 52, we can clearly observe the distortions created by three phase line to ground fault which caused the Phase-to-phase voltage of generation side to decrease, but in the load side phase-to-phase voltage does not get affected due to symmetrical fault as shown in Figure 53, because ANN tuned DVR is used in the system which isolates the load while three phase line to ground fault occurs in the system (Hasan et al., 2020).

In Figure 54 we can clearly observe that at the time of fault ANN tuned DVR inject phase voltages in load while isolating it from generation side, which assures continues phase voltage supply to the load without distortions (Daniar et al., 2023).

In Figure 55 and Figure 56 phase current of generation side and phase current of load side is shown in per unit system (Alibegovic & Freiland Paulsson, 2021).

Figure 51. Simulation results at constant irradiance of 1000 w/m², constant temperature of 25°C and wind speed 12m/s respectively, waveform of phase-to-phase RMS voltage of load and RMS phase current of load

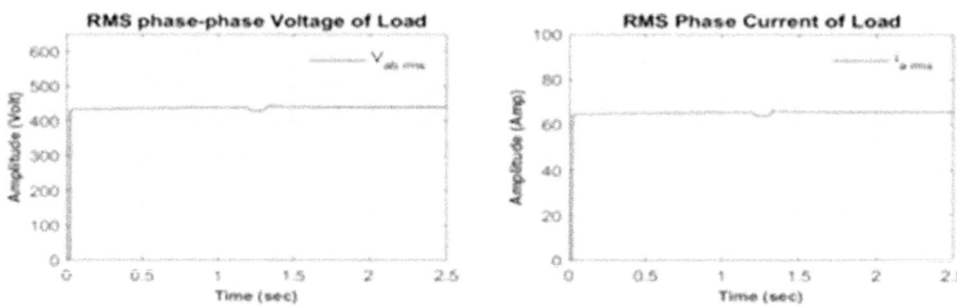

Figure 52. Simulation results at constant irradiance of 1000 w/m², constant temperature of 25°C and wind speed 12m/s respectively, waveform of phase-to-phase voltage of generation side in PU

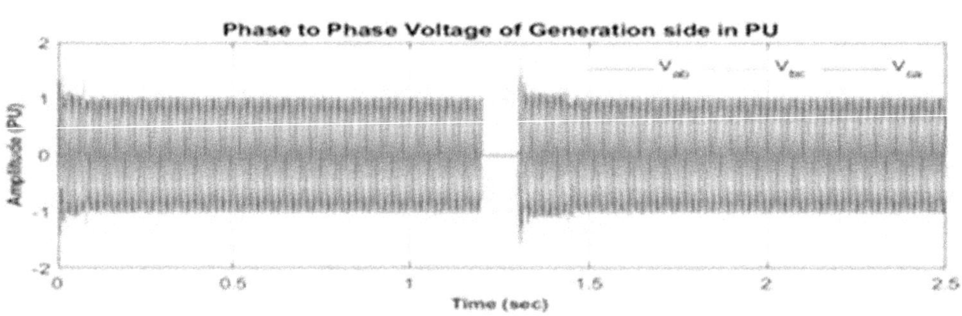

Figure 53. Simulation results at constant irradiance of 1000 w/m², constant temperature of 25°C and wind speed 12m/s respectively, waveform of phase-to-phase voltage of load side in PU

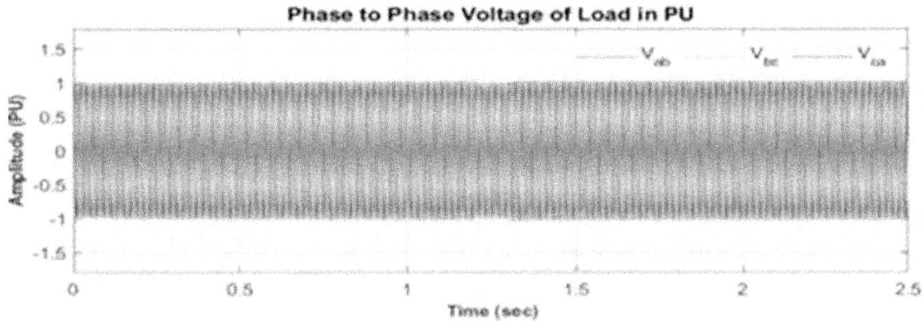

Figure 54. Simulation results at constant irradiance of 1000 w/m², constant temperature of 25°C and wind speed 12m/s respectively, waveform of injected phase-to-phase voltage in PU

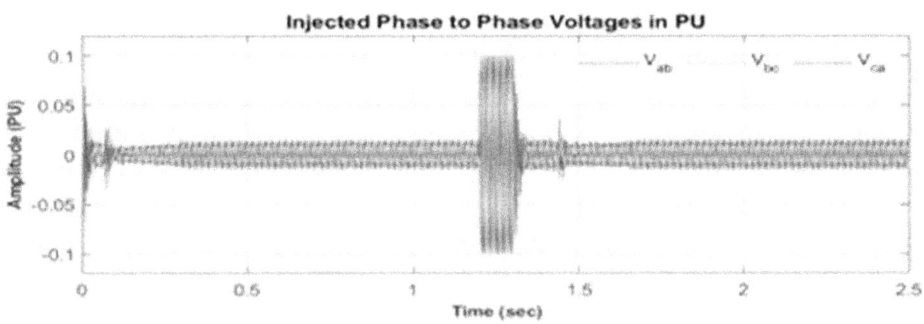

Figure 55. Simulation results at constant irradiance of 1000 w/m², constant temperature of 25°C and wind speed 12m/s respectively, waveform of phase current of generation side in PU

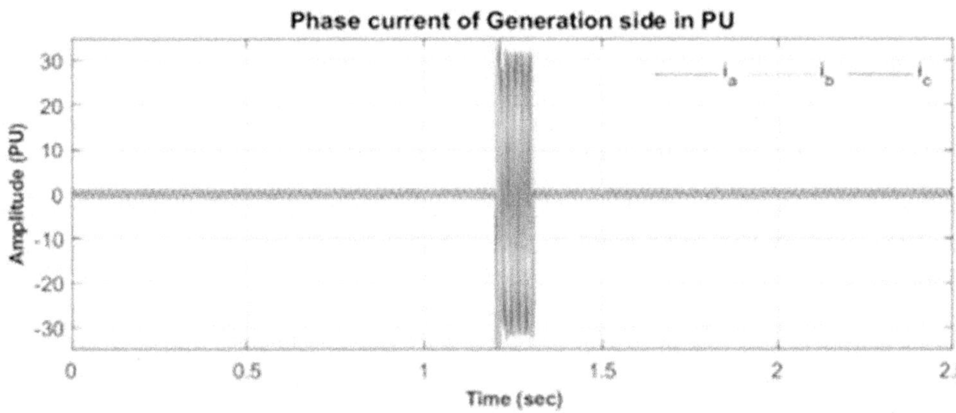

Figure 56. Simulation results at constant irradiance of 1000 w/m2, constant temperature of 25°C and wind speed 12m/s respectively, waveform of phase current of load side in PU

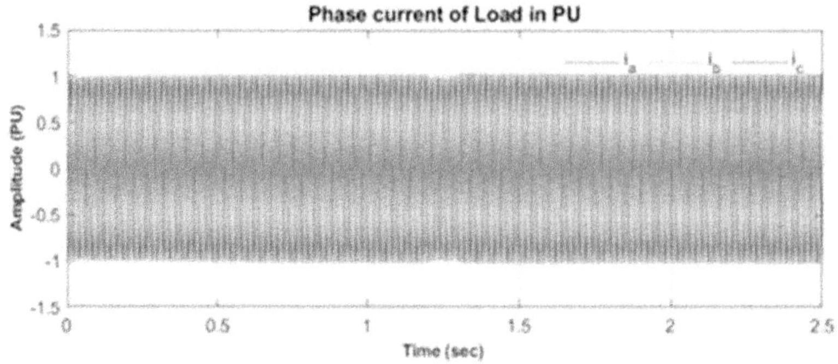

CONCLUSION

In this paper, the main objective of this thesis is to improve the power quality of the hybrid micro grid system. This is done by solving the voltage sagging and swelling problem. It also reduces total harmonic distortion due to asymmetrical conditions. Using an artificial neural network, an efficient control strategy is proposed here. This allows better voltage regulation on DC bus (±3.01V that fulfils the standards ± 5% range) than the conventional control strategy. ANN-enabled HESS control quickly responds to load and generation changes resulting in power smoothing with the added advantage of an increase in battery life span by reducing instantaneous current stress. A FLC-based VSI inverter limits voltage output to desired levels and reduces harmonic distortion within IEEE 519 limits.

REFERENCES

Ahmed, S., Benoudjafer, C., & Benachaiba, C. (2018). Mppt technique for standalone hybrid pv-wind using fuzzy controller. In Artificial Intelligence in Renewable Energetic Systems: Smart Sustainable Energy Systems (pp. 185-196). Springer International Publishing.

Al Hadi, F. M., Aly, H. H., & Little, T. (2023). Harmonic Forecasting of Wind and Solar Hybrid Model Driven by DFIG and PMSG using ANN and ANFIS. *IEEE Access : Practical Innovations, Open Solutions*.

Alexander, A., & Thathan, M. (2015). Modelling and analysis of modular multilevel converter for solar photovoltaic applications to improve power quality. *IET Renewable Power Generation*, *9*(1), 78–88.

Alibegovic, L., & Freiland Paulsson, E. (2021). *Harmonic Emissions in Microgrids*. CODEN: LUTEDX/TEIE.

Arezki, S., & Boudour, M. (2014, September). Improvement of power quality for hybrid PV-FC power supply system. In *2014 16th International Power Electronics and Motion Control Conference and Exposition* (pp. 725-730). IEEE.

Arul, P. G., Ramachandaramurthy, V. K., & Rajkumar, R. K. (2015). Control strategies for a hybrid renewable energy system: A review. *Renewable & Sustainable Energy Reviews*, *42*, 597–608.

Aworo, O. (2019). *Resilient Microgrids Using a State Controller* [Doctoral dissertation]. University of Pittsburgh.

Bahloul, I., Bouzid, M. B. K., El Khil, S. K., & Champenois, G. (2023). Robust Novel Indicator To Distinguish between an Inter-Turn Short Circuit Fault and Load Unbalance in PMSG. *IEEE Transactions on Industry Applications*.

Barrett, M. (2020, July). The early detection of stator winding shorted turns in permanent magnet three phase generators by analysis of its phase voltages. In *2020 IEEE 14th International Conference on Compatibility, Power Electronics and Power Engineering (CPE-POWERENG)* (Vol. 1, pp. 317-323). IEEE.

Bellini, A., Bifaretti, S., Iacovone, V., & Cornaro, C. (2009, September). Simplified model of a photovoltaic module. In *2009 Applied Electronics* (pp. 47-51). IEEE.

Bento, A., Luís, R., Pinto, S., & Silva, J. F. (2019, October). A novel multilevel T-Type indirect matrix converter for three-phase open-end AC loads. In *IECON 2019-45th Annual Conference of the IEEE Industrial Electronics Society* (Vol. 1, pp. 6575-6580). IEEE.

Bhukya, M. N., Kota, V. R., & Depuru, S. R. (2019). A simple, efficient, and novel standalone photovoltaic inverter configuration with reduced harmonic distortion. *IEEE Access : Practical Innovations, Open Solutions*, *7*, 43831–43845.

Calik, H., Dabakoglu, A., & Oguz, Y. (2021). Investigation of Dynamic Behaviour of Double Feed Induction Generator and Permanent Magnet Synchronous Generator Wind Turbines in Failure Conditions. *International Journal of Renewable Energy Research*, *11*, 721–729.

Chatterjee, S., Misbahuddin, M., Vamsi, P., & Ahmed, M. H. (2023, May). Power Quality Improvement and Fault Diagnosis of PV System By Machine Learning Techniques. In *2023 International Conference on Signal Processing, Computation, Electronics, Power and Telecommunication (IConSCEPT)* (pp. 1-6). IEEE.

Chauhan, A., & Saini, R. P. (2014). A review on Integrated Renewable Energy System based power generation for stand-alone applications: Configurations, storage options, sizing methodologies and control. *Renewable & Sustainable Energy Reviews*, *38*, 99–120.

Chawda, G. S., Shaik, A. G., Shaik, M., Padmanaban, S., Holm-Nielsen, J. B., Mahela, O. P., & Kaliannan, P. (2020). Comprehensive review on detection and classification of power quality disturbances in utility grid with renewable energy penetration. *IEEE Access : Practical Innovations, Open Solutions*, *8*, 146807–146830.

Chettibi, N., Mellit, A., Sulligoi, G., & Pavan, A. M. (2016). Adaptive neural network-based control of a hybrid AC/DC microgrid. *IEEE Transactions on Smart Grid*, *9*(3), 1667–1679.

Chishti, F., Murshid, S., & Singh, B. (2019). LMMN-based adaptive control for power quality improvement of grid intertie wind–PV system. *IEEE Transactions on Industrial Informatics*, *15*(9), 4900–4912.

Daniar, A., Khan, S. A., & Gardner, M. C. (2023). A Passive Inter Turn Short Circuit Fault Mitigation Technique for Electric Machines. *IEEE Transactions on Energy Conversion*.

Dida, A. (2017). *Contribution à la Commande Neuro-Floue de la Machine Asynchrone à Double Alimentation Utilisé dans un Système Eolien* [Doctoral dissertation]. Université Mohamed Khider-Biskra.

Fekry, H. M., Eldesouky, A. A., Kassem, A. M., & Abdelaziz, A. Y. (2020). Power management strategy based on adaptive neuro fuzzy inference system for AC microgrid. *IEEE Access : Practical Innovations, Open Solutions*, *8*, 192087–192100.

Fouad, K., Boulouiha, H. M., Allali, A., Taibi, A., & Denai, M. (2018). Multivariable control of a grid-connected wind energy conversion system with power quality enhancement. *Energy Systems*, *9*(1), 25–57.

Gawande, P. N., & Dambhare, S. S. (2016, May). A novel algorithm to detect symmetrical faults during power swing for a double circuit multi-terminal transmission line. In *2016 IEEE/PES Transmission and Distribution Conference and Exposition (T&D)* (pp. 1-7). IEEE.

Habib, H. U. R., Wang, S., Elkadeem, M. R., & Elmorshedy, M. F. (2019). Design optimization and model predictive control of a standalone hybrid renewable energy system: A case study on a small residential load in Pakistan. *IEEE Access : Practical Innovations, Open Solutions*, *7*, 117369–117390.

Hasan, S., Muttaqi, K. M., & Sutanto, D. (2020). Detection and characterization of time-variant nonstationary voltage sag waveforms using segmented Hilbert–Huang transform. *IEEE Transactions on Industry Applications*, *56*(4), 4563–4574.

Khasim, S. R., Dhanamjayulu, C., Padmanaban, S., Holm-Nielsen, J. B., & Mitolo, M. (2021). A novel asymmetrical 21-level inverter for solar PV energy system with reduced switch count. *IEEE Access : Practical Innovations, Open Solutions*, *9*, 11761–11775.

Kumar, V., Pandey, A. S., & Sinha, S. K. (2016, March). Grid integration and power quality issues of wind and solar energy system: A review. In *2016 International conference on emerging trends in electrical electronics & sustainable energy systems (ICETEESES)* (pp. 71-80). IEEE.

Li, G., Liang, J., Ma, F., Ugalde-Loo, C. E., & Liang, H. (2018). Analysis of single-phase-to-ground faults at the valve-side of HB-MMCs in HVDC systems. *IEEE Transactions on Industrial Electronics*, *66*(3), 2444–2453.

Li, L., Wu, M., Wu, S., Li, J., & Song, K. (2019). A three-phase to single-phase AC-DC-AC topology based on multi-converter in AC electric railway application. *IEEE Access : Practical Innovations, Open Solutions*, *7*, 111539–111558.

Liu, J., Zhao, C., & Xie, Z. (2021). Power and current limiting control of wind turbines based on PMSG under unbalanced grid voltage. *IEEE Access : Practical Innovations, Open Solutions*, *9*, 9873–9883.

Moawwad, A., El Moursi, M. S., & Xiao, W. (2016). Advanced fault ride-through management scheme for VSC-HVDC connecting offshore wind farms. *IEEE Transactions on Power Systems*, *31*(6), 4923–4934.

Nam-In, T., & Kumsuwan, Y. (2020, June). PWM Techniques for a 13-Level Asymmetrical Dual Inverter Configuration Without the Overcharging of Floating Capacitor. In *2020 17th International Conference on Electrical Engineering/Electronics, Computer, Telecommunications and Information Technology (ECTI-CON)* (pp. 535-538). IEEE.

Rai, H. M., Trivedi, A., & Shukla, S. (2013). ECG signal processing for abnormalities detection using multi-resolution wavelet transform and Artificial Neural Network classifier. *Measurement*, *46*(9), 3238–3246.

Roy, P., He, J., Zhao, T., & Singh, Y. V. (2022). Recent advances of wind-solar hybrid renewable energy systems for power generation: A review. *IEEE Open Journal of the Industrial Electronics Society*, *3*, 81–104.

Roy, R. B., Rokonuzzaman, M., Amin, N., Mishu, M. K., Alahakoon, S., Rahman, S., ... Pasupuleti, J. (2021). A comparative performance analysis of ANN algorithms for MPPT energy harvesting in solar PV system. *IEEE Access : Practical Innovations, Open Solutions*, *9*, 102137–102152.

Sadeghkhani, I., Golshan, M. E. H., Guerrero, J. M., & Mehrizi-Sani, A. (2016). A current limiting strategy to improve fault ride-through of inverter interfaced autonomous microgrids. *IEEE Transactions on Smart Grid*, *8*(5), 2138–2148.

Sahoo, S. K., Somkun, S., Velu, S. K., Padhan, S., Jeyaraj, B. E., & Puppala, R. (2023). *Recent Advances in Renewable Energy Automation and Energy Forecasting*. Academic Press.

Swapna, V., & Gayatri, M. T. L. (2021, November). Power Quality Enhancement using Hybrid Energy Storage based Dynamic Voltage Restorer for Grid Integrated Distributed Generation. In *2021 IEEE Bombay Section Signature Conference (IBSSC)* (pp. 1-5). IEEE.

Tanaka, T., Ma, K., Wang, H., & Blaabjerg, F. (2018). Asymmetrical reactive power capability of modular multilevel cascade converter based STATCOMs for offshore wind farm. *IEEE Transactions on Power Electronics*, *34*(6), 5147–5164.

Tata, H., Madeti, S. R. K., & Sai Veerraju, M. (2022, August). Power Quality Improvement of a Hybrid Renewable Energy Systems Using Dynamic Voltage Restorer with PI Controller. In *International Conference on Recent Advances in Mechanical Engineering Research and Development* (pp. 549-559). Singapore: Springer Nature Singapore.

Xue, C., Wang, J., & Li, Y. (2022). Model predictive control for grid-tied multi-port system with integrated PV and battery storage. *IEEE Transactions on Smart Grid*, *13*(6), 4596–4609.

Zanib, N., Batool, M., Riaz, S., & Nawaz, F. (2022). Performance analysis of renewable energy based distributed generation system using ANN TUNED UPQC. *IEEE Access : Practical Innovations, Open Solutions*, *10*, 110034–110049.

Zhang, Q., He, J., Xu, Y., Hong, Z., Chen, Y., & Strunz, K. (2021). Average-value modeling of direct-driven PMSG-based wind energy conversion systems. *IEEE Transactions on Energy Conversion*, *37*(1), 264–273.

Zhang, S., Du, P., Chen, C., Zhong, W. D., & Alphones, A. (2019). Robust 3D indoor VLP system based on ANN using hybrid RSS/PDOA. *IEEE Access : Practical Innovations, Open Solutions*, *7*, 47769–47780.

Zichang, L., Yadong, L., Yingjie, Y., Peng, W., & Xiuchen, J. (2020). An identification method for asymmetric faults with line breaks based on low-voltage side data in distribution networks. *IEEE Transactions on Power Delivery*, *36*(6), 3629–3639.

Chapter 12
Optimization in the Selection and Use of Audiovisual Resources for Superior Education:
A Process Proposal

Luis A. Millan-Tudela
https://orcid.org/0000-0003-3669-9960
University of Alicante, Spain

Bartolomé Marco-Lajara
https://orcid.org/0000-0001-8811-9118
University of Alicante, Spain

Javier Martínez-Falcó
https://orcid.org/0000-0001-9004-5816
University of Alicante, Spain

Eduardo Sánchez-García
https://orcid.org/0000-0002-8524-4906
University of Alicante, Spain

ABSTRACT

The use of audiovisual content (short self-elaborated videos, movie extracts, and similar) in teaching has been adopted since the digitalization in education facilities took place. However, the inclusion of these materials should be assessed to check for the fit with the learning goals established in each learning environment. This chapter proposes a methodology to validate the adequacy and the level of congruence of a given audiovisual content with those objectives, providing a deep analysis on the requirements that the selection of materials should meet.

DOI: 10.4018/978-1-6684-9601-5.ch012

INTRODUCTION

Lecture teaching, widely extended in many higher education systems, is usually dominated by the unidirectionality of the professor's discourse to the students. In essence, lecture teaching is identified as the classic formula of classroom education, consisting of the elaboration of some theoretical materials, according to the contents of the subject in question, which are explained to the student. However, this approach, in its most elementary form, usually generates problems in learning by not including elements that promote discussion, as well as by not offering intervals of disconnection from the theoretical contents that allow students to assimilate them.

This type of teaching, however, is indispensable as part of formal education systems, and should not be understood as something opposed to other methodologies, but complementary to them. This is why numerous academics from different fields have proposed tools that lead to the enrichment of this type of teaching. For example, Brame (2016) proposes "instructional activities involving students in doing things and thinking about what they are doing". Baños and Farré (2011) propose the inclusion of short case studies in the subject of Pharmacology. Caballero Miguez and Garza Gil (2012) defend "cooperative work and participatory content development" in the field of Economics as an innovative approach. In a similar vein, Martí et al. (2007) advocate project-based learning (PBL), presenting the results obtained in the area of Computer Graphics. Other authors, however, do not share this complementary view. Based on empirical results, Fernández et al. (2010) propose a transition from lectures to active methodologies and continuous evaluation, while Fernández Jambrina (2013) postulates the replacement of face-to-face teaching by online sessions for the first stages of training at the university level. Similarly, Luján-Mora (2013) stresses the use of Massive Open Online Courses (MOOCs) as a substitute for lectures. In a more general sense, Oye et al. (2012) defend e-learning as it "allows users to fruitfully gather knowledge and education both by synchronous and asynchronous methodologies to effectively face the need to rapidly acquire up to date know-how within productive environments".

Although proposals such as the above can be grouped according to the main objective of complementing or substituting lectures, it is evident that the existence of new technologies offers an enriching support for learning, regardless of the modality and area of knowledge. Among these, one of the easiest proposals to implement is the inclusion of audiovisual content that helps to exemplify, deepen, and discuss the theoretical contents of lectures. However, the authors of this chapter share the idea that the use of these resources should not be indiscriminate but should be aligned with a series of precepts that guarantee their adequate use. In this sense, a decision system is proposed that allows the teacher to make an adequate and effective selection of the audiovisual contents to be included.

For this purpose, the research is divided into the following sections. Section 2 sets out the objectives pursued in this work, both in terms of the development of the decision process and its contrast. Section 3 presents both the design of the process and the statistical tools that will allow its evaluation in later stages of the research. Section 4 contains a summary of the results of this work, as well as a series of conclusions that allow to connect them with the previous development. Finally, in Section 5 a series of discussions is developed. This discussion, that act as a complement to the conclusions, facilitate a better design of the sessions and teaching materials based on the use of audiovisual supports and contents through the proposed process.

OBJECTIVES

The general objective (GO) of this research is the development of a sequential process that, through the verification and filtering based on pre-established criteria, makes it possible to assess the suitability of audiovisual materials for use in the teaching of a given subject.

In order to achieve this, a series of specific objectives (SO) are also established. Through these, a greater degree of specificity is achieved and, consequently, a greater ease in defining the actions to be carried out. These specific objectives are listed and explained below.

SO_1: establishment of filtering criteria. To ensure that the audiovisual contents complementary to lectures have an affinity, as well as an effective and efficient impact on student learning, they must be aligned both with the contents to be taught and with the skills (both soft and hard) to be developed in the students according to the studies being taught.

SO_2: selection of the contrasting methodology. Although the details of this specific objective are out of the scope of this work, it must be statistically corroborated that the process developed generates a real benefit in the learning process. For this purpose, the statistical technique to be used to verify whether the process is truly effective must be determined.

SO_3: devising a validation scale. In support of SO_2, a series of dimensions to be measured should be established as a mean to obtain data on the results perceived by the students through the collection of primary information.

METHODOLOGY

Process Development

One of the main conditions for the inclusion of materials in teaching is that they should be aligned both with the contents to be taught and with the skills to be developed in students. Indeed, the materials must be related to the contents, as it is understood that they should be complementary to them. However, the range of skills to be developed can go unnoticed, being a key element in higher education nowadays.

Currently, phenomena such as globalization and digitalization (among others) are causing what is known as a VUCA environment (Volatile, Uncertain, Complex and Ambiguous) (Guerras Martín & Navas López, 2022). Being deeply studied in social sciences, the existence of this type of environment leads to greater demands for flexibility, adaptation, and resilience by people, also in their careers. As a consequence, the different higher education institutions are developing a series of regulations for the development of skills in students that, although not directly related to the regulated contents to be taught, allow the development of a series of skills that reinforce their professional aptitudes. Therefore, the contents to be assessed through the process developed must include an assessment of these skills.

It should be noted that, although it is highly advisable to link the materials to the skills, it is possible that some audiovisual materials facilitate the assimilation of contents without addressing any of the target skills, so that this last point should be considered as part of a more positive, but not exclusive, assessment.

Once the previous points have been assessed, the next question to be answered is the characteristics of the format. In this regard, the duration of the content (in the case of recordings) should be evaluated first. Although no threshold is established at this stage of the research, the professor must assess whether the length of the materials is, on the one hand, sufficient to allow an effective transmission of knowledge and, on the other hand, adequately concise to avoid distracting the students. If the materials are too long, the teacher should plan to (1) edit them, or (2) eliminate them if this is not possible.

In addition, there is the matter of available resources. It must be assessed whether the material with potential use based on the above criteria is easily accessible with the means available in the classroom, as well as for students outside of it.

This process can be represented in the form of a decision tree, as shown in Figure 1.

Statistical Assessment Tool

In order to check whether the proposed process implies an improvement in the design and development of the teaching sessions, after its implementation, information

Figure 1. Proposed process for the optimization of audiovisual content
Source: Own elaboration

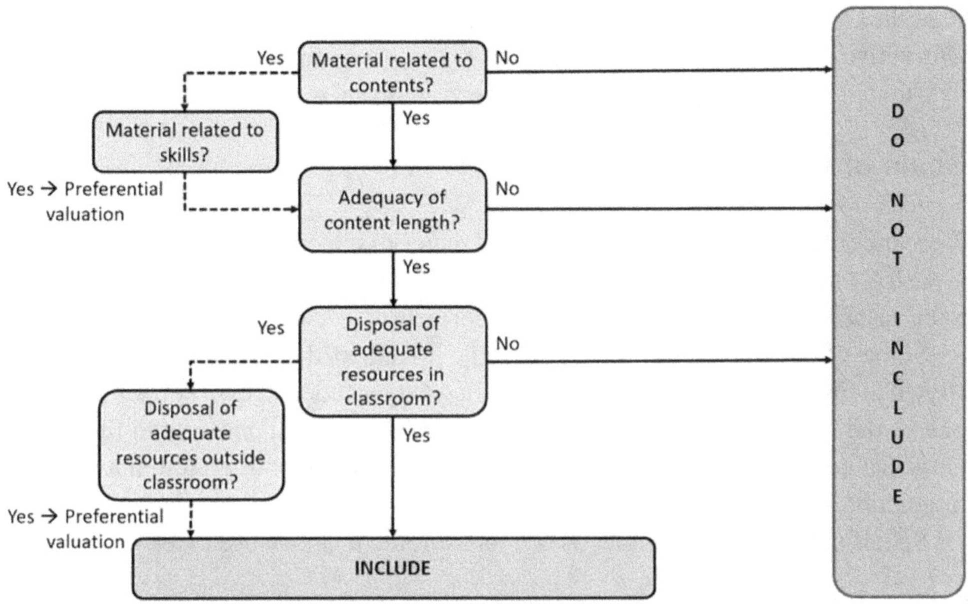

must be collected from the students and then compared with the results obtained in the subjects in question.

Among the different tools to be applied, the most appropriate are those known as structural equation models (SEM) and, within these, those that apply the partial least squares (PLS) algorithm. This is due to the concurrence of several circumstances.

The first is that student consultation is based on student opinion, which implies the measurement of perceptions, attitudes and beliefs. For this purpose, PLS-SEM is a suitable element that allows the assessment of these types of latent variables (i.e., not directly measurable) through a series of manifest variables (Hair et al., 2022), which usually take the form of questionnaires. Furthermore, as indicated by these authors, PLS-SEM works properly with constructs that can be shaped both reflectively and formatively, which allows the validation model to be modified by other researchers more easily in the face of new scientific evidence that may require it.

Another reason is the ability to model complex relationships, so that a multitude of connections between constructs can be represented in order to maximize the explained variance of the endogenous latent variable (Hair et al., 2022), which in this case is student performance based on their perceptions of the audiovisual materials used.

Finally, an issue of utmost importance is that PLS-SEM works well with small sample sizes and non-normally distributed variables (Hair et al., 2022). This is crucial since the data must be obtained directly from the students present in the teaching sessions, which means that the sample will be quantitatively limited. In addition, the fact of being able to deal with non-normal samples offers a greater degree of security regarding its proper application, since, unlike other types of research, there are no alternatives for data collection in the event of non-compliance with the normally distributed data assumption after the collection of the responses.

Scale of Validation

Once the criteria to be met by the process of material selection have been determined, as well as the statistical methodology for verifying their effectiveness, the dimensions to be measured for this purpose must be defined.

Some of these dimensions are directly linked to the different testing stages shown in Figure 1. However, others are aimed at the students' own perception, since one of the objectives in the implementation of audiovisual contents in the master class is a positive appreciation by the students, so that their attention, interest, and interaction are increased.

Specifically, the dimensions whose measurement is proposed are:

- Affinity
- Format
- Accessibility
- Attractiveness
- Performance

Below, a justification for the need to evaluate each of these dimensions is provided. It should be noted, however, that all of them should be measured in a reflective manner, since, due to their nature, it is quite likely that there will be a high overlap between the different manifest variables (indicators) belonging to the same dimension.

Dimension 1: Affinity

In this context, affinity refers to the degree of alignment of the audiovisual materials used with both the contents and the target skills, which involves evaluating the first phase of the model developed. In this sense, it should be considered that it is essential that the association of the materials with the training objectives is made not only by the professors, but also by the students, since they are the final addressees. Therefore, it is essential to check the suitability of such contents, since it is possible that their use does not generate the mental links estimated by the teacher.

For this purpose, and as in the case of the decision-making process, both contents and skills must be considered. Since these are different in each subject and training plan, it is crucial to make them known to the students prior to their evaluation. In fact, the items used in the questionnaire should explicitly specify that the questions are addressed to what happened in the target subject, since sometimes there is an overlapping of contents and skills between subjects of the same training plan with their reinforcement as the main objective.

Dimension 2: Format

The second phase of the proposed process refers to the appropriateness of the format, understanding as such the duration-storyline binomial. Specifically, it is possible that an excessive (or insufficient) duration may lead to distraction (or non-comprehension) and, consequently, to the non-achievement of learning. Since audiovisual materials can either remain in their original format or be modified to improve their appropriateness, it is important to verify that the version offered to the students allows them to consolidate learning to a greater extent than the version not used.

To this end, it is necessary to collect information on the perception of the duration of the audiovisual content, both individually and as a whole (to detect possible saturation), as well as the quality of the content in terms of the storyline. This last point should be developed for two main reasons: (1) the original contents may be confusing in themselves, and (2) possible editions of the contents may generate a disconnection between the different parts of which they are composed.

Dimension 3: Accessibility

The last point linked to a phase of the designed process is the accessibility of the contents. Here, different aspects must be checked. Thus, in addition to the possibility of being used both in the classroom and on private equipment, the presence of subtitles, whether the video and/or audio quality is adequate, as well as its availability on a permanent basis, among others, must be added.

This is because the materials must remain accessible in a variety of situations and circumstances, so they must have different features that allow the transmission of the inherent knowledge regardless of the user, which in turn facilitates the creation of an inclusive learning environment. In addition, many of the materials used in learning sessions are used by students in their professional careers (especially in the early stages), so it is essential that they allow access to the same information regardless of the time elapsed since their use in the academic context.

Dimension 4: Attractiveness

Once the data on the different aspects directly linked to the designed process have been collected, a series of tests must be carried out on the students' opinion regarding the contents. As previously indicated, one of the premises on which the inclusion of audiovisual materials is based is their capacity to capture the students' attention, so that an effective transmission of the contents is more likely to be achieved. Therefore, it is not enough that they meet the requirements set for the selection process, but they must generate an attraction effect. Consequently, the quality of attractiveness must be duly assessed.

In later stages of the research, the specific aspects (manifest variables) to be measured must be defined. However, the closeness to the socio-cultural reality of the students, the curiosity generated in them, and the degree of originality perceived can be listed as essential. As can be seen, attractiveness is not limited to a single quality, but should encompass different perspectives that allow a holistic view of the students' global perception to be obtained.

Dimension 5: Performance

Since the ultimate objective of this project is to improve learning, some measure must be used to assess whether the impact of the methodology developed is positive based on statistical significance criteria. For this purpose, performance measurement is considered to be the most efficient way to achieve it.

However, its valuation cannot be established exclusively on the basis of academic results in the form of academic grading. Although this is an indicator with a high potential for application, learning is not reflected solely through the evaluation that each student finally obtains. As mentioned in previous sections, both generic and specific (or transversal) skills are added to the contents of the subject, and many times they are not generally included in the students' grade. Taking into account the complementary role of audiovisual materials in lectures, specific indicators should be used so that, in addition to the formal results (grades), allow developing a measurement that is closer to the reality of everything assimilated by the students. This entails assessing their perceptions of other types of learning that, indirectly, have been obtained as a result of the design of the sessions proposed.

RESULTS AND CONCLUSION

By means of the present research, a decision process that makes it possible to evaluate the appropriateness of audiovisual content in lectures has been developed.

In addition, the use of a PLS-SEM model has been justified in order to assess the effect of the proposed process on learning after its implementation. Finally, the concrete elements to be measured, i.e., the dimensions to be considered in order to assess the degree of success of the proposal, have been defined.

All these results, moreover, have the advantage of being based on a generic model of action, which allows them to be applied to different fields of knowledge, regardless of their nature. What is more, the fact that no thresholds are established within the process allows it to be adapted to the needs of each learning situation, and some thresholds may be established in subsequent research based on this proposal.

However, the work has some limitations. For example, it is possible that not setting validity thresholds, while facilitating its implementation in modalities with very different characteristics, may lead to a certain laxity in the process. Furthermore, the fact that the validity has not been statistically tested (since data must be obtained after a pilot implementation) means that the results obtained should be interpreted as provisional, pending the completion of the next stages of the research, which have nevertheless been outlined in the most basic terms in order to offer a roadmap.

DISCUSSION

Education is one of the fundamental pillars of societies. Indeed, their advancement has been linked to the learning of their members, whether formal or informal (e.g., experiential). However, this must be a two-way process. That is, developments in the environment itself (social, economic, technological, etc.) can offer a series of tools to optimise the learning process, and the latter, therefore, can offer tools to promote these developments.

In the area of education, the development of technology in recent decades has led to a process of transformation at the teaching level. In the field of formal higher education, a multitude of prestigious institutions are developing open training courses in the form of MOOCs with a worldwide scope and asynchronously, establishing collaborative teaching networks with live (synchronous) interventions between teachers from different regions and thus enabling an enriching exchange of knowledge and experience.

This evolution of technology, together with research in education, has led to the creation and development of teaching methodologies that are now widely applied. In addition to the aforementioned, simple tools such as project-based learning, gamification or collaborative learning (all of them listed in the section "Introduction") can be highlighted. These techniques are often framed within the continuous assessment modality, which allows the compartmentalisation of content to facilitate its assimilation. However, many of them seem to be contrary to lecture-based teaching (understood as a traditional modality) when the latter has its own advantages.

Among its virtues is the fact that it is a standardised type of teaching, which allows students to maintain a certain consistency in their study techniques between subjects. In addition, lecture-based teaching is often developed through the use of similar content regardless of the professor, which provides equal opportunities for students to access knowledge. However, it is also characterised by rigidity and monotony on many occasions. Precisely, the use of new technologies can help to exploit its strengths and avoid its weaknesses. In essence, starting from lecture-based sessions (the core of education), professors can use the different teaching innovation methodologies that have been developed in a complementary way, in order to adapt this base to their own style and thus achieve a better learning result.

Among the different proposals, the one presented in this paper aims to serve as an auxiliary tool for lectures, helping to optimise the design of teaching materials, reinforcing the transmission of contents, and facilitating the acquisition of skills. Adequate design of materials is essential, as they are an indispensable part of the communication channel between teacher and student. Therefore, their creation should not be based on fully systematised processes that affect the interaction between the parties, but neither should it be based exclusively on perceptions, as important elements may be omitted. In addition, many of the materials developed are subsequently used by students in their careers, so that their enrichment must be accompanied by content sufficiency, i.e., they must have all the necessary information to be able to perform the most basic professional tasks, as well as to facilitate the search for more advanced information.

Finally, it should be mentioned that the use of methodologies such as those shown in "Introduction", together with tools such as the one proposed, should be re-evaluated on the basis of incipient technological developments. The first half of the 2020s is being marked by an exponential development of artificial intelligence, to such an extent that there are freely available programmes that make it possible to automate the writing of papers, the preparation of presentations, the translation of documents, the processing of data and even the interpretation of content. These tools, if used improperly, interfere with the main objective of learning based on the criteria of autonomy, effort, and critical thinking, just mentioning some. However, their use can open the door to new ways of knowledge acquisition, as they offer a capacity for data processing that, on most occasions, is not attainable by human capabilities.

Within these new possibilities, it is worth noting that outsourcing certain activities to tools such as those based on artificial intelligence can lead to significant time savings in the development of both materials and teaching sessions themselves. Given the limitations that often arise in terms of time available in the classroom, as well as professors' working hours, the use of these tools in the development of the most automatic and monotonous tasks allows redirecting efforts towards the development of the key elements in learning. In other words, to the creation of enriching learning sessions that address the critical concepts to be acquired by students (regarding both the subject and the skills to be developed) and that encourage their autonomy. To this end, the process proposed in this chapter should be understood as a tool for this purpose, as it allows for the evaluation of audiovisual content, both in its original form and in its editing by both individuals and artificial intelligence programmes. Thus, the use of these tools makes it possible to focus attention on the proposed process, thus contributing to a much more in-depth and thoughtful evaluation of the audiovisual materials included in the documents shared with the students and, therefore, to an educational proposal that is both effective and more efficient than what has been achieved so far.

REFERENCES

Baños, J. E., & Farré, M. (2011). Dinamización de la clase magistral en medicina: Diez ejemplos de minicasos utilizados en la docencia de la farmacología. *Educación Médica, 14*(2), 105–112.

Brame, C. (2016). Active learning. Vanderbilt University Center for Teaching.

Caballero Miguez, G., & Garza Gil, M. D. (2012). Innovando la docencia superior en Economía: trabajo cooperativo y elaboración participativa de contenidos. REDU: Revista De Docencia Universitaria.

Fernández, L., Yepes, M., & Fernández, L. (2010). Docencia de Electrónica Física: Un cambio desde las clases magistrales a metodologías activas y evaluación continua. *Docencia de Electrónica Física: Un cambio desde las clases magistrales a metodologías activas y evaluación continua,* 179-190.

Fernández Jambrina, L. (2013). Docencia no presencial como alternativa a la clase magistral en los primeros cursos de ingeniería. Academic Press.

Guerras Martín, L. Á., & Navas López, J. E. (2022). *La dirección estratégica de la Empresa: Teoría y aplicaciones* (6th ed.). Thomson Reuters.

Hair, J. F., Hult, G. T. M., Ringle, C. M., & Sarstedt, M. (2022). *A primer on Partial Least Squares Structural Equation Modeling (PLS-SEM)* (3rd ed.). SAGE Publications Inc.

Luján-Mora, S. (2013). De la clase magistral tradicional al MOOC: Doce años de evolución de una asignatura sobre programación de aplicaciones web. *REDU. Revista De Docencia Universitaria*, *11*, 279–300. doi:10.4995/redu.2013.5557

Martí, E., Julià, C., & Gil, D. (2007). *PBL en la docencia de Gráficos por Computador: Una comparativa con la clase magistral.* JAC.

Oye, N. D., Salleh, M., & Iahad, N. A. (2012). E-learning methodologies and tools. *International Journal of Advanced Computer Science and Applications*, *3*(2).

KEY TERMS AND DEFINITIONS

Artificial Intelligence: Software that, through emulating human reasoning, develop certain tasks that may be impossible or extremely time-consuming for human beings.

Contents: Explicit knowledge that must be acquired by students through a certain academic subject in the context of their education, and that has been explicated through a formal contract.

MOOC: Acronym for Massive Online Open Course, a learning modality that consists of attending to a course that is entirely developed on the Internet, with non-restriction access, and that may be certified for free or by paying a fixed amount.

Project-Based Learning: Methodology that consists of the acquisition of knowledge indirectly through the development of projects.

Skills: In the context of the chapter, human aptitudes that are needed in order to develop tasks in a work environment but are not directly related with the technical knowledge required for the position in question.

Compilation of References

Abbasi, M. H., Hosseinlou, M. H., & Fadaki, S. M. J. (2020). An investigation of Bus Rapid Transit system (BRT) based on economic and air pollution analysis. *Case Studies on Transport Policy*, *8*(2), 553–563. doi:10.1016/j.cstp.2019.11.008

Abbas, M. K., Hassan, Q., Tabar, V. S., Tohidi, S., Jaszczur, M., Abdulrahman, I. S., & Salman, H. M. (2023). Techno-economic analysis for clean hydrogen production using solar energy under varied climate conditions. *International Journal of Hydrogen Energy*, *48*(8), 2929–2948. doi:10.1016/j.ijhydene.2022.10.073

About Resource Smart Schools. (2019). Available online: https://www.sustainability.vic.gov.au/School/About-ResourceSmartSchools

Acar-Erdol, T., & Gözütok, F. D. (2019). Toplumsal cinsiyet eşitliği eğitim programının değerlendirilmesi [Evaluation of gender equality curriculum]. *İlköğretim Onlne, 18*(4), 1497-1519.

Acar-Erdol, T., & Gözütok, F. D. (2017). Ortaöğretim öğrencileri için toplumsal cinsiyet eşitliği öğretim programı ihtiyaç analizi:Bir Anadolu lisesi örneği [Needs assessment for gender equality curriculum among high school students: An example of an Anatolian High School]. *Eğitim ve Bilim*, *42*(190), 39–65. doi:10.15390/EB.2017.6429

Adewopo, J. B., Solano-Hermosilla, G., Colen, L., & Micale, F. (2021). Using crowd-sourced data for real-time monitoring of food prices during the COVID-19 pandemic: Insights from a pilot project in northern Nigeria. *Global Food Security*, *29*, 100523. Advance online publication. doi:10.1016/j.gfs.2021.100523 PMID:34178595

Aguirre, J. M., & Garcia, L. M. (2021). Integrating sustainability education into teacher preparation programs: A systematic literature review. *Sustainability*, *13*(9), 4813.

Ahanchian, M. R., & Soleimani, E. (2017). The perception of faculty members on being professional in Iran's higher education system. *IRPHE*, *23*(2), 1–23.

Ahmad, M., & Khiyal, M. S. H. (2023). Assessment of land administration in Pakistan and the potential role of volunteered geographic information. In Handbook of Research on Driving Socioeconomic Development With Big Data. https://doi.org/ doi:10.4018/978-1-6684-5959-1.ch014

Ahmad, M., Khayal, M. S. H., & Tahir, A. (2022). Analysis of Factors Affecting Adoption of Volunteered Geographic Information in the Context of National Spatial Data Infrastructure. *ISPRS International Journal of Geo-Information, 11*(2). Advance online publication. doi:10.3390/ijgi11020120

Ahmed, S., Benoudjafer, C., & Benachaiba, C. (2018). Mppt technique for standalone hybrid pv-wind using fuzzy controller. In Artificial Intelligence in Renewable Energetic Systems: Smart Sustainable Energy Systems (pp. 185-196). Springer International Publishing.

Ahmed, A. A., & Ahmed, A. E. A. A. (2021). Experimenting the strategy of climbing the plateau in teaching history to develop cognitive achievement, meta cognitive skills and mental alertness among first-grade secondary students. *Journal of Scientific Research in Education, 22*(2), 431–461.

Al Hadi, F. M., Aly, H. H., & Little, T. (2023). Harmonic Forecasting of Wind and Solar Hybrid Model Driven by DFIG and PMSG using ANN and ANFIS. *IEEE Access : Practical Innovations, Open Solutions.*

Alaa, M. (2021). Experimenting the Strategy of Climbing the Plateau in Teaching History to Develop Cognitive Achievement, Meta-cognitive Skills and Mental Alertness among First-grade Secondary Students. *Journal of Scientific Research in Education, 22*(2), 431–461.

Alamir, N., Kamel, S., Megahed, T. F., Hori, M., & Abdelkader, S. M. (2023). Developing Hybrid Demand Response Technique for Energy Management in Microgrid Based on Pelican Optimization Algorithm. *Electric Power Systems Research, 214*, 108905. doi:10.1016/j.epsr.2022.108905

Al-Anbaki, A., & Al-Karkhi, A. (2015). The effect of the hill-climbing strategy on the achievement of fifth-grade literary students in modern and contemporary European history. *Diyala Journal*, (65), 343–371.

Al-Azawi, R., Al-Faliti, F., & Al-Blushi, M. (2016). Educational gamification vs. game based learning: Comparative study. *International Journal of Innovation, Management and Technology, 7*(4), 132–136. doi:10.18178/ijimt.2016.7.4.659

Alexander, A., & Thathan, M. (2015). Modelling and analysis of modular multilevel converter for solar photovoltaic applications to improve power quality. *IET Renewable Power Generation, 9*(1), 78–88.

Alibegovic, L., & Freiland Paulsson, E. (2021). *Harmonic Emissions in Microgrids.* CODEN: LUTEDX/TEIE.

Ambusaidi, A., Al-Dairiya, H., & Al-Rashidi, H. (2023). *Sustainable Development in School Environment.* Al Maseera House for Publication and Distribution.

Anđelković, A. S., & Bajatović, D. (2020). Integration of weather forecast and artificial intelligence for a short-term city-scale natural gas consumption prediction. *Journal of Cleaner Production, 266*, 122096. doi:10.1016/j.jclepro.2020.122096

Appiah-Otoo, I., Song, N., Acheampong, A. O., & Yao, X. (2022). Crowdfunding and renewable energy development: What does the data say? *International Journal of Energy Research*, *46*(2), 1837–1852. doi:10.1002/er.7301

Aratemur-Çimen, C., & Bayhan, S. (2018). *Değişen ders kitaplarında sekülerizm ve toplumsal cinsiyet eşitliği araştırması*. Karşılaştırmalı Eğitim Derneği.

Arbia, G., Solano-Hermosilla, G., Nardelli, V., Micale, F., Genovese, G., Amerise, I. L., & Adewopo, J. (2023). From mobile crowdsourcing to crowd-trusted food price in Nigeria: Statistical pre-processing and post-sampling. *Scientific Data*, *10*(1), 446. doi:10.103841597-023-02211-1 PMID:37438443

Arezki, S., & Boudour, M. (2014, September). Improvement of power quality for hybrid PV-FC power supply system. In *2014 16th International Power Electronics and Motion Control Conference and Exposition* (pp. 725-730). IEEE.

Arul, P. G., Ramachandaramurthy, V. K., & Rajkumar, R. K. (2015). Control strategies for a hybrid renewable energy system: A review. *Renewable & Sustainable Energy Reviews*, *42*, 597–608.

Ashfaq, A., Kamran, M., Rehman, F., Sarfaraz, N., Ilyas, H. U., & Riaz, H. H. (2022, March). Role of Artificial Intelligence in Renewable Energy and its Scope in Future. In *2022 5th International Conference on Energy Conservation and Efficiency (ICECE)* (pp. 1-6). IEEE. 10.1109/ICECE54634.2022.9758957

Ashwin, P. & McVitty, D. (2015). The Meanings of Student Engagement: Implications for Policies and Practices. In Curaj, A., Matei, L., Pricopie, R., Salmi, J., Scott, P. (eds) The European Higher Education Area. Springer, Cham. https://doi.org/10.1007/978-3-319-20877-0_23

Asiama, K., Bennett, R., & Zevenbergen, J. (2017). Participatory Land Administration on Customary Lands: A Practical VGI Experiment in Nanton, Ghana. *ISPRS International Journal of Geo-Information*, *6*(7), 186. Advance online publication. doi:10.3390/ijgi6070186

Aslan, G. (2015). Öğretmen adaylarının toplumsal cinsiyet algılarına ilişkin metaforik bir çözümleme [A metaphoric analysis regarding gender perceptions of preservice]. *Eğitim ve Bilim*, *40*, 363–384. doi:10.15390/EB.2015.2930

Association for the Advancement of Sustainability in Higher Education. (n.d.). *About*. Retrieved July 25, 2023, from https://www.aashe.org/

Aworo, O. (2019). *Resilient Microgrids Using a State Controller* [Doctoral dissertation]. University of Pittsburgh.

Bahloul, I., Bouzid, M. B. K., El Khil, S. K., & Champenois, G. (2023). Robust Novel Indicator To Distinguish between an Inter-Turn Short Circuit Fault and Load Unbalance in PMSG. *IEEE Transactions on Industry Applications*.

Bai, S., Hew, K. F., & Huang, B. (2020). Does gamification improve student learning outcome? Evidence from a meta-analysis and synthesis of qualitative data in educational contexts. *Educational Research Review*, *30*, 100322. doi:10.1016/j.edurev.2020.100322

Baldini, N., Grimaldi, R., & Sobrero, M. (2006). Institutional changes and the commercialization of academic knowledge: A study of Italian universities' patenting activities between 1965 and 2002. *Research Policy*, *35*(4), 518–532. doi:10.1016/j.respol.2006.01.004

Baños, J. E., & Farré, M. (2011). Dinamización de la clase magistral en medicina: Diez ejemplos de minicasos utilizados en la docencia de la farmacología. *Educación Médica*, *14*(2), 105–112.

Barrett, M. (2020, July). The early detection of stator winding shorted turns in permanent magnet three phase generators by analysis of its phase voltages. In *2020 IEEE 14th International Conference on Compatibility, Power Electronics and Power Engineering (CPE-POWERENG)* (Vol. 1, pp. 317-323). IEEE.

Barth, M., Rieckmann, M., Robèrt, K. H., Fischer, D., Karatzoglou, B., & Lozano, R. (2021). Advancing education for sustainable development in higher education: Reflections on the role of universities. *Sustainability Science*, *17*(1), 1–15. PMID:33442425

Barton, A., & Sakwa, L. N. (2012). The representation of gender in English textbooks in Uganda. *Pedagogy, Culture & Society*, *20*(2), 173–190. doi:10.1080/14681366.2012.669394

Barzehkar, M., Parnell, K. E., Soomere, T., Dragovich, D., & Engström, J. (2021). Decision support tools, systems and indices for sustainable coastal planning and management: A review. *Ocean and Coastal Management*, *212*, 105813. Advance online publication. doi:10.1016/j.ocecoaman.2021.105813

Bayhan, S., & Aratemur-Çimen, C. (2019). 2017 müfredat reformu sonrası ders kitaplarında toplumsal cinsiyet eşitliği [Gender equality in the textbooks of 2017 curricula]. *Eğitim Bilim Toplum Dergisi*, *17*(67), 62-88.

Bayhan, S. (2020). *2005'ten 2020'ye öğretim programları toplumsal cinsiyet eşitliği ve sekülerizm araştırması sonuç raporu*. Karşılaştırmalı Eğitim Derneği.

Bazzocchi, F., Ciacci, C., & Di Naso, V. (2021). Evaluation of environmental and economic sustainability for the building envelope of low-carbon schools. *Sustainability (Basel)*, *13*(4), 1702. doi:10.3390u13041702

Bellini, A., Bifaretti, S., Iacovone, V., & Cornaro, C. (2009, September). Simplified model of a photovoltaic module. In *2009 Applied Electronics* (pp. 47-51). IEEE.

Bell, L., & Stevenson, H. (2015). Towards an analysis of the policies that shape public education: Setting the context for school leadership. *Management in Education*, *29*(4), 146–150. doi:10.1177/0892020614555593

Ben Rhaiem, M. A., Selmi, M., Farah, I. R., & Bouzeghoub, A. (2022). A big spatiotemporal streaming data architecture for smart city crisis monitoring using VGI. *Proceedings - 2022 2nd International Conference of Smart Systems and Emerging Technologies, SMARTTECH 2022*. 10.1109/SMARTTECH54121.2022.00035

Bento, A., Luís, R., Pinto, S., & Silva, J. F. (2019, October). A novel multilevel T-Type indirect matrix converter for three-phase open-end AC loads. In *IECON 2019-45th Annual Conference of the IEEE Industrial Electronics Society* (Vol. 1, pp. 6575-6580). IEEE.

Bevir, M. (2012). *Governance: A very short introduction*. Tehran: Kargadan Pub.

Beynaghi, A., Trencher, G., Moztarzadeh, F., Mozafari, M., Maknoon, R., & Leal Filho, W. (2016). Future sustainability scenarios for universities: Moving beyond the United Nations Decade of Education for Sustainable Development. *Journal of Cleaner Production*, *112*, 3464–3478. doi:10.1016/j.jclepro.2015.10.117

Bharadwaj, S., Ballare, S., Rohit, & Chandel, M. K. (2017). Impact of congestion on greenhouse gas emissions for road transport in Mumbai metropolitan region. *Transportation Research Procedia*, *25*, 3542–3555. doi:10.1016/j.trpro.2017.05.282

Bhukya, M. N., Kota, V. R., & Depuru, S. R. (2019). A simple, efficient, and novel standalone photovoltaic inverter configuration with reduced harmonic distortion. *IEEE Access : Practical Innovations, Open Solutions*, *7*, 43831–43845.

Biemmi, I. (2015). Gender in schools and culture: Taking stock of education in Italy. *Gender and Education*, *27*(7), 812–827. doi:10.1080/09540253.2015.1103841

Blake, J., Sterling, S., & Kagawa, F. (2013). Getting it together. Interdisciplinarity and sustainability in the higher education institution. *Pedagogic Research Institute and Observatory*, *4*, 1–71.

Bland, D., & Atweh, B. (2007). Students as researchers: Engaging students' voices in PAR. *Educational Action Research*, *15*(3), 337–349. doi:10.1080/09650790701514259

Boeve-de Pauw, J., Ceulemans, K., Delvaux, E., Hens, Z., & Lievens, J. (2021). How educators' perceptions of sustainability relate to their teaching practices: A study on the role of education for sustainable development in teacher education. *Environmental Education Research*, *27*(3), 313–327.

Bolstad, R. (2021). Schools' Carbon Footprint Pilot Project: Final Report for Auckland Council. New Zealand Council for Educational Research.

Bovill, C., Cook-Sather, A., & Felten, P. (2011) Students as co-creators of teaching approaches, course design and curricula: implications for academic developers. International Journal for Academic Development, 16 (2). pp. 133-145. ISSN 1360-144X

Bovill, C., Cook-Sather, A., Felten, P., Millard, L., & Moore-Cherry, N. (2016). Addressing potential challenges in co-creating learning and teaching: Overcoming resistance, navigating institutional norms and ensuring inclusivity in student–staff partnerships. *Higher Education*, *71*(2), 195–208. doi:10.100710734-015-9896-4

Brame, C. (2016). Active learning. Vanderbilt University Center for Teaching.

Brans, J. P. (2002). Or, Ethics and Decisions: The Oath of Prometheus. *European Journal of Operational Research*, *140*(2), 191–196. doi:10.1016/S0377-2217(02)00062-0

British Educational Research Association (BERA). (2018). *Ethical Guidelines for Educational Research* (4th ed.). British Educational Research Association.

British Psychological Association (BPS). (2018). *Code of Ethics and Conduct*. British Psychological Association.

Broman, G. I., & Robèrt, K. H. (2017). A framework for strategic sustainable development. *Journal of Cleaner Production, 140*, 17–31. doi:10.1016/j.jclepro.2015.10.121

Brooks, R., Te Riele, K., & Maguire, M. (2014). Ethics and education research. *Sage (Atlanta, Ga.)*.

Brown, V. A., Deane, P. M., Harris, J. A., & Russell, J. Y. (2010). Towards a just and sustainable future. In V. A. Brown, J. A. Harris, & J. Y. Russell (Eds.), *Tackling Wicked Problems: Through the Transdisciplinary Imagination* (pp. 3–15). Earthscan.

Burgess, D. J., & Borgida, E. (1999). Who women are, who women should be: Descriptive and prescriptive gender stereotyping in sex discrimination. *Psychology, Public Policy, and Law*, *5*(3), 665–692. doi:10.1037/1076-8971.5.3.665

Burns, H. (2011). Teaching for transformation: (Re) designing sustainability courses based on ecological principles. *Journal of Sustainability Education, 2*.

Burns, J. P., & Zhiren, Z. (2012). Performance Management in the Government of the People's Republic of China: Accountability and Control in the Implementation of Public Policy. *OECD Journal on Budgeting*.

Caballero Miguez, G., & Garza Gil, M. D. (2012). Innovando la docencia superior en Economía: trabajo cooperativo y elaboración participativa de contenidos. REDU: Revista De Docencia Universitaria.

Cairney & Weible. (2018). Practical lessons from policy theories. *Policy and Politics*.

Calik, H., Dabakoglu, A., & Oguz, Y. (2021). Investigation of Dynamic Behaviour of Double Feed Induction Generator and Permanent Magnet Synchronous Generator Wind Turbines in Failure Conditions. *International Journal of Renewable Energy Research*, *11*, 721–729.

Campbell, E. (2001). Let right be done: Trying to put ethical standards into practice. *Journal of Education Policy*, *16*(5), 395–411. doi:10.1080/02680930110071011

Candel, J., & Biesbroek, R. (2016). Toward a processual understanding of policy integration. *Policy Sciences*, *49*(3), 211–231. doi:10.100711077-016-9248-y

Carbon Trust. (n.d.). *Carbon Neutral*. Retrieved from: https://www.carbontrust.com/what-we-do/assurance-and-labelling/carbon-neutral-verification

Carinci, S. (2007). Inclusion of gender in curriculum and instructional teaching practices and its impact on the adult learner. *Western Region Research Conference on the Education of Adults (WRRCEA)*.

CEİD. (2022). *Türkiye'de Toplumsal Cinsiyet Eşitliğini İzleme Raporu 2021-2022*. Cinsiyet Eşitliği İzleme Derneği.

Cejudo, G. M., & Michel, C. L. (2019). *Instruments for policy integration*. Public Administration Department.

CEPES. (2011). *From Gender Studies to Gender IN Studies: Case Studies on Gender-Inclusive Curriculum in Higher Education*. UNESCO.

Chalkley, B. (2006). Education for sustainable development: Continuation. *Journal of Geography in Higher Education*, *30*(2), 235–236. doi:10.1080/03098260600717307

Chalk, S. G., & Miller, J. F. (2006). Key challenges and recent progress in batteries, fuel cells, and hydrogen storage for clean energy systems. *Journal of Power Sources*, *159*(1), 73–80. doi:10.1016/j.jpowsour.2006.04.058

Chami, R., Cosimano, T., Fullenkamp, C., & Oztosun, S. (2019). Nature's solution to climate change. *Finance and Development Magazine*, *56*, 34–38.

Chan, L. S. (2023). Transition from fossil fuel propelled transport to electrified mass transit railway system-Experience from Hong Kong. *Energy Policy*, *173*, 113372. doi:10.1016/j.enpol.2022.113372

Chatterjee, L., & Reynolds, J. (2008). *Carbon Emissions from Schools: Where they arise and how to reduce them*. https://www.sdcommission.org.uk/data/files/publications/Publish_Schools_Carbon_Strategy.pdf

Chatterjee, S., Misbahuddin, M., Vamsi, P., & Ahmed, M. H. (2023, May). Power Quality Improvement and Fault Diagnosis of PV System By Machine Learning Techniques. In *2023 International Conference on Signal Processing, Computation, Electronics, Power and Telecommunication (IConSCEPT)* (pp. 1-6). IEEE.

Chauhan, A., & Saini, R. P. (2014). A review on Integrated Renewable Energy System based power generation for stand-alone applications: Configurations, storage options, sizing methodologies and control. *Renewable & Sustainable Energy Reviews*, *38*, 99–120.

Chawda, G. S., Shaik, A. G., Shaik, M., Padmanaban, S., Holm-Nielsen, J. B., Mahela, O. P., & Kaliannan, P. (2020). Comprehensive review on detection and classification of power quality disturbances in utility grid with renewable energy penetration. *IEEE Access : Practical Innovations, Open Solutions*, *8*, 146807–146830.

Chawla, L., Watson, C., & Gelinas, J. (2021). Preparing teachers for sustainability education: Embedding a field-based approach. *The Journal of Environmental Education*, *52*(1), 26–38.

Chen, X., & Kwan, M. P. (2015). Contextual uncertainties, human mobility, and perceived food environment: The uncertain geographic context problem in food access research. American Journal of Public Health. doi:10.2105/AJPH.2015.302792

Chen, X., & Yang, X. (2014). Does food environment influence food choices? A geographical analysis through "tweets." *Applied Geography (Sevenoaks, England)*, *51*, 82–89. doi:10.1016/j.apgeog.2014.04.003

Chettibi, N., Mellit, A., Sulligoi, G., & Pavan, A. M. (2016). Adaptive neural network-based control of a hybrid AC/DC microgrid. *IEEE Transactions on Smart Grid*, *9*(3), 1667–1679.

Chishti, F., Murshid, S., & Singh, B. (2019). LMMN-based adaptive control for power quality improvement of grid intertie wind–PV system. *IEEE Transactions on Industrial Informatics*, *15*(9), 4900–4912.

Christensen, T., & Lægreid, P. (2007). The Whole-of-Government Approach to Public Sector Reform. *JSTOR Collection,* 1059-1066.

Codish, D., & Ravid, G. (2014). Personality based gamification: Educational gamification for extroverts and introverts. *Proceedings of the 9th CHAIS Conference for the Study of Innovation and Learning Technologies: Learning in the Technological Era, 1,* 36-44.

CONAHCYT. (2023). *Consejo nacional de humanidades ciencia y tecnologias*. Obtenido de Consejo nacional de humanidades ciencia y tecnologias: https://conahcyt.mx/conahcyt/areas-del-conahcyt/uasr/desarrollo-regional/fondos/programa-de-estimulos-a-la-innovacion/

Corbett, D. C., & Rostow, W. W. (1960). The Stages of Economic Growth: A Non-Communist Manifesto. *International Journal (Toronto, Ont.)*, *16*(1), 83. Advance online publication. doi:10.2307/40198523

Corcoran, P. B., Weakland, J. P., & Wals, A. E. (2021). *Higher education for sustainability: Cases, challenges, and opportunities from across the curriculum*. Routledge.

Cramwinckel, J. (2019). *The role of global open geospatial data in measuring SDG indicator 11.7. 1: Public open spaces*. Wageningen University.

CrowdWater. (2023). *CrowdWater*. https://crowdwater.ch/en/start/

CWU. (2016). *Philly Fresh Food Map (Powered by OpenStreetMap)*. http://www.gis.cwu.edu/phillyfood/

Dai, D., & Wang, F. (2011). Geographic disparities in accessibility to food stores in southwest Mississippi. *Environment and Planning. B, Planning & Design*, *38*(4), 659–677. Advance online publication. doi:10.1068/b36149

Dale, A., & Newman, L. (2005). Sustainable development, education and literacy. *International Journal of Sustainability in Higher Education*, *6*(4), 351–362. doi:10.1108/14676370510623847

Daniar, A., Khan, S. A., & Gardner, M. C. (2023). A Passive Inter Turn Short Circuit Fault Mitigation Technique for Electric Machines. *IEEE Transactions on Energy Conversion*.

Dasgupta, K., Roy, P. K., & Mukherjee, V. (2020). Power flow based hydro-thermal-wind scheduling of hybrid power system using sine cosine algorithm. *Electric Power Systems Research, 178*, 106018. doi:10.1016/j.epsr.2019.106018

Davies, J., & Brember, I. (1995). Attitudes to school and the curriculum in year 2, year 4 and year 6: Changes over four years. *European Conference on Educational Research.*

Deng, X., Chen, W., Zhou, Q., Zheng, Y., Li, H., Liao, S., & Biljecki, F. (2023). Exploring spatiotemporal pattern and agglomeration of road CO2 emissions in Guangdong, China. *The Science of the Total Environment, 871*, 162134. doi:10.1016/j.scitotenv.2023.162134 PMID:36775171

Department for Education (DfE). (2022). *Sustainability and climate change: a strategy for the education and children's services systems*. Policy paper published on 21 April 2022. Retrieved from Sustainability and climate change: a strategy for the education and children's services systems. www.gov.uk

Deterding, S., Dixon, D., Khaled, R., & Nacke, L. (2011). From game design elements to gamefulness: Defining "gamification". *Proceedings of the 15th International Academic MindTrek Conference: Envisioning Future Media Environments,* 9-15.

Deutschland.de. (2023, July 19). https://www.deutschland.de/es/desarrollo-sostenible

Dicheva, D., Dichev, C., Agre, G., & Angelova, G. (2015). Gamification in education: A systematic mapping study. *Journal of Educational Technology & Society, 18*(3), 75–88.

Dida, A. (2017). *Contribution à la Commande Neuro-Floue de la Machine Asynchrone à Double Alimentation Utilisé dans un Système Eolien* [Doctoral dissertation]. Université Mohamed Khider-Biskra.

Dincer, I. (2000). Renewable energy and sustainable development: A crucial review. *Renewable & Sustainable Energy Reviews, 4*(2), 157–175. doi:10.1016/S1364-0321(99)00011-8

Dingus, T. A., Guo, F., Lee, S., Antin, J. F., Perez, M., Buchanan-King, M., & Hankey, J. (2016). Driver crash risk factors and prevalence evaluation using naturalistic driving data. *Proceedings of the National Academy of Sciences of the United States of America, 113*(10), 2636–2641. doi:10.1073/pnas.1513271113 PMID:26903657

Doh, J., & Stumpf, S. (2005). Toward a framework of responsible leadership and governance. In Handbook on Responsible Leadership and Governance in Global Business. Edward Elgar. doi:10.4337/9781845425562.00007

Dos Santos, E. B., da Costa Maynard, D., Zandonadi, R. P., Raposo, A., & Botelho, R. B. A. (2022). Sustainability recommendations and practices in school feeding: a systematic review. *Kompass Nutrition & Dietetics, 2*(2), 83-102.doi:10.1159/000526379

Downey, D. B., Kuhfeld, M., & van Hek, M. (2022). Schools as a relatively standardizing institution: The case of gender gaps in cognitive skills. *Sociology of Education, 95*(2), 89–109. doi:10.1177/00380407211070319

Duffy, J., Warren, K., & Walsh, M. (2001). Classroom interactions: Gender of teacher, gender of student, and classroom subject. *Sex Roles*, *45*(9/10), 579–593. doi:10.1023/A:1014892408105

ECOES. (2023). *Espacio en comun de educacion superior*. Obtenido de http://www.ecoes.unam.mx/presentacion.html

Edwards, N. (2013). Bad Pharma. *Journal of the Royal Society of Medicine*, *106*(2), 72–73. doi:10.1177/0141076813477618

El Financiero. (2022). *Nueva ruta del Trolebús Línea 5 en Reforma: horario, costo y estaciones*. Available at: https://www.elfinanciero.com.mx/cdmx/2022/12/27/nueva-ruta-del-trolebus-linea-5-en-reforma-horario-costo-y-estaciones/

Enge, M. R. (2014). Gender equality in public higher educationinstitutions of Ethiopia: The case of science, technology, engineering, and mathematics. *Discourse and Communication for Sustainable Education*, *5*(1), 3–21. doi:10.2478/dcse-2014-0001

Erden, F. T. (2009). A course on gender equity in education: Does it affect gender role attitudes of preservice teachers? *Teaching and Teacher Education*, *25*(3), 409–414. doi:10.1016/j.tate.2008.11.001

ERG. (2023). *Türkiye'de Kapsayıcı ve Nitelikli Eğitim için Öncelikler: 14 Mayıs 2023 Seçimlerinde Göreve Gelecek Hükümete ve Meclise Öneriler*. Eğitim Reformu Girişimi.

European Commission. (2019). *A European Green Deal*. Retrieved from https://ec.europa.eu/info/publications/communication-european-green-deal_en

European Commission. (2020). *EU climate action*. Retrieved from https://ec.europa.eu/clima/policies/strategies_en

European Environment Agency. (2023, July 19). *European environment agency*. Retrieved from European environment agency: https://www.eea.europa.eu/themes/sustainability-transitions/sustainable-development-goals-and-the/country-profiles/belgium-country-profile-sdgs-and

Farrel, T. M., Sutton, J. E., Clark, D. E., Horner, W. R., Morris, K. I., Finison, K. S., Menchen, G. E., & Cohen, K. H. (1996). Moose-Motor Vehicle collisions. *Archives of Surgery*, *131*(4), 377–381. doi:10.1001/archsurg.1996.01430160035005 PMID:8615722

Farr-Wharton, G., Choi, J. H. J., & Foth, M. (2014). Food talks back: Exploring the role of mobile applications in reducing domestic food wastage. *Proceedings of the 26th Australian Computer-Human Interaction Conference, OzCHI 2014*. 10.1145/2686612.2686665

Fast, V., & Rinner, C. (2018). Toward a participatory VGI methodology: Crowdsourcing information on regional food assets. *International Journal of Geographical Information Science*, *32*(11), 2209–2224. doi:10.1080/13658816.2018.1480784

Fawzy, S., Osman, A. I., Doran, J., & Rooney, D. W. (2020). Strategies for mitigation of climate change: A review. *Environmental Chemistry Letters*, *18*(6), 2069–2094. doi:10.100710311-020-01059-w

Fazal, S., Haque, M. E., Arif, M. T., Gargoom, A., & Oo, A. M. T. (2023). Grid integration impacts and control strategies for renewable based microgrid. *Sustainable Energy Technologies and Assessments*, *56*, 103069. doi:10.1016/j.seta.2023.103069

Fekry, H. M., Eldesouky, A. A., Kassem, A. M., & Abdelaziz, A. Y. (2020). Power management strategy based on adaptive neuro fuzzy inference system for AC microgrid. *IEEE Access : Practical Innovations, Open Solutions*, *8*, 192087–192100.

Fernández Jambrina, L. (2013). Docencia no presencial como alternativa a la clase magistral en los primeros cursos de ingeniería. Academic Press.

Fernández, L., Yepes, M., & Fernández, L. (2010). Docencia de Electrónica Física: Un cambio desde las clases magistrales a metodologías activas y evaluación continua. *Docencia de Electrónica Física: Un cambio desde las clases magistrales a metodologías activas y evaluación continua,* 179-190.

Ferrer-Estévez, M., & Chalmeta, R. (2021). Integrating sustainable development goals in educational institutions. *International Journal of Management Education*, *19*(2), 1–19. doi:10.1016/j.ijme.2021.100494

Food and nutrition in numbers: 2014. (2015). *Choice Reviews Online, 52*(12). doi:10.5860/CHOICE.191448

Foreign, Commonwealth, and Development Office (FCDO). (2022). *Addressing the climate, environment, and biodiversity crises in and through girls' education.* FCDO Position Paper. Available at: https://assets.publishing.service.gov.uk/government/uploads/system/uploads/attachment_data/file/1122368/Addressing_the_climate__environment__and_biodiversity_crises_in_and_through_girls__education.pdf

Fort, E., Bouffard, E., Charnay, P., Bernard, M., Boisson, D., Laumon, B., Hours, M. (2011). Return to work following road accidents: Factors associated with late work resumption. *Journal of Rehabilitation Medicine, 43*(4), 283-291. doi:. doi:113372

Fouad, K., Boulouiha, H. M., Allali, A., Taibi, A., & Denai, M. (2018). Multivariable control of a grid-connected wind energy conversion system with power quality enhancement. *Energy Systems*, *9*(1), 25–57.

Frawley, T. (2005). Gender bias in the classroom: Current controversies and implications for teachers. *Childhood Education*, *81*(4), 221–227. doi:10.1080/00094056.2005.10522277

Friedlingstein, P., Jones, M. W., O'sullivan, M., Andrew, R. M., Hauck, J., Peters, G. P., Peters, W., Pongratz, J., Sitch, S., Le Quéré, C., Bakker, D. C. E., Canadell, J. G., Ciais, P., Jackson, R. B., Anthoni, P., Barbero, L., Bastos, A., Bastrikov, V., Becker, M., ... Zaehle, S. (2019). Global carbon budget 2019. *Earth System Science Data*, *11*(4), 1783–1838. doi:10.5194/essd-11-1783-2019

Gajda, A., Bojko, A., & Stoecker, E. (2022). The vicious circle of stereotypes: Teachers' awareness of and responses to students' gender-stereotypical behaviour. *PLoS One*, *17*(6), 1–24. doi:10.1371/journal.pone.0269007 PMID:35704605

Gamarra, A. R., Istrate, I. R., Herrera, I., Lago, C., Lizana, J., & Lechon, Y. (2017). Energy and water consumption and carbon footprint of school buildings in hot climate conditions. Results from life cycle assessment. *Journal of Cleaner Production*, *195*, 1326–1337. doi:10.1016/j.jclepro.2018.05.153

Gao, S., Li, L., Li, W., Janowicz, K., & Zhang, Y. (2014). Constructing gazetteers from volunteered Big Geo-Data based on Hadoop. *Computers, Environment and Urban Systems*. Advance online publication. doi:10.1016/j.compenvurbsys.2014.02.004

Gaudet, J., & Lapointe, C. (2013). What could an equitable instructional design achieve? An experiment with trainers. *New Horizons in Adult Education and Human Resource Development*, *16*(3), 12–26. doi:10.1002/nha3.10149

Gawande, P. N., & Dambhare, S. S. (2016, May). A novel algorithm to detect symmetrical faults during power swing for a double circuit multi-terminal transmission line. In *2016 IEEE/PES Transmission and Distribution Conference and Exposition (T&D)* (pp. 1-7). IEEE.

Geng, K., Brands, D. K., Verhoef, E. T., & Wang, Y. (2023). The impact of tradable rush hours permits on peak demand: Evidence from an on-campus field experiment. *Transportation Research Part D, Transport and Environment*, *120*, 103775. doi:10.1016/j.trd.2023.103775

Ghofrani, A., Zaidan, E., & Abulibdeh, A. (2022). Simulation and impact analysis of behavioral and socioeconomic dimensions of energy consumption. *Energy*, *240*, 122502. doi:10.1016/j.energy.2021.122502

Gilardi, A., Mateu, J., Borgoni, R., & Lovelace, R. (2020). Multivariate hierarchical analysis of car crashes data considering a spatial network lattice. *Journal of the Royal Statistical Society. Series A, (Statistics in Society)*. Advance online publication. doi:10.48550/arxiv.2011.12595

Gobierno de Mexico. (2023, July 19). Retrieved from Gobierno de Mexico: https://e.economia.gob.mx/guias/conocer-los-objetivos-de-desarrollo-sostenible-y-el-plan-de-acciones-en-mexico/

Goodchild, M. F. (2007). Citizens as sensors: The world of volunteered geography. GeoJournal. doi:10.100710708-007-9111-y

Gottschalk, J. (2021). *Writing Strategies for Talent Development: From Struggling to Gifted Learners, Grades 3–8*. Routledge. doi:10.4324/9781003089247

Government and Ministries. (2023, July 19). *Prime Minister's office*. Retrieved from Prime Minister's office: https://vnk.fi/en/sustainable-development

Gramk, C. (2021). *Green fiscal policies, An armoury of instruments to recover growth sustainably*. United Nations- ECLAC Office Brasilia.

Gray, C., & Leith, H. (2004). Perpetuating gender stereotypes in the classroom: A teacher perspective. *Educational Studies*, *30*(1), 3–17. doi:10.1080/0305569032000159705

Graymore, M. L. M., Wallis, A. M., & Richards, A. J. (2009). An Index of Regional Sustainability: A GIS-based multiple criteria analysis decision support system for progressing sustainability. *Ecological Complexity*, *6*(4), 453–462. Advance online publication. doi:10.1016/j.ecocom.2009.08.006

Green Peace. (2023). *Climate Justice and Social Justice: Two Sides of the Same Coin.* Available at: https://www.greenpeace.org/international/story/58334/climate-justice-and-social-justice-two-sides-of-the-same-coin/

Grigorieva, O., & Nikulshin, A. H. (2022). Electric buses on the streets of Moscow: experience, problems, prospects. *X International Scientific Siberian Transport Forum, 63*, 670-675. 10.1016/j.trpro.2022.06.061

Groh, F. (2012). Gamification: State of the art definition and utilization. *Institute of Media Informatics Ulm University, 39*(31).

Growcom, A. (2008). *Vegetable Industry Carbon Footprint Scoping Study.* https://www.daf.qld.gov.au/__data/assets/pdf_file/0003/59025/Hort-Fruit-Drought-Carbon-Report1.pdf

Guerras Martín, L. Á., & Navas López, J. E. (2022). *La dirección estratégica de la Empresa: Teoría y aplicaciones* (6th ed.). Thomson Reuters.

Habiba, U., & Abedin, Md. (2015). Introduction and Overview of Food Security and Risk Reduction Issues. doi:10.1007/978-4-431-55411-0_1

Habib, H. U. R., Wang, S., Elkadeem, M. R., & Elmorshedy, M. F. (2019). Design optimization and model predictive control of a standalone hybrid renewable energy system: A case study on a small residential load in Pakistan. *IEEE Access : Practical Innovations, Open Solutions*, *7*, 117369–117390.

Hair, J. F., Hult, G. T. M., Ringle, C. M., & Sarstedt, M. (2022). *A primer on Partial Least Squares Structural Equation Modeling (PLS-SEM)* (3rd ed.). SAGE Publications Inc.

Hák, T., Janoušková, S., & Moldan, B. (2016). Sustainable Development Goals: A need for relevant indicators. *Ecological Indicators*, *60*, 565–573. doi:10.1016/j.ecolind.2015.08.003

Hasan, S., Muttaqi, K. M., & Sutanto, D. (2020). Detection and characterization of time-variant nonstationary voltage sag waveforms using segmented Hilbert–Huang transform. *IEEE Transactions on Industry Applications*, *56*(4), 4563–4574.

Hassan, Q., Abbas, M. K., Tabar, V. S., Tohidi, S., Sameen, A. Z., & Salman, H. M. (2023). Techno-economic assessment of battery storage with photovoltaics for maximum self-consumption. *Energy Harvesting and Systems*, (0).

Hassan, Q., Sameen, A. Z., Salman, H. M., Al-Jiboory, A. K., & Jaszczur, M. (2023). *The role of renewable energy and artificial intelligence towards environmental sustainability and net zero*. Academic Press.

Hassan, Q., Tabar, V. S., Sameen, A. Z., Salman, H. M., & Jaszczur, M. (2023). A review of green hydrogen production based on solar energy: Techniques and methods. *Energy Harvesting and Systems*, (0).

Hassan, Q., Abdulateef, A. M., Hafedh, S. A., Al-samari, A., Abdulateef, J., Sameen, A. Z., Salman, H. M., Al-Jiboory, A. K., Wieteska, S., & Jaszczur, M. (2023). Renewable energy-to-green hydrogen: A review of main resources routes, processes and evaluation. *International Journal of Hydrogen Energy*, *48*(46), 17383–17408. doi:10.1016/j.ijhydene.2023.01.175

Hayles, C., & de la Harpe, B. (2010). Can Greening the Curriculum Influence Students' Perceptions and Awareness of Sustainability? In *W089-Special Track 18th CIB World Building Congress May 2010 Salford, United Kingdom* (p. 48).Academic Press.

He, B., Ye, L., Pei, M., Lu, P., Dai, B., Li, Z., & Wang, K. (2022). A combined model for short-term wind power forecasting based on the analysis of numerical weather prediction data. *Energy Reports*, *8*, 929–939. doi:10.1016/j.egyr.2021.10.102

Henderson, K., & Loreau, M. (2023). A model of Sustainable Development Goals: Challenges and opportunities in promoting human well-being and environmental sustainability. *Ecological Modelling*, *475*, 110164. doi:10.1016/j.ecolmodel.2022.110164

Heris, M. S., Piralaiy, E., & Fard, A. M. (2013). A study of status of the research ethics compliance among physical education academic members of Islamic Azad University in reigns 2 & 13. *European Journal of Experimental Biology*, *3*(1), 568–571.

Herron, T. L., & Gilbertson, D. L. (2004). Ethical principles vs. ethical rules: The moderating effect of moral development on audit independence judgments. *Business Ethics Quarterly*, *14*(3), 499–523. doi:10.5840/beq200414320

Holder-Webb, L., & Cohen, J. (2012). The cut and paste society: Isomorphism in codes of ethics. *Journal of Business Ethics*, *107*(4), 485–509. doi:10.100710551-011-1060-1

Holst, J. (2023). Towards coherence on sustainability in education: A systematic review of Whole Institution Approaches. *Sustainability Science*, *18*(2), 1015–1030. doi:10.100711625-022-01226-8

Hou, H., Liu, C., Wang, Q., Wu, X., Tang, J., Shi, Y., & Xie, C. (2022). Review of load forecasting based on artificial intelligence methodologies, models, and challenges. *Electric Power Systems Research*, *210*, 108067. doi:10.1016/j.epsr.2022.108067

Howlett, M. P., & Saguin, K. (2019). *Policy capacity for policy integration: Implications for the sustainable development goals*. Lee Kuan Yew School of Public Policy Research Paper, (18-06).

Howlett, M., Vince, J., & del Rio, P. (2017). Policy integration and multi-level governance: Dealing with the vertical dimension of policy mix designs. *Politics and Governance*, *5*(2), 69–78. doi:10.17645/pag.v5i2.928

Huang, B., & Wang, J. (2020). Big spatial data for urban and environmental sustainability. *Geo-Spatial Information Science*, *23*(2), 125–140. Advance online publication. doi:10.1080/10095020.2020.1754138

INEGI. (2017). *Encuesta origen-destino en hogares de la Zona Metropolitan del Valle de Mexico* [Origin-Destination survey in households of the metropolitan Zone of the Valley of Mexico]. INEGI.

INEGI. (2023). *Marco geoestadístico*. Available at: https://www.inegi.org.mx/temas/mg/#Descargas

Intergovernmental Panel on Climate Change (IPCC). (2018). *Global warming of 1.5°C.* Retrieved from https://www.ipcc.ch/sr15/download/

Intergovernmental Panel on Climate Change (IPCC). (2022). *Climate Change: A Threat to Human Wellbeing and Health of the Planet. Taking Action Now Can Save Our Future*. Available at: https://www.ipcc.ch/2022/02/28/pr-wgii-ar6/

International Energy Agency (IEA). (2022). *Global Energy and Climate Model documentation.* Retrieved from https://iea.blob.core.windows.net/assets/2db1f4ab-85c0-4dd0-9a57-32e542556a49/GlobalEnergyandClimateModelDocumentation2022.pdf

International Labour Organization. (2021). *Guidelines for a just transition towards environmentally sustainable economies and societies for all*. ILO.

Iphofen, R. (2020). An introduction to research ethics and scientific integrity. Handbook of research ethics and scientific integrity, 3-13.

Ishikawa, S. (2017). Status and Problems of Humanities Education with Respect to Pharmacy Education. Yakugaku zasshi. *Journal of the Pharmaceutical Society of Japan*, *137*(4), 399–405. PMID:28381714

Jafari, E. (2022). Academic dishonesty in e-learning, a threat to the integrity of the medical education system under the COVID-19 epidemic. *Indian Journal of Medical Education*, *22*(23), 158–169.

Jafari, E., & Alamolhoda, J. (2021). Lived Experience of Faculty Members of Ethics in Virtual Education. *Technology. Knowledge and Learning*, *2021*, 1–9.

Janvier, A., & Farlow, B. (2015). The ethics of neonatal research: An ethicist's and a parents' perspective. *Seminars in Fetal & Neonatal Medicine*, *20*(6), 436–441. doi:10.1016/j.siny.2015.10.003 PMID:26497942

Jha, S. K., Bilalovic, J., Jha, A., Patel, N., & Zhang, H. (2017). Renewable energy: Present research and future scope of Artificial Intelligence. *Renewable & Sustainable Energy Reviews*, *77*, 297–317. doi:10.1016/j.rser.2017.04.018

Jones, P., Selby, D., & Sterling, S. (2010). *Sustainability Education: Perspectives and Practice Across Higher Education. Earthscan.* Routledge.

Jordan, M. E., Bartlett, R., Cribb, G., & Dadey, N. (2021). Strengthening higher education educators' capacity to teach sustainability: A collective case study. *International Journal of Sustainability in Higher Education*, *22*(3), 508–525.

Jurado Rota, J., Pérez Albert, M. Y., & Serrano Giné, D. (2019). Visitor monitoring in protected areas: An approach to Natura 2000 sites using Volunteered Geographic Information (VGI). *Geografisk Tidsskrift - Danish Geografisk Tidskrift, 119*(1), 69–83. Advance online publication. doi:10.1080/00167223.2019.1573409

Kalogirou, S. A. (2001). Artificial neural networks in renewable energy systems applications: A review. *Renewable & Sustainable Energy Reviews, 5*(4), 373–401. doi:10.1016/S1364-0321(01)00006-5

Kapp, K. M. (2012). *The gamification of learning and instruction: Game-based methods and strategies for training and education*. John Wiley & Sons.

Karacasu, M., Er, A., Bilgic, S., & Barut, H. B. (2011). Variations in traffic accidents on seasonal, monthly, daily and hourly basis: Eskisehir Case. *Procedia: Social and Behavioral Sciences, 20*, 167–775. doi:10.1016/j.sbspro.2011.08.085

Keller, G. (2001). The new demographics of higher education. *Review of Higher Education, 24*(3), 219–235. doi:10.1353/rhe.2001.0004

Khan, O., Parvez, M., Kumari, P., Parvez, S., & Ahmad, S. (2023). The future of pharmacy: How AI is revolutionizing the industry. *Intelligent Pharmacy, 1*(1), 32–40. doi:10.1016/j.ipha.2023.04.008

Khare, R., Villuri, V. G. K., & Chaurasia, D. (2021). Urban sustainability assessment: The evaluation of coordinated relationship between BRTS and land use in transit-oriented development mode using DEA model. *Ain Shams Engineering Journal, 12*(1), 107–117. doi:10.1016/j.asej.2020.08.012

Khasim, S. R., Dhanamjayulu, C., Padmanaban, S., Holm-Nielsen, J. B., & Mitolo, M. (2021). A novel asymmetrical 21-level inverter for solar PV energy system with reduced switch count. *IEEE Access : Practical Innovations, Open Solutions, 9*, 11761–11775.

Kitagawa, F., & Oba, J. (2010). Managing differentiation of higher education system in Japan: Connecting excellence and diversity. *Higher Education, 59*(4), 507–524. doi:10.100710734-009-9262-5

Koch, T. (2000). We live in the city not in a study. *ORMS Today, 27*(2), 16–17.

Kollmayer, M., Schultes, M.-T., Lüftenegger, M., Finsterwald, M., Spiel, C., & Schober, B. (2020). REFLECT – A teacher training program to promote gender equality in schools. *Frontiers in Education, 5*(136), 1–8. doi:10.3389/feduc.2020.00136

Kostas, M. (2023). Textbooks, students and teachers talk around gender: A new materialist approach to children's agency. *Teaching and Teacher Education, 125*, 104052. Advance online publication. doi:10.1016/j.tate.2023.104052

Koyuncu Şahin, M., Esen Çoban, A., & Korkmaz, A. (2018). Toplumsal cinsiyet eşitliği ve Türk eğitim sistemindeki yeri: Okul öncesi öğretmen adaylarının gözünden. *Uluslararası Bilimsel Araştırmalar Dergisi, 3*(2), 735–752. doi:10.21733/ibad.457232

Kumar, V., Pandey, A. S., & Sinha, S. K. (2016, March). Grid integration and power quality issues of wind and solar energy system: A review. In *2016 International conference on emerging trends in electrical electronics & sustainable energy systems (ICETEESES)* (pp. 71-80). IEEE.

Kumar, K., Rao, R. S., Kaiwartya, O., Kaiser, S., & Padmanaban, S. (2022). *Sustainable Developments by Artificial Intelligence and Machine Learning for Renewable Energies*. Elsevier.

Kumlu, E., & Çomoğlu, H. İ. (2021). Gender issues in english language teacher education: Cindrella's awakening through comparative literature. *Educational Policy Analysis and Strategic Research, 16*(1), 99–102. doi:10.29329/epasr.2020.334.6

Kuruvilla, M. (2014). Gender studies course at UG/PG levels and gender awareness training to teachers. *Higher Education for the Future, 1*(1), 29–38. doi:10.1177/2347631113518393

Lawrence, H., Robertson, C., Feick, R., & Nelson, T. (2020). The spatial-comprehensiveness (S-COM) index: Identifying optimal spatial extents in volunteered geographic information point datasets. *ISPRS International Journal of Geo-Information, 9*(9), 497. Advance online publication. doi:10.3390/ijgi9090497

Leal Filho, W. (2010). Teaching sustainable development at university level: Current trends and future needs. *Journal of Baltic Science Education, 9*(4), 273–284.

Leng, J., Ruan, G., Jiang, P., Xu, K., Liu, Q., Zhou, X., & Liu, C. (2020). Blockchain-empowered sustainable manufacturing and product lifecycle management in industry 4.0: A survey. *Renewable & Sustainable Energy Reviews, 132*, 110112. doi:10.1016/j.rser.2020.110112

Liao, X., Wu, G., Yang, L., & Barth, M. J. (2023). A real-world data-driven approach for estimating environmental impacts of traffic accidents. *Transportation Research Part D, Transport and Environment, 117*, 103664. doi:10.1016/j.trd.2023.103664

Li, G., Liang, J., Ma, F., Ugalde-Loo, C. E., & Liang, H. (2018). Analysis of single-phase-to-ground faults at the valve-side of HB-MMCs in HVDC systems. *IEEE Transactions on Industrial Electronics, 66*(3), 2444–2453.

Li, L., Wu, M., Wu, S., Li, J., & Song, K. (2019). A three-phase to single-phase AC-DC-AC topology based on multi-converter in AC electric railway application. *IEEE Access : Practical Innovations, Open Solutions, 7*, 111539–111558.

Liu, J., Zhao, C., & Xie, Z. (2021). Power and current limiting control of wind turbines based on PMSG under unbalanced grid voltage. *IEEE Access : Practical Innovations, Open Solutions, 9*, 9873–9883.

Lubicz-Nawrocka, T. (2018). Students as partners in learning and teaching: The benefits of co-creation of the curriculum. *International Journal for Students as Partners, 2*(1), 47–63. Advance online publication. doi:10.15173/ijsap.v2i1.3207

Lubicz-Nawrocka, T., & Owen, J. (2022). Curriculum Co-creation in a Postdigital World: Advancing Networked Learning and Engagement. Postdigit Sci Educ 4, 793–813. https://doi.org/10.1007/s42438-022-00304-5

Luckin, R., Holmes, W., Griffiths, M., & Forcier, L. B. (2016). *Intelligence Unleashed: An argument for AI in education*. Academic Press.

Lu, J., Li, B., Li, H., & Al-Barakani, A. (2021). Expansion of city scale, traffic modes, traffic congestion, and air pollution. *Cities (London, England)*, *108*, 1–15. doi:10.1016/j.cities.2020.102974

Luján-Mora, S. (2013). De la clase magistral tradicional al MOOC: Doce años de evolución de una asignatura sobre programación de aplicaciones web. *REDU. Revista De Docencia Universitaria*, *11*, 279–300. doi:10.4995/redu.2013.5557

Lu, Z., & Zhou, S. (2018). *Integration of large scale wind energy with electrical power systems in China*. John Wiley & Sons. doi:10.1002/9781118910054

Magacho, G., Espagne, E., Godin, A., Mantes, A., & Yilmaz, D. (2023). Macroeconomic exposure of developing economies to low-carbon transition. *World Development*, *167*, 106231. doi:10.1016/j.worlddev.2023.106231

Mansour, S., Alahmadi, M., & Abulibdeh, A. (2021, August). 2022. Spatial assessment of audience accessibility to historical monuments and museums in Qatar during the 2022 FIFA World Cup. *Transport Policy*, *127*, 116–129. doi:10.1016/j.tranpol.2022.08.022

Marshall, S., & Smith, J. (1987). Sex differences in learning mathematics: A longitudinal study with item and error analyses. *Journal of Educational Psychology*, *79*(4), 372–383. doi:10.1037/0022-0663.79.4.372

Martí, E., Julià, C., & Gil, D. (2007). *PBL en la docencia de Gráficos por Computador: Una comparativa con la clase magistral*. JAC.

Mâță, L., & Boghian, I. (2019). Perception of Teachers in Higher Education towards Ethical Issues of Information Technology Use. *Romanian Journal for Multidimensional Education*, *11*(1), 156–169. doi:10.18662/rrem/183

Mercier-Laurent, E. (2021). Can Artificial Intelligence Effectively Support Sustainable Development? In *Artificial Intelligence for Knowledge Management: 8th IFIP WG 12.6 International Workshop, AI4KM 2021, Held at IJCAI 2020, Yokohama, Japan, January 7–8, 2021, Revised Selected Papers 8* (pp. 144-159). Springer International Publishing.

Mesbahi, M., & Abbaszadeh, A. (2013). System model of professional behavior in supreme education (Qualitative research). *Bioethics Journal*, *3*(10), 11–34.

Mexico. (2023). *Creación del Consejo Nacional de Ciencia y Tecnología*. Obtenido de Creación del Consejo Nacional de Ciencia y Tecnología: https://www.gob.mx/epn/articulos/creacion-del-consejo-nacional-de-ciencia-y-tecnologia#:~:text=El%20Consejo%20Nacional%20de%20Ciencia,personalidad%20jur%C3%ADdica%20y%20patrimonio%20propio

Mexico. (2023). *Programa de Mejoramiento del Profesorado, PROMEP*. Obtenido de Programa de Mejoramiento del Profesorado, PROMEP: https://www.gob.mx/sep/documentos/programa-de-mejoramiento-del-profesorado-promep-eed

Mezirow, J. (2000). *Learning as Transformation*. Jossey-Bass.

Microsoft. (2021). *Microsoft to be carbon negative by 2030*. Retrieved from https://blogs.microsoft.com/blog/2020/01/16/microsoft-will-be-carbon-negative-by-2030/

Ministry of Education, British Columbia. (n.d.). *Sustainable Schools: Best Practices Guide*. https://www2.gov.bc.ca/assets/gov/education/kindergarten-to-grade-12/teach/teaching-tools/environmental-learning/sustbestpractices.pdf

Ministry of Education. (2023). *Green Schools Project*. Sultanate of Oman Ministry of Energy and Minerals-Oman. Retrieved from https://www.mem.gov.om/en-US/Sectors/Electricity/Pages/National-Energy-Strategy.aspx

Ministry of National Education (MEB). (2019). *PISA 2018 Türkiye ön raporu*. Türkiye Cumhuriyeti Milli Eğitim Bakanlığı.

Mirzaie, S. A., & Qarakhani, M. (2013). The Research Ethics In Iranian social Science. *Iranian Journal of Sociology*, *15*(2), 3–23.

Mishra, B. K. (2023). *Handbook of Research on Applications of AI, Digital Twin, and Internet of Things for Sustainable Development*. IGI. doi:10.4018/978-1-6684-6821-0

MIT. (2021). *Why do people call climate change an "existential treat"?* Available at: https://climate.mit.edu/ask-mit/why-do-some-people-call-climate-change-existential-threat

Moawwad, A., El Moursi, M. S., & Xiao, W. (2016). Advanced fault ride-through management scheme for VSC-HVDC connecting offshore wind farms. *IEEE Transactions on Power Systems*, *31*(6), 4923–4934.

Mobasheri, A. (2017). A rule-based spatial reasoning approach for openstreetmap data quality enrichment; case study of routing and navigation. *Sensors (Basel)*, *17*(11), 2498. Advance online publication. doi:10.339017112498 PMID:29088125

Mogensen, F., & Schnack, K. (2022). The power of learning communities: Supporting sustainability educators through networks of practice. *International Journal of Sustainability in Higher Education*, *23*(1), 10–25.

Mohammed, S., Alkhereibi, A. H., Abulibdeh, A., Jawarneh, R. N., & Balakrishnan, P. (2023). GIS-based spatiotemporal analysis for road traffic crashes; in support of sustainable transportation planning. *Transportation Research Interdisciplinary Perspectives*, *20*, 100836. doi:10.1016/j.trip.2023.100836

Molina, M. (2023). *¿Notas la diferencia? Detectan nueva hora pico un la Ciudad de México*. Available at: https://www.debate.com.mx/cdmx/Notas-la-diferencia-Detectan-nueva-hora-pico-en-la-Ciudad-de-Mexico-20230227-0199.html

Moller-Jensen, M. (2021). Frictions of everyday mobility: Traffic, transport and gendered confrontations on the roads of Accra. *Mobilities*, *16*(4), 461–475. doi:10.1080/17450101.2021.1917969

Mora, A., Riera, D., Gonzalez, C., & Arnedo-Moreno, J. (2015). A literature review of gamification design frameworks. *2015 7th International Conference on Games and Virtual Worlds for Serious Applications (VS-Games),* 1-8. 10.1109/VS-GAMES.2015.7295760

Moradi, M., Khatooni, M., Jahani Hashemi, H., Zeighami, R., & Sheikh, M. (2013). Relationship between professional commitment and job satisfaction in Qazvin's pubic educational hospital nurses. *Medical Ethics (Burlington, Mass.), 7*(24), 55–78.

Mousavi, S., Mohsenpour, M., Borhani, F., & Ebadi, M. (2015). Ethical Sensitivity of nurses and nursing students working in Aja University of Medical Sciences. *Quarterly Journal of Medical Ethics, 31*(9), 127–143.

Mulholland, E. (2018). *The Implementation of the 2030 Agenda and the SDGs in Europe: Overview and Updates*. ESDN.

Nadae, J., Carvalho, & Monteiro, M. (2019). Integrated management systems as a driver for sustainability: the review and analysis of the literature and the proposition of the conceptual framework. São Paulo: Associação Brasileira de Engenharia de Produção.

Nam-In, T., & Kumsuwan, Y. (2020, June). PWM Techniques for a 13-Level Asymmetrical Dual Inverter Configuration Without the Overcharging of Floating Capacitor. In *2020 17th International Conference on Electrical Engineering/Electronics, Computer, Telecommunications and Information Technology (ECTI-CON)* (pp. 535-538). IEEE.

Nara, B. B., Lengoiboni, M., & Zevenbergen, J. (2020). Implications of customary land rights inequalities for food security: A study of smallholder farmers in Northwest Ghana. *Land (Basel), 9*(6), 178. Advance online publication. doi:10.3390/land9060178

Nasir, O., Javed, R. T., Gupta, S., Vinuesa, R., & Qadir, J. (2023). Artificial intelligence and sustainable development goals nexus via four vantage points. *Technology in Society, 72*, 102171. doi:10.1016/j.techsoc.2022.102171

National Governance Association. (2022). *Environmental sustainability: a whole school approach A guide for governing boards*. https://www.nga.org.uk/getmedia/42a15743-7a17-4dfb-b15a-804e5f26b5a2/nga-environmental-sustainability-20211104.pdf

Neves, P., & do Céu, M. (2018). On (scientific) integrity: Conceptual clarification. *Medicine, Health Care, and Philosophy, 21*(2), 181–187. doi:10.100711019-017-9796-8 PMID:28905288

Nicholas, P. K., & Breakey, S. (2017). Climate Change, Climate Justice, and Environmental Health: Implications for the Nursing Profession. *Journal of Nursing Scholarship, 49*(6), 606–616. doi:10.1111/jnu.12326 PMID:28749596

Nieto, B. (2020). *Teaching materials for the subject Business Finance in the Master of Business Administration*. Unpublished manuscript.

Nutley, T., McNabb, S., & Salentine, S. (2013). Impact of a decision-support tool on decision making at the district level in Kenya. *Health Research Policy and Systems, 11*(1), 34. Advance online publication. doi:10.1186/1478-4505-11-34 PMID:24011028

O'Mathúna, D., & Siriwardhana, C. (2017). Research ethics and evidence for humanitarian health. *Lancet*, *390*(10109), 2228–2229. doi:10.1016/S0140-6736(17)31276-X PMID:28602560

Odell, P., Rauland, V., & Murcia, K. (2020). Schools: An untapped opportunity for a carbon neutral future. *Sustainability, 13*(1), 1-24. doi:10.3390/su13010046

OECD. (2018). *Education policy implementation: A literature review and proposed framework.* Directorate for Education and Skills.

OECD. (2019). *PISA 2018 Results* (Vol. 2). Where All Students Can Succeed.

Ohman, J. (2008). *Values and Democracy in Education for Sustainable Development.* Liber.

Okenwa-Emgwa, L., & Straus, E. (2018). Higher education as a platform for capacity building to address violence against women and promote gender equality: The Swedish example. *Public Health Reviews*, *39*(31), 1–6. doi:10.118640985-018-0108-5 PMID:30473904

Okonkwo, E. C., Wole-Osho, I., Bamisile, O., Abid, M., & Al-Ansari, T. (2021). Grid integration of renewable energy in Qatar: Potentials and limitations. *Energy*, *235*, 121310. doi:10.1016/j.energy.2021.121310

Oman Observer. (2019). *Oman inaugurates 500MW solar project in Ibri.* Retrieved from https://www.omanobserver.om/article/1070002/business/oman-inaugurates-500mw-solar-project-in-ibri

Omoiee Milan, M., Mehdinezhad, V., & Yaghoubi, N. (2011). Assessing Factors Affecting the Tendency to Use Electronic Learning Systems among Faculty Members. *Interdisciplinary Journal of Virtual Learning in Medical Sciences*, *2*(3), 28–38.

Öngören, S. (2019). Öğretmen adaylarının toplumsal cinsiyet algılarına göre okul öncesi öğretmenliğine ilişkin görüşleri [Teacher candidates' views on preschool teaching according to social gender perceptions]. *OPUS International Journal of Society Researches*, *11*(18), 1773–1796. doi:10.26466/opus.516102

Owen, G. (2013). *Game theory.* Emerald Group Publishing.

Oye, N. D., Salleh, M., & Iahad, N. A. (2012). E-learning methodologies and tools. *International Journal of Advanced Computer Science and Applications*, *3*(2).

Öz-Aydin, S. (2016). The Awareness of Turkish High School Students about Carbon Footprint and the Effects of the High School Biology Curriculum on This Awareness. *International Electronic Journal of Environmental Education*, *6*(2), 29–53. doi:10.18497/iejee-green.56932

Pandey, G. (2018). Challenges and future prospects of agri-nanotechnology for sustainable agriculture in India. *Environmental Technology & Innovation*, *11*, 299–307. doi:10.1016/j.eti.2018.06.012

Pang, Y., Pan, L., Zhang, J., Chen, J., Dong, Y., & Sun, H. (2022). Integrated sizing and scheduling of an off-grid integrated energy system for an isolated renewable energy hydrogen refueling station. *Applied Energy*, *323*, 119573. doi:10.1016/j.apenergy.2022.119573

Panwar, N. L., Kaushik, S. C., & Kothari, S. (2011). Role of renewable energy sources in environmental protection: A review. *Renewable & Sustainable Energy Reviews*, *15*(3), 1513–1524. doi:10.1016/j.rser.2010.11.037

Patton, M. Q. (2002). *Qualitative research and evaluation methods* (3rd ed.). Sage Publisher.

Patton, M. Q. (2014). *Qualitative research & evaluation methods: Integrating theory and practice*. Sage publications.

Peidro Payá, D. (2016). *Teaching materials for the subject Business in the Degree of Mechanical Engineering*. Unpublished manuscript.

Pollock, E. R., Young, M. D., Lubans, D. R., Coffey, J. E., Hansen, V., & Morgan, P. J. (2021). Understanding the impact of a teacher education course on attitudes towards gender equity in physical activity and sport: An exploratory mixed methods evaluation. *Teaching and Teacher Education*, *105*, 1–15. doi:10.1016/j.tate.2021.103421

Praveen, P., Khan, A., Verma, A. R., Kumar, M., & Peoples, C. (2023). Smart Village Infrastructure and Rural Communities. In Smart Village Infrastructure and Sustainable Rural Communities (pp. 1-15). IGI Global. doi:10.4018/978-1-6684-6418-2.ch001

Preston, D. (1997). Can business ethics really exist? *Computers & Society*, *27*(1), 6–9. doi:10.1145/382069.382070

Priestley, M., Edwards, R., Priestley, A., & Miller, K. (2012). Teacher agency in curriculum making: Agents of change and spaces for maneuver. *Curriculum Inquiry*, *43*(2), 191–214. doi:10.1111/j.1467-873X.2012.00588.x

PRODEP. (2023). *Programa para el Desarrollo Profesional Docente, para el Tipo Medio Superior*. Obtenido de Programa para el Desarrollo Profesional Docente, para el Tipo Medio Superior: http://prodep.desarrolloprofesionaldocente.sems.gob.mx/

Prosser, J. (Ed.). (1998). *Image based research: a sourcebook for qualitative researchers*. Falmer Press.

Qin, Y., Xu, Z., Wang, X., & Skare, M. (2023). Artificial Intelligence and Economic Development: An Evolutionary Investigation and Systematic Review. *Journal of the Knowledge Economy*, 214–219. doi:10.100713132-023-01183-2

Qiu, Y., Zhao, X., Fan, D., Li, S., & Zhao, Y. (2022). Disaggregating population data for assessing progress of SDGs: methods and applications. International Journal of Digital Earth, 15(1). doi: 10.1080/17538947.2021.2013553

Quinn, S., & Yapa, L. (2016). OpenStreetMap and Food Security: A Case Study in the City of Philadelphia. *The Professional Geographer*, *68*(2), 271–280. Advance online publication. doi: 10.1080/00330124.2015.1065547

Rai, H. M., Trivedi, A., & Shukla, S. (2013). ECG signal processing for abnormalities detection using multi-resolution wavelet transform and Artificial Neural Network classifier. *Measurement*, *46*(9), 3238–3246.

Raj, A., Milton, A., & Ekstrand, M. D. (2021). Pink for princesses, blue for superheroes: The need to examine gender stereotypes in kids' products in search and recommendations. arXiv preprint arXiv:2105.09296.

Randstad. (2021). Sustainability at Work 2021: Toward 2030 Together. Available at: https://www.randstad.com/s3fs-media/rscom/public/2021-09/sustainability-at-work-2021.pdf

Ran, J., & Nedovic-Budic, Z. (2018). Designing an Information Infrastructure for Policy Integration of Spatial Planning and Flood Risk Management. *International Journal of E-Planning Research*, *7*(1), 53–85. doi:10.4018/IJEPR.2018010104

Raza, A., Zhong, M., Akuh, R., & Safdar, M. (2023). Public transport equity with the concept of time-dependent accessibility using geostatistical methods, Lorenz curves, and Gini coefficients. *Case Studies on Transport Policy*, *11*, 100956. doi:10.1016/j.cstp.2023.100956

Reamer, F. G. (2017). Evolving ethical standards in the digital age. *Australian Social Work*, *70*(2), 148–159. doi:10.1080/0312407X.2016.1146314

Republic, Ministry of the Environment of the Czech. (2023, July 19). *Ministry of the Environment of the Czech Republic*. Retrieved from Ministry of the Environment of the Czech Republic: https://www.mzp.cz/en

Resnik, D. B. (2020). What is ethics in research and why it is important. National Institute of Environmental Health Sciences, 8(8).

Rilla, C. (2013). *School garden design as catalyst for environmental education and community engagement: Los Angeles unified school district case studies*. University of Southern California.

Rissanen, M., & Löfström, E. (2014). Students' research ethics competences and the university as a learning environment. *International Journal for Educational Integrity*, *10*(2). Advance online publication. doi:10.21913/IJEI.v10i2.1004

Rodrigues, R., Pietzcker, R., Fragkos, P., Price, J., McDowall, W., Siskos, P., Fotiou, T., Luderer, G., & Capros, P. (2022). Narrative-driven alternative roads to achieve mid-century CO2 net neutrality in Europe. *Energy*, *239*, 121908. doi:10.1016/j.energy.2021.121908

Rohit, A. K., Devi, K. P., & Rangnekar, S. (2017). An overview of energy storage and its importance in Indian renewable energy sector: Part I–Technologies and Comparison. *Journal of Energy Storage*, *13*, 10–23. doi:10.1016/j.est.2017.06.005

Rollason, E., Bracken, L. J., Hardy, R. J., & Large, A. R. G. (2018). The importance of volunteered geographic information for the validation of flood inundation models. *Journal of Hydrology (Amsterdam)*, *562*, 267–280. Advance online publication. doi:10.1016/j.jhydrol.2018.05.002

Roy, P., He, J., Zhao, T., & Singh, Y. V. (2022). Recent advances of wind-solar hybrid renewable energy systems for power generation: A review. *IEEE Open Journal of the Industrial Electronics Society*, *3*, 81–104.

Roy, R. B., Rokonuzzaman, M., Amin, N., Mishu, M. K., Alahakoon, S., Rahman, S., ... Pasupuleti, J. (2021). A comparative performance analysis of ANN algorithms for MPPT energy harvesting in solar PV system. *IEEE Access : Practical Innovations, Open Solutions*, *9*, 102137–102152.

Sadeghi-Niaraki, A., Jelokhani-Niaraki, M., & Choi, S. M. (2020). *A volunteered geographic information-based environmental decision support system for waste management and decision making*. Sustainability. doi:10.3390u12156012

Sadeghkhani, I., Golshan, M. E. H., Guerrero, J. M., & Mehrizi-Sani, A. (2016). A current limiting strategy to improve fault ride-through of inverter interfaced autonomous microgrids. *IEEE Transactions on Smart Grid*, *8*(5), 2138–2148.

Sadker, D. (1999). Gender Equity: Still Knocking at the Classroom Door. *Educational Leadership*, *80-83*. Advance online publication. doi:10.1080/10665680003301l2

Sahoo, S. K., Somkun, S., Velu, S. K., Padhan, S., Jeyaraj, B. E., & Puppala, R. (2023). *Recent Advances in Renewable Energy Automation and Energy Forecasting*. Academic Press.

Saint-Germain, P. M. (2007). *Faculty Integrating Technology*. Retrieved from home.csulb.edu: https://home.csulb.edu/~msaintg/ppa590/intro.htm

Sanchez, E., van Oostendorp, H., Fijnheer, J. D., & Lavoué, E. (2020). *Gamification. Encyclopedia of Education and Information Technologies*. Springer.

Santos-Reyes, J., Avalos-Bravo, V., & Padilla-Pérez, D. (2017). Bus Rapid Transit Systems Road Safety: A Case Study of Mexico City. In H. Yaghoubi (Ed.), *Urban Transport Systems* (pp. 125–144). INTECH Publishers Press. doi:10.5772/65583

Santos-Reyes, J., Padilla-Pérez, D., & Beard, A. N. (2015). Modeling Critical Infrastructure Interdependency: The Case of the Mexico City Metro Transport System. *Human and Ecological Risk Assessment*, *21*(5), 1428–1444. doi:10.1080/10807039.2014.957956

Santos-Reyes, J., Padilla-Pérez, D., & Beard, A. N. (2019). Transport Infrastructure interdependency: Metro's failure propagation in the road transport system in Mexico City. *Sustainability (Basel)*, *11*(17), 4757. doi:10.3390u11174757

Sarwar, S., Idris, Z. M., & Ali, S. M. (2018). Paid academic writing services: A perceptional study of business students. *International Journal of Experiential Learning & Case Studies*, *3*(1), 73–83. doi:10.22555/ijelcs.v3i1.2078

Schrag, Z. M. (2010). *Ethical imperialism: Institutional review boards and the social sciences, 1965–2009*. JHU Press. doi:10.1353/book.471

Schrank, D., Eisele, B., & Lomax, T. (2019). *Urban mobility report 2019*. Available at: https://static.tti.tamu.edu/tti.tamu.edu/documents/umr/archive/mobility-report-2019.pdf

Science-Based Targets initiative (SBTi). (2021). *Science-Based Targets: Scaling Urgent Corporate Climate Action Worldwide.* Retrieved from https://sciencebasedtargets.org/resources/files/SBTiProgressReport2021.pdf

Scott, G., Tilbury, D., Sharp, L., & Deane, L. (2012). Turnaround Leadership for Sustainability in Higher Education. Australian Government, OLT, Sydney.

Scott, G., & Rajabifard, A. (2017). Sustainable development and geospatial information: A strategic framework for integrating a global policy agenda into national geospatial capabilities. *Geo-Spatial Information Science*, *20*(2), 59–76. doi:10.1080/10095020.2017.1325594

Secretaria de Gobernación. (2019). *Plan Nacional De Desarrollo 2019-2024.* Presidencia de la Republica.

See, L., McCallum, I., Fritz, S., Perger, C., Kraxner, F., Obersteiner, M., Baruah, U. D., Mili, N., & Kalita, N. R. (2013). Mapping Cropland in Ethiopia Using Crowdsourcing. *International Journal of Geosciences*, *4*(6), 6–13. Advance online publication. doi:10.4236/ijg.2013.46A1002

See, L., Mooney, P., Foody, G., Bastin, L., Comber, A., Estima, J., Fritz, S., Kerle, N., Jiang, B., Laakso, M., Liu, H. Y., Milèinski, G., Nikšieč, M., Painho, M., Podör, A., Olteanu-Raimond, A. M. R., & Rutzinger, M. (2016). Crowdsourcing, citizen science or volunteered geographic information? The current state of crowdsourced geographic information. *ISPRS International Journal of Geo-Information*, *5*(5), 55. Advance online publication. doi:10.3390/ijgi5050055

Sen, G., Chau, H. W., Tariq, M. A. U. R., Muttil, N., & Ng, A. W. (2022). Achieving sustainability and carbon neutrality in higher education Institutions: A review. *Sustainability (Basel)*, *14*(1), 222. doi:10.3390u14010222

SEP. (2023). *Secretaria de Educcion Publica.* Obtenido de https://becas.sep.gob.mx/

Şerban, A. C., & Lytras, M. D. (2020). Artificial intelligence for smart renewable energy sector in europe—Smart energy infrastructures for next generation smart cities. *IEEE Access : Practical Innovations, Open Solutions*, *8*, 77364–77377. doi:10.1109/ACCESS.2020.2990123

Shang, W.-L., & Lv, Z. (2023). Low carbon technology for carbon neutrality in sustainable cities: A survey. *Sustainable Cities and Society*, *92*, 104489. doi:10.1016/j.scs.2023.104489

Sharmin, S., Yabar, H., & Richards, D. (2023). Green energy optimization in dinajpur, Bangladesh: A path to net neutrality. *Sustainability (Basel)*, *15*(2), 1336. doi:10.3390u15021336

Shohel, M. M. C., & Howes, A. J. (2011). Models of Education for Sustainable Development and Nonformal Primary Education in Bangladesh. *Journal of Education for Sustainable Development*, *5*(1), 129–139. doi:10.1177/097340821000500115

Sleem, H., Abdelhai, R. A. A., Al-Abdallat, I., Al-Naif, M., Gabr, H. M., Kehil, E. T., Sadiq, B. B., Yousri, R., Elsayed, D., Sulaiman, S., & Silverman, H. (2010). Development of an accessible self-assessment tool for research ethics committees in developing countries. *Journal of Empirical Research on Human Research Ethics; JERHRE*, *5*(3), 85–98. doi:10.1525/jer.2010.5.3.85 PMID:20831423

Sleurs, W. (2008). *Competencies for ESD (Education for Sustainable Development).* Teachers. A Framework to Integrate ESD in the Curriculum of Teacher Training Institute.

Smith, E. (2008). Pitfalls and Promises: The Use of Secondary Data Analysis in Educational Research. *British Journal of Educational Studies*, *56*(3), 323–339. doi:10.1111/j.1467-8527.2008.00405.x

Soden, R., & Palen, L. (2014). From Crowdsourced Mapping to Community Mapping: The Post-earthquake Work of OpenStreetMap Haiti. *COOP 2014 - Proceedings of the 11th International Conference on the Design of Cooperative Systems, 27-30 May 2014, Nice (France).* 10.1007/978-3-319-06498-7_19

Solano-Hermosilla, G., Barreiro-Hurle, J., Adewopo, J. B., & Gorrín-González, C. (2022). Increasing engagement in price crowdsourcing initiatives: Using nudges in Nigeria. *World Development*, *152*, 105818. Advance online publication. doi:10.1016/j.worlddev.2022.105818 PMID:35370345

Spanish National Commission for Markets and Competition. (2014). *El programa de clemencia de la CNMC.* https://www.cnmc.es/ambitos-de-actuacion/competencia/programa-de-clemencia

Spencer, R., Rehman, S., & Kirk, S. (2015). Understanding gender norms, nutrition, and physical activity in adolescent girls: A scoping review. *The International Journal of Behavioral Nutrition and Physical Activity*, *12*(6), 1–10. doi:10.118612966-015-0166-8 PMID:25616739

Stake, J., & Hoffmann, F. (2001). Changes in student social attitudes, activism, and personal confidence in higher education: The role of women's studies. *American Educational Research Journal*, *38*(2), 411–436. doi:10.3102/00028312038002411

Staufenberg, J. (2022). Focus feature: decarbonising college estates. *FE Week.* feweek.co.uk

Steel, B. S. (1996). Thinking globally and acting locally?: Environmental attitudes, behavior and activism. *Journal of Environmental Management*, *47*(1), 27–36. doi:10.1006/jema.1996.0033

Sterling, S. (2001). Sustainable education: Re-visioning learning and change. *Schumacher Briefings, 38.*

Sterling, S. (2021). Sustainability education: What should be the core content of teacher education courses in sustainability? *Environmental Education Research*, *27*(2), 153–167.

Stern, M. J., & Powell, R. B. (2022). Educating for sustainability in the anthropocene. *The Journal of Environmental Education*, *53*(1), 1–15.

Stibbe, A. (2021). Ecolinguistics and the role of educators in addressing the climate and ecological crisis. *Environmental Education Research*, *27*(6), 887–903.

Stokel-Walker, C. (2015). What exactly is 'game theory'? *BBC News.* https://www.bbc.com/news/magazine-31503875

Stream Tracker. (2023). *Why Stream Tracker?* https://www.streamtracker.org/

Superior. (2023). *ANUIES*. Obtenido de ANUIES: http://www.anuies.mx/anuies/acerca-de-la-anuies/mision-vision-y-objetivos-estrategicos

Sutrop, M., Parder, M. L., & Juurik, M. (2020). Research Ethics Codes and Guidelines. Handbook of Research Ethics and Scientific Integrity, 67-89.

Swapna, V., & Gayatri, M. T. L. (2021, November). Power Quality Enhancement using Hybrid Energy Storage based Dynamic Voltage Restorer for Grid Integrated Distributed Generation. In *2021 IEEE Bombay Section Signature Conference (IBSSC)* (pp. 1-5). IEEE.

Tanaka, T., Ma, K., Wang, H., & Blaabjerg, F. (2018). Asymmetrical reactive power capability of modular multilevel cascade converter based STATCOMs for offshore wind farm. *IEEE Transactions on Power Electronics*, *34*(6), 5147–5164.

Tata, H., Madeti, S. R. K., & Sai Veerraju, M. (2022, August). Power Quality Improvement of a Hybrid Renewable Energy Systems Using Dynamic Voltage Restorer with PI Controller. In *International Conference on Recent Advances in Mechanical Engineering Research and Development* (pp. 549-559). Singapore: Springer Nature Singapore.

The Bedford College Group (TBCG). (2021). *Bedford College Sustainability Policy*. Available at: bedfordcollegegroup.ac.uk

The Bedford College Group (TBCG). (2022). *Skills for a sustainable future: strategic plan 2022-2027*. Author.

The Sustainability Consortium. (n.d.). *About*. Retrieved July 25, 2023, from https://sustainabilityconsortium.org/

The United Nations Framework Convention on Climate Change (UNFCCC). (n.d.). *What is carbon neutrality?* Retrieved from https://enb.iisd.org/negotiations/un-framework-convention-climate-change-unfccc#:~:text=The%20UN%20Framework%20Convention%20on,interference%20with%20the%20climate%20system.%E2%80%9D

Thomas, I. (2009). Critical thinking, transformative learning, sustainable education, and problem-based learning in universities. *Journal of Transformative Education*, *7*(3), 245–264. doi:10.1177/1541344610385753

Tilbury, D. (2007). Education for sustainability: A review of practice. *International Journal of Sustainability in Higher Education*, *8*(4), 500–510.

Tilbury, D., & Mochizuki, Y. (2021). Sustainability education and education for sustainable development: Their role in learning, decision making and professional capacity. *Sustainability Science*, *16*(1), 41–52.

Tlili, A. H., Huang, R., Chang, T.-W., Nascimbeni, F., & Burgos, D. (2019). Open educational resources and practices in China: A systematic literature review. *Sustainability (Basel)*, *11*(18), 4867. doi:10.3390u11184867

Torre-Sierra, A. D., & Guichot-Reina, V. (2022). The influence of school textbooks on the configuration of gender identity: A study on the unequal representation of women and men in the school discourse during the Spanish democracy. *Teaching and Teacher Education*, *117*, 1–13. doi:10.1016/j.tate.2022.103810

Tracy, D. M., & Lane, M. B. (1999). Gender-equitable teaching behaviors: Preservice teachers' awareness and implementation. *Equity & Excellence in Education*, *32*(3), 93–104. doi:10.1080/1066568990320311

Trolebús. (2023). *Trolebus – CDMX*. Available at: https://www.ste.cdmx.gob.mx/red-de-servicio/lineas-de-trolebus

Tronchin, L., Manfren, M., & Nastasi, B. (2018). Energy efficiency, demand side management and energy storage technologies–A critical analysis of possible paths of integration in the built environment. *Renewable & Sustainable Energy Reviews*, *95*, 341–353. doi:10.1016/j.rser.2018.06.060

TÜİK. (2022). *Türkiye Aile Yapısı Araştırması 2021*. Türkiye İstatistik Kurumu.

UK Government. (2020). *Ten Point Plan for a Green Industrial Revolution*. Retrieved from https://assets.publishing.service.gov.uk/government/uploads/system/uploads/attachment_data/file/936567/10_POINT_PLAN_BOOKLET.pdf

UN General Assembly. (2015). *Transforming our world: the 2030 agenda for sustainable development, 21 October 2015*. A/RES/70/1.

UN Sustainable Development Group. (2023, July 19). *India*. Retrieved from India: https://unsdg.un.org/es/un-in-action/india

UN. (2015). General Assembly. *Transforming our world: the 2030 Agenda for Sustainable Development*. https://sdgs.un.org/2030agenda

UN. (2023). https://sdgs.un.org/goalsadresinden alındı

UNDP. (2023). *Sustainable Development Goals*. https://www.undp.org/sustainable-development-goals

UNECE. (2021). *Sustainable Use of Natural Resources*. United Nations.

UNESCO. (2015). *A guide for gender equality in teacher education policy and practices*. https://unesdoc.unesco.org/ark:/48223/pf0000231646

UNESCO. (2017). *Education for Sustainable Development Goals: Learning Objectives*. United Nations Educational, Scientific and Cultural Organization.

UNESCO. (2019). *From access to empowerment: UNESCO strategy for gender equality in and through education 2019-2025*. United Nationas Educational, Scientific and Cultural Organization., doi:10.54675/IJGQ3826

UNESCO. (2019). Retrieved from Artificial intelligence in education: https://www.unesco.org/en/digital-education/artificial-intelligence

UNESCO. (2021). *Interagency Consultation Meeting on Co-operation and Co-ordination in Environmental Education, 3rd, Paris, 1990. Contacto, 4.*

UNESCO. (2022). *Education for Sustainable Development: A Roadmap*. https://en.unesco.org/themes/education-sustainabledevelopment

UNESCO. (2022). Education in Latin America and the Caribbean at a crossroads, Regional monitoring report SDG4 - Education 2030. France. UNESCO.

UNESCO. (2023). *Education transforms lives*. Retrieved from UNESCO EDUCATION: https://www.unesco.org/en/education

UNESCO. (2023). *Gender equality and education.* UNESCO Priority Gender Equality. https://www.unesco.org/en/gender-equality/education

UNESCO. (n.d.). *Education for sustainable development.* Retrieved July 25, 2023, from https://www.unesco.org/en/education-sustainable-development

UN-GGIM. (2012). *Monitoring sustainable development: Contribution of geospatial information to the rio+ 20 processes*. United Nations New York.

UNICEF. (2023). *Gender equality in education.* UNICEF. https://www.unicef.org/education/girls-education

United Nations Educational, Scientific and Cultural Organization (UNESCO). (2005). *United Nations Decade of Education for Sustainable Development 2005–2014: UNESCO International Implementation Scheme*. UNESCO.

United Nations Framework Convention on Climate Change (UNFCCC). (1992). *What is action for climate empowerment?* UNFCCC.

United Nations Framework Convention on Climate Change (UNFCCC). (2013). *Youth in action on climate change: Inspirations from around the world.* Available at: https://unfccc.int/resource/docs/publications/publication_youth_2013.pdf

United Nations. (2018). *Working together: Integration, institutions and the sustainable development goals*. Department of Economic and Social Affairs.

United Nations. (2023a). *Sustainable Development Goals.* Retrieved from United Nations Web site: https://www.un.org/sustainabledevelopment/education/

United Nations. (2023b). *Sustainable Development Goals: 17 Goals to Transform our World.* Retrieved from https://www.un.org/en/exhibits/page/sdgs-17-goals-transform-world

UN-SDG (UN-Sustainable Development Goals). (2023). *Goals 13: Take urgent action to combat climate change and its impacts*. Available at: https://www.un.org/sustainabledevelopment/climate-change/

Unterhalter, E., North, A., Arnot, M., Lloyd, C., Moletsane, L., Murphy-Graham, E., . . . Saito, M. (2014). Interventions to enhance girls' education and gender equality. Education Rigorous Literature Review. Department for International Development.

Vallance, R. J. (2004, November). Formation in research ethics: Developing a teaching approach for the social sciences. *AARE Conference 28th Nove2nd Dec The University of Melbourne.*

Vare, P., Rieckmann, M., & Escrigas, C. (2022). Assessing the impact of sustainability education for teachers: A systematic review. *Environmental Education Research*, *28*(1), 20–34.

Vaughan, L. (2011). Hunger In America: Utilizing GIS and Volunteered Geographic Information (VGI) to Improve Food Security. *139st APHA Annual Meeting and Exposition 2011.*

Vonthron, S., Perrin, C., & Soulard, C. T. (2023). Mapping the food environment: The reliability of volunteered geographical information and institutional data sources in France. *Transactions in GIS*, *27*(1), 134–151. Advance online publication. doi:10.1111/tgis.13013

Wals, A. E. J. (2007). Rethinking sustainability education: Towards a transformative learning approach. *Journal of Education for Sustainable Development*, *1*(1), 7–19.

Wang, Y., Li, Y., Liu, C., & Li, Z. (2020). Is traffic accident related to air pollution? A case report from an island of Taihu Lake, China. *Atmospheric Pollution Research*, *11*(5), 1028–1033. doi:10.1016/j.apr.2020.02.018

Wbcsd. (2023, July 19). *Brazil - CEBDS*. Retrieved from Brazil - CEBDS: https://sdghub.com/brazil-cebds/

Wei, Y. M., Chen, K., Kang, J. N., Chen, W., Wang, X. Y., & Zhang, X. (2022). Policy and management of carbon peaking and carbon neutrality: A literature review. *Engineering (Beijing)*, *14*, 52–63. doi:10.1016/j.eng.2021.12.018

Weng, J. (2015). Brain as an Emergent Finite Automaton: A Theory and Three Theorems. *International Journal of Intelligence Science*, *5*(2), 112–131. doi:10.4236/ijis.2015.52011

Werbach, K. (Re)defining gamification: A process approach. *Persuasive Technology: 9th International Conference, 9,* 266-272.

Wessells, M. (2013). Reflections on Ethical and Practical Challenges of Conducting Research with Children in War Zones: Toward a Grounded Approach. *Research methods in conflict settings: A view from below*, 81.

WHO. (2020). *Road Traffic Injuries*. https://www.who.int/news-room/fact-sheets/detail/road-traffic-injuries

Widener, M. J., & Li, W. (2014). Using geolocated Twitter data to monitor the prevalence of healthy and unhealthy food references across the US. *Applied Geography (Sevenoaks, England)*, *54*, 189–197. Advance online publication. doi:10.1016/j.apgeog.2014.07.017

Wiek, A., Withycombe, L., & Redman, C. L. (2011). Key competencies in sustainability: A reference framework for academic program development. *Sustainability Science*, *6*(2), 203–218. doi:10.100711625-011-0132-6

Wiki. (2020). *Food Security*. https://wiki.openstreetmap.org/wiki/Food_security

Williams, J. H., DeBenedictis, A., Ghanadan, R., Mahone, A., Moore, J., Morrow, W. R. III, ... Torn, M. S. (2019). The technology path to deep greenhouse gas emissions cuts by 2050: The pivotal role of electricity. *Science*, *365*(6458), 71–76. doi:10.1126cience.1208365 PMID:22116030

Wolek, M., Wolanski, M., Bartłomiejczyk, M., Wyszomirski, O., Grzelec, K., & Hebel, K. (2021). Ensuring sustainable development of urban public transport: A case study of the trolleybus system in Gdynia and Sopot (Poland). *Journal of Cleaner Production*, *279*, 123807. doi:10.1016/j.jclepro.2020.123807

Women, U. N. (2022). *Progress on the sustainable development goals: The gender snapshot 2022*. https://www.unwomen.org/en/digital-library/publications/2022/09/progress-on-the-sustainable-development-goals-the-gender-snapshot-2022

World Bank. (2021). *CO2 emissions (kt)*. Retrieved from https://data.worldbank.org/indicator/EN.ATM.CO2E.KT?locations=CN

World Health Organization. (2015). *Global health ethics key issues: Global network of WHO collaborating centres for bioethics*. World Health Organization.

Wu, X., Liao, B., Su, Y., & Li, S. (2023). Multi-objective and multi-algorithm operation optimization of integrated energy system considering ground source energy and solar energy. *International Journal of Electrical Power & Energy Systems*, *144*, 108529. doi:10.1016/j.ijepes.2022.108529

Xiao, L., Wang, J., Dong, Y., & Wu, J. (2015). Combined forecasting models for wind energy forecasting: A case study in China. *Renewable & Sustainable Energy Reviews*, *44*, 271–288. doi:10.1016/j.rser.2014.12.012

Xinhua. (2020). *China aims to hit peak emissions before 2030, carbon neutrality before 2060*. Retrieved from http://www.xinhuanet.com/english/2020-09/23/c_139388712.htm

Xue, C., Wang, J., & Li, Y. (2022). Model predictive control for grid-tied multi-port system with integrated PV and battery storage. *IEEE Transactions on Smart Grid*, *13*(6), 4596–4609.

Xu, M., Shuai, B., Wang, X., Liu, H., & Zhou, H. (2023). Analysis of the accessibility of connecting transport at high-speed rail stations from the perspective of departing passengers. *Transportation Research Part A, Policy and Practice*, *173*, 103714. doi:10.1016/j.tra.2023.103714

Yalcin, G. (2013). Non-spatial analysis for the road traffic accidents. *Procedia: Social and Behavioral Sciences*, *92*, 1033–2038. doi:10.1016/j.sbspro.2013.08.795

Yar, D., Ulaş, S., & Vefikuluçay-Yılmaz, D. (2022). Üniversite öğrencilerinin cinsiyet rollerinin benlik saygısı ve yaşam doyumu üzerindeki etkisi. *IBAD Sosyal Bilimler Dergisi*, (13), 324–343. doi:10.21733/ibad.1143175

Yıldırım, A., & Şimşek, H. (2018). *Sosyal Bilimlerde Nitel Araştırma Yöntemleri* (11 b). Seçkin Yayıncılık.

Zabala, I., & García, M. (2008). Historia de la Educación Ambiental desde su discusión y análisis en los congresos internacionales. *Revista de Investigacion, 32*(63), 201–218.

Zanib, N., Batool, M., Riaz, S., & Nawaz, F. (2022). Performance analysis of renewable energy based distributed generation system using ANN TUNED UPQC. *IEEE Access : Practical Innovations, Open Solutions, 10*, 110034–110049.

Zeug, H., Zeug, G., Bielski, C., Solano-Hermosilla, G., & M'barek, R. (2017). *Innovative food price collection in developing countries. Focus on crowdsourcing in Africa*. Academic Press.

Zhang, J., Tong, Z., Ji, Z., Gong, Y., & Sun, Y. (2022). Effects of Climate Change Knowledge on Adolescents' Attitudes and Willingness to Participate in Carbon Neutrality Education. *International Journal of Environmental Research and Public Health, 19*(17), 10655. doi:10.3390/ijerph191710655

Zhang, M. X. (2022). Globalization, Green Economy and Environmental Challenges: State of the Art Review for Practical Implications. *Frontiers in Environmental Science, 10.*

Zhang, D., Chen, X., & Yao, H. (2015). Development of a prototype web-based decision support system for watershed management. *Water (Basel), 7*(2), 780–793. Advance online publication. doi:10.3390/w7020780

Zhang, G. (2020). Spatial and temporal patterns in volunteer data contribution activities: A case study of eBird. *ISPRS International Journal of Geo-Information, 9*(10), 597. Advance online publication. doi:10.3390/ijgi9100597

Zhang, G. (2022). Detecting and Visualizing Observation Hot-Spots in Massive Volunteer-Contributed Geographic Data across Spatial Scales Using GPU-Accelerated Kernel Density Estimation. *ISPRS International Journal of Geo-Information, 11*(1), 55. Advance online publication. doi:10.3390/ijgi11010055

Zhang, L., Wen, J., Li, Y., Chen, J., Ye, Y., Fu, Y., & Livingood, W. (2021). A review of machine learning in building load prediction. *Applied Energy, 285*, 116452. doi:10.1016/j.apenergy.2021.116452

Zhang, Q., He, J., Xu, Y., Hong, Z., Chen, Y., & Strunz, K. (2021). Average-value modeling of direct-driven PMSG-based wind energy conversion systems. *IEEE Transactions on Energy Conversion, 37*(1), 264–273.

Zhang, S., Du, P., Chen, C., Zhong, W. D., & Alphones, A. (2019). Robust 3D indoor VLP system based on ANN using hybrid RSS/PDOA. *IEEE Access : Practical Innovations, Open Solutions, 7*, 47769–47780.

Ziakopoulos, A., & Yannis, G. (2020, February). A review of spatial approaches in road safety. *Accident; Analysis and Prevention, 135*, 105323. doi:10.1016/j.aap.2019.105323 PMID:31648775

Zichang, L., Yadong, L., Yingjie, Y., Peng, W., & Xiuchen, J. (2020). An identification method for asymmetric faults with line breaks based on low-voltage side data in distribution networks. *IEEE Transactions on Power Delivery*, *36*(6), 3629–3639.

About the Contributors

Javier Martínez-Falcó is Assistant Professor in the Department of Management at the University of Alicante. In the field of research, he focuses his interest on issues related to the Strategic Management of the Company, specifically in the areas of Knowledge Management and Corporate Sustainability of wine companies, on which he has written several publications in the form of articles, book chapters and contributions to conferences. He has also participated in several national research projects and teaches Strategic Management on the ADE, TADE and DADE degree courses.

Bartolomé Marco-Lajara is Professor of Business Organization at the University of Alicante (Spain). He holds a PhD in Economics and his doctoral dissertation focused in strategic alliances management. His research interests are on Strategic Management and Tourism Management. He is author of several books, book chapters and international articles related to teaching methodology and the areas above mentioned. On the other hand, he is member of the Tourism Research Institute at the University of Alicante since its foundation. He has participated in many public- and private-funded projects, being the main researcher in many of them. He was assistant dean of the Economics Faculty at the University of Alicante during eight years, holding now the position of Head of the Department of Business Organization at this University.

Eduardo Sánchez-García (PhD) is an assistant professor in the Department of Business Organization at the University of Alicante (Spain). His research is in the field of strategic management of the company. Specifically, he focuses on the study of the territorial agglomeration of companies in clusters and the absorptive capacity, and its influence on the innovative performance and entrepreneurial orientation of companies. He is co-author of several papers presented at conferences, both national and international, as well as several articles and book chapters related to his field of research.

Luis A. Millán;Tudela is Assistant Professor in the Department of Management at University of Alicante. MBA and Bachelor of Mechanical Engineering, his research focuses on business longevity, and more specifically on the factors that influence the survival of companies. This activity is developed within the research group "Dirección Estratégica, conocimiento e innovación en una economía global (DECI-GLOBAL)". He also teaches strategic management in the Degree of Business Administration.

* * *

Munir Ahmad is a Ph.D. in Computer Science with over 23 years of extensive experience in spatial data development, management, processing, visualization, and quality control. He has dedicated expertise in open data, crowdsourced data, volunteered geographic information, and spatial data infrastructure. A seasoned professional with extensive knowledge in the field, he has served as a trainer for the latest spatial technologies. With a passion for research and over 25 publications in the same field, he obtained his Ph.D. degree in Computer Science from Preston University Pakistan, in 2022. He is committed to advancing the industry and sharing knowledge through his expertise and experience. #SpatialData #GIS #GeoTech

Jamileh Alamolhoda is a professor at faculty of education and psychology, Shahid Beheshti University, Tehran, Iran.

Huda M. Aldayri is a cultural affairs specialist at the Omani National Commission for Education, Culture and Science. She holds a PhD in social studies curricula and teaching methods from Sultan Qaboos University. She has published a number of scientific papers on sustainable development issues in a number of peer-reviewed scientific journals. She is a member of the Green Schools and Ecotourism Project, member. And she is a rapporteur in the strategic research committee in the education sector.

Abdullah K. Ambusaidi is a Professor of Science Education. He received his Ph.D. from the Centre of Science Education at Glasgow University/UK. Currently, he is the undersecretary for education at the Ministry of Education/Oman. He worked before at Sultan Qaboos University and held several positions, including Director of the Center for Humanities Research and Dean of Postgraduate Studies. He has supervised over 80 master's and doctorate degrees in science education and is an external and internal examiner for more than 80 master's and doctoral dissertations. He has more than 150 published and accepted research papers in well-known international journals. He is a consultant for many educational projects in Oman, including TAMAM Project (school-based development) and Green Schools Initiative. He is also a leading researcher of various educational projects, such as the school garden project, ecotourism project, health education project, and the Omani Heritage Project in the school curricula.

E. Ivette Cota-Rivera received a M.Eng. from the Autonomous University of Baja California (UABC, Universidad Autonoma de Baja California). In 2016, she received her Ph.D. in Energy Engineering and is currently working as a professor in the Energy Engineering program at the Polytechnic University of Baja California (UPBC, Universidad Politecnica de Baja California). Her research interests include energy efficiency and sustainability, artificial intelligence, energy science and technology, as well as programming.

İpek Derman graduated from Hacettepe University, Department of Biology Education in 2009. She received her MS in 2013 and earned a doctoral degree in 2019 in the Division of Curriculum and Instruction from Hacettepe University. She earned doctoral degrees in the same Division from Hacettepe University in 2019. In 2011 she was hired as a research assistant by the Department of Educational Sciences at Hacettepe University. Her field of study includes curriculum development, teacher training, learning theories, teaching methods, science education, higher education, qualitative, quantitative, and mixed research methods. She has published numerous national and international articles in the field of education. She writes and presents extensively on curriculum and instruction areas.

Esmaeil Jafari is Assistant Professor at Faculty of Education and Psychology, Shahid Beheshti University, Tehran, Iran.

Fabian N. Murrieta-Rico received B.Eng. and M.Eng. degrees from Instituto Tecnológico de Mexicali (ITM) in 2004 and 2013 respectively. In 2017, he received his PhD in Materials Physics at Centro de Investigación Científica y Educación Superior de Ensenada (CICESE). He has worked as an automation engineer, systems designer, as a university professor, and as postdoctoral researcher at Facultad de Ingeniería, Arquitectura y Diseño from Universidad Autónoma de Baja California (UABC) and at the Centro de Nanociencias y Nanotecnología from Universidad Nacional Autónoma de México (CNyN-UNAM), currently he works as professor at the Universidad Politécnica de Baja California. His research has been published in different journals and presented at international conferences since 2009. He has served as reviewer for different journals, some of them include IEEE Transactions on Industrial Electronics, IEEE Transactions on Instrumentation, Measurement and Sensor Review. His research interests are focused on the field of time and frequency metrology, synthesis and characterization of zeolite, and highly sensitive chemical detectors.

Jesús Heriberto Orduño-Osuna has obtained a Bachelor's degree in Mechatronics Engineering from Universidad Politécnica de Baja California. Currently, he is concurrently pursuing Master's degrees in Computer Science and Applied Mathematics at Universidad Internacional de La Rioja in Mexico. Professionally, he holds the esteemed position of a professor at his alma mater. In this role, he has the privilege of imparting his extensive knowledge and experience to enthusiastic engineering and technology students.

Jaime Santos-Reyes is a lecturer at the National Polytechnic Institute, Mexico, whose main research interests are seismic risk, safety management systems, critical infrastructures modelling, accident and risk analysis, reliability engineering. He obtained a PhD from Heriot-Watt University, UK, in 2001. He has spent 3 years working as a research associate at Heriot-Watt University, Edinburgh, Scotland, UK.

Mohamed A. Shahat is an Assistant Professor of science education at Sultan Qaboos University (SQU), Oman, and an Associate Professor at Aswan University (AU), Egypt. He earned his Ph.D. in physics education from the University of Duisburg-Essen, Germany. He attained a post-doctoral fellowship at the University of Duisburg-Essen. He has supervised and examined many masters and doctorate degrees in science education. Currently, he is leading various national and international research projects, including STEM and entrepreneurial education and students' learning engagement at SQU. He participated in multiple national and international conferences and published extensive articles in national and international journals. He published chapters in regional and globally edited books. He is a member of different national and international journal editorial boards. He has also been awarded other teaching, research, and reviewing awards.

M. Mahruf C. Shohel is currently working at the University of Roehampton. He is an academic researcher with special interests in education, childhood studies, international development, technology-enhanced learning and social science research methods. He has written extensively on development issues in the Global South and conducted research on disadvantaged children including socioeconomically deprived children, street children, sex worker's children and displaced refugee children. Currently, Dr Shohel is engaged in the fields of education in emergencies, education for sustainable development and global citizenship, emerging technologies in education, students' learning journeys and their engagement, and teaching and learning in higher education.

Sonia Singh, PhD., is an Associate Professor of OB & HRM at JAIN (Deemed-to-be University), Bengaluru, India. She completed her doctoral degree in Management and holds double Masters in Science & Management Program. She is a University rank holder and is UGC NET qualified. She has an overall experience of seventeen years in Academics as well as Industry and has worked with some of the world's leading organizations. Her key competent areas are HR Business Partnering, Talent Acquisition & Retention and Learning & Development function. She has to her credit two Patents and authored several research papers and book chapter in Journals of repute. She is contributing her knowledge and experience as a member of Research Advisory Committee in reputed Universities and Editorial board member of International peer-reviewed Journals. Her research interests include Organizational Psychology, Agile HR, Diversity & Inclusion, Sustainable HRM and Leadership Studies.

Preksha Yadav, PhD, an alumna of IIM-Indore, possesses extensive experience of over fourteen years in academia, research, and administration. Currently serving as an Assistant Professor at JAIN (Deemed-to-be University), she holds a UGC-NET qualification. Additionally, she has served as a corporate trainer, imparting valuable knowledge on stress management, leadership, interpersonal skills, and life skills to employees in both public and private sectors. Dr. Yadav's research interests span multiple disciplines, including Sustainability, Green HRM, Digital Wellbeing, Mental Health, and Online Learning, among others. Her scholarly contributions include more than 15 publications, encompassing research papers, articles, and book chapters. She has actively participated in over 20 national and international conferences held across the country. She has obtained 2 patents and authored 1 book.

Index

H

I

M

O

P

Q

R

S

T

U

V

W

www.ingramcontent.com/pod-product-compliance
Lightning Source LLC
LaVergne TN
LVHW080642011026
849160LV00014B/1294